T3-BNT-663

# New Comprehensive Biochemistry

Volume 40

*General Editor*

G. BERNARDI
*Paris*

ELSEVIER

Amsterdam · Boston · Heidelberg · London · New York · Oxford
Paris · San Diego · San Francisco · Singapore · Sydney · Tokyo

# Emergent Collective Properties, Networks and Information in Biology

J. Ricard

*Institut Jacques Monod,*
*CNRS, Universités Paris VI et Paris VII,*
*2 Place Jussieu,*
*75251 Paris Cedex 05, France*

ELSEVIER

Amsterdam · Boston · Heidelberg · London · New York · Oxford
Paris · San Diego · San Francisco · Singapore · Sydney · Tokyo

| ELSEVIER B.V. | ELSEVIER Inc. | ELSEVIER Ltd | ELSEVIER Ltd |
| --- | --- | --- | --- |
| Radarweg 29, P.O. Box 211, | 525 B Street, Suite 1900 | The Boulevard, Langford Lane | 84 Theobalds Road |
| 1000 AE Amsterdam | San Diego CA 92101-4495 | Kidlington Oxford OX5 1GB | London WC1X 8RR |
| The Netherlands | USA | UK | UK |



First edition 2006

Library of Congress Cataloging in Publication Data
A catalog record is available from the Library of Congress.

British Library Cataloguing in Publication Data

Emergent collective properties, networks and information in biology. - (New comprehensive biochemistry ; v. 40)
1. Biochemistry 2. Principal components analysis 3. Reduction (Chemistry)
I. Ricard, Jacques, 1929–
572.3′3

ISBN-10: 0-444-52159-3
ISBN-13: 978-0-444-52159-0

ISBN: 0-444-80303-3 (Series)
ISSN: 0167-7306

∞ The paper used in this publication meets the requirements of ANSI/NISO Z39.48-1992 (Permanence of Paper).
Printed in The Netherlands.

# Preface

Classical Science, i.e., the scientific activities that have sprung up in Europe since the seventeenth century, relies upon a principle of reduction which is the very basis of the analytic method developed by Descartes. This principle consists, for instance, in deconstructing a complex system, and studying its component sub-systems independently with the hope it will then be possible to understand the logic of the overall system. For centuries, this analytic approach has been extremely fruitful and has led to most achievements of classical science.

Today, molecular biology can still be considered an excellent example of this analytic approach of the real world. Most molecular biologists thought in the 1960s and 1970s that all the properties of living organisms were already present, in potential state, in the structure of biomacromolecules such as nucleic acids and proteins. Thus, for instance, there is little doubt that the project aimed at deciphering large genomes was based, either explicitly or implicitly, on the belief that knowledge of the genome is sufficient to predict and explain most functional properties of living systems, including man. If this idea were correct, no emergence of a novel property (i.e., a property present in the system but not present in its components) were to be expected. In fact, this view has been accepted for decades, at least tacitly.

More recently however, it became increasingly obvious that the global properties of a system cannot always be predicted from the independent study of the corresponding sub-systems. This paradigmatic change became evident in 1999 when a special issue of the journal *Science,* entitled "Beyond Reductionism", appeared. In this issue, a number of scientists working in fields as diverse as fundamental physics, chemistry, biology, and social sciences reach the same conclusion, namely that important results cannot be understood if one sticks to the idea that reduction is sufficient to understand the real world. If, however, one accepts the idea that emergence of global properties of a system out of the interactions between local component sub-systems is real, it is essential to understand the physical nature of emergence and to express this idea, not in metaphysical, but in quantitative scientific terms.

The concept of network as a mathematical description of a set of states, or events, linked according to a certain topology has been developed recently and has led to a novel approach of real world. This approach is no doubt important in the field of biology. In fact, biological systems can be considered networks. Thus, for instance, an enzyme-catalyzed reaction is a network that links, according to a certain topology, the various states of the protein and of its complexes with the substrates and products of the chemical reaction. Connections between neurons, social relations in animal and human populations are also examples of networks. Hence there is little doubt that the concept of network transgresses the boundaries between traditional scientific disciplines.

A very important concept in modern science is that of information. Originally this concept was formulated by Shannon in the context of the communication of a message between a source and a destination. According to Shannon's theory, transfer of a message in a communication channel requires a specific association of signs which contributes to the mathematical expression of the so-called mutual information of the system. In this perspective, cell information can be thought of as the ability of a system to associate in a specific manner the molecular signs. If such a specific association of signs is an essential requirement for the existence of information, most biochemical networks should possess information for they usually involve specific association of molecular signals. Enzymes for instance associate in a specific manner two or three substrates. Information, in this case, is not related to the communication of a message but rather to the organization of a network. It is therefore of interest to know whether Shannon's theory can be used as such, or has to be modified, in order to describe in quantitative terms the organization of a given system. One can consider that, from this point of view, three possible types of networks can be thought of. First, one can imagine that the properties of the network are the properties of its component sub-systems. The properties of the overall network can then be reduced to those of its components. Second, the network has lesser degrees of freedom than the set of its nodes, but its global properties are qualitatively novel. Then the system behaves as an *integrated whole*. Last, the network has more degrees of freedom and qualitatively novel properties. It can then be considered *emergent* for it possesses more information than the set of its components. An important question is to know the physical constraints that generate these different types of behavior. Although it is relatively simple to study the properties of networks under thermodynamic equilibrium conditions, there is little doubt that, in the cell, they constitute open systems. Hence it is of interest to know whether departure from thermodynamic equilibrium results in a change of information content of systems and how the multiplicity of pathways leading to the same node of a network affects information.

Sets of enzyme reactions form networks that, as we shall see later, may possibly contain information. If this view were confirmed, this would imply that information linked with network topology is superimposed to the genetic information required for enzyme synthesis. In this perspective, the total information of a cell would be larger than its genetic information. Robustness of networks is an important parameter that contributes to define their activity and one may wonder whether there exists a relationship between network information and robustness.

Simple statistical mechanics of networks requires that the concepts of activity and concentration be valid. This is usually not the case in living cells as the number of molecules of a given chemical species is usually too small to allow one to disregard the influence of stochastic fluctuations of the number of molecules in a given region of space. It is therefore of interest to take account of the potential influence of molecular noise on networks dynamics. This matter raises another puzzling question: how is it possible to explain that elementary processes are subjected to molecular noise whereas the biological functions that rely upon these elementary processes appear to be strictly deterministic?

This book aims at answering these questions. It presents the conditions required for the reduction of the properties of a biological system to those of its components; the mathematical background required to study the organization of biological networks; the main properties of biological networks; the mathematical analysis of communication in living systems; the statistical mechanics of network organization, integration, and emergence; the mechanistic causes of network information, integration, and emergence; the information content of metabolic networks; the role of functional connections in biochemical networks; the information flow in protein edifices; the quantitative and systemic approach of gene networks; the importance of stochastic fluctuations in network function and dynamics.

Although these topics are biological in essence, they are treated in a physical perspective for it has now become possible to use physical *concepts*, and not only physical *techniques*, to understand some aspects of the internal logic of biological events. This book is based on a theoretical study of simple model networks for two reasons. First, because it appears that complexity is not complication and that complex events, such as emergence, can already be detected and studied with apparently simple model systems. Second, because apparently simple model networks can be studied analytically in a rigorous way, without having recourse to blind computer simulation. Indeed such models are far too simple to be a true *description* of real biochemical networks but they nevertheless offer a rigorous *explanation* of important biochemical events. In the same vein, Figures have often been presented as simple schemes in order to make it plain what a phenomenon is, without reference to specific numerical data.

It is a real pleasure to thank my colleague and friend Dick D'Ari who has been kind enough to read and correct the manuscript of this book and who has spent hours discussing its content with me. Brigitte Meunier has been extremely helpful on several occasions. Last, I have a special debt to my wife Käty who, in spite of the burden imposed on both of us, has always encouraged me to write this book.

Jacques Ricard
*Paris*

# Contents

Preface ........ v

Other volumes in the series ........ xv

*Chapter 1.* Molecular stereospecific recognition and reduction in cell biology ........ 1

1. The concepts of reduction, integration, and emergence ........ 2
2. Stereospecific recognition under thermodynamic equilibrium conditions as the logical basis for reduction in biology ........ 3
3. Most biological systems are not in thermodynamic equilibrium conditions ........ 5
   - 3.1. Simple enzyme reactions cannot be considered equilibrium systems ........ 5
   - 3.2. Complex enzyme reactions cannot be described by equilibrium models ........ 8
     - 3.2.1. Steady-state rate and induced fit ........ 8
     - 3.2.2. Steady state and pre-equilibrium ........ 11
     - 3.2.3. Pauling's principle and the constancy of catalytic rate constant along the reaction coordinate ........ 12
4. Coupled scalar–vectorial processes in the cell occur under nonequilibrium conditions ........ 13
   - 4.1. Affinity of a diffusion process ........ 13
   - 4.2. Carriers and scalar–vectorial couplings ........ 14
5. Actin filaments and microtubules are nonequilibrium structures ........ 19
6. The mitotic spindle is a dissipative structure ........ 22
7. Interactions with the environment, nonequilibrium, and emergence in biological systems ........ 23

References ........ 24

*Chapter 2.* Mathematical prelude: elementary set and probability theory ........ 27

1. Set theory ........ 27
   - 1.1. Definition of sets ........ 27
   - 1.2. Operations on sets ........ 28
   - 1.3. Relations and graphs ........ 30
   - 1.4. Mapping ........ 32

2. Probabilities ........ 32
   2.1. Axiomatic definition of probability and fundamental theorems ........ 32
   2.2. Properties of the distribution function and the Stieltjes integral ........ 36
3. Probability distributions ........ 39
   3.1. Binomial distribution ........ 39
   3.2. The Poisson distribution ........ 42
   3.3. The Laplace–Gauss distribution ........ 44
4. Moments and cumulants ........ 45
   4.1. Moments ........ 45
      4.1.1. Monovariate moments ........ 46
      4.1.2. Bivariate moments ........ 48
   4.2. Cumulants and characteristic functions ........ 51
5. Markov processes ........ 52
6. Mathematics as a tool for studying the principles that govern network organization, information, and emergence ........ 54

References ........ 54

*Chapter 3.* Biological networks ........ 57

1. The concept of network ........ 58
2. Random networks ........ 59
3. Percolation as a model for the emergence of organization in a network ........ 63
4. Small-world and scale-free networks ........ 70
   4.1. Metabolic networks are fundamentally different from random graphs ........ 70
   4.2. Properties of small-world networks ........ 72
   4.3. Scale-free networks ........ 74
5. Attack tolerance of networks ........ 75
6. Towards a general science of networks ........ 77

References ........ 79

*Chapter 4.* Information and communication in living systems ........ 83

1. Components of a communication system ........ 84
2. Entropy and information ........ 85
3. Communication and mapping ........ 92
4. The subadditivity principle ........ 93
5. Nonextensive entropies ........ 96
6. Coding ........ 97
   6.1. Code words of identical length ........ 97
   6.2. Code words of variable length ........ 98
7. The genetic code and the Central Dogma ........ 102

8. Accuracy of the communication channel between DNA and proteins ..... 105
References ..... 107

*Chapter 5.* Statistical mechanics of network information, integration, and emergence ..... 109

1. Information and organization of a protein network ..... 111
   1.1. Subsets of the protein network ..... 111
   1.2. Definition of self- and mutual information of integration ..... 114
2. Subadditivity and lack of subadditivity in protein networks ..... 117
3. Emergence and information of integration of a network ..... 119
4. The physical nature of emergence in protein networks ..... 124
5. Reduction and lack of reduction of biological systems ..... 125
6. Information, communication, and organization ..... 129

References ..... 131

*Chapter 6.* On the mechanistic causes of network information, integration, and emergence ..... 133

1. Information and physical interaction between two events ..... 133
2. Emergence and topological information ..... 136
3. Organization and the different types of mutual information ..... 140
   3.1. Mutual information and negative correlation ..... 141
   3.2. Mutual information and physical interaction between two binding processes ..... 141
   3.3. Mutual information and network topology ..... 141
4. Organization and negative mutual information ..... 142

References ..... 144

*Chapter 7.* Information and organization of metabolic networks ..... 145

1. Mutual information of individual enzyme reactions ..... 146
2. Relationships between enzyme network organization and catalytic efficiency ..... 148
3. Mutual information of integration and departure from pseudo-equilibrium ..... 151
4. Mutual information of integration of multienzyme networks ..... 154
   4.1. Metabolic networks as networks of networks ..... 154
   4.2. Robustness of multienzyme networks ..... 155
   4.3. Regular multienzyme networks ..... 157
   4.4. Fuzzy-organized multienzyme networks ..... 161
   4.5. Topological information of regular and fuzzy-organized networks ..... 162

5. Enzyme networks and Shannon communication–information theory ........ 163
References ........ 165

*Chapter 8.* Functional connections in multienzyme complexes: information, and generalized microscopic reversibility ........ 167

1. Network connections and mutual information of integration in multienzyme complexes........ 168
   1.1. Linear networks........ 168
   1.2. Functions of connection........ 171
2. Generalized microscopic reversibility ........ 176
3. Connections in a network displaying generalized microscopic reversibility........ 178
4. Mutual information of integration and reaction rate........ 178
5. Possible functional advantages of physically associated enzymes ........ 180
References ........ 181

*Chapter 9.* Conformation changes and information flow in protein edifices........ 185

1. Phenomenological description of equilibrium ligand binding and nonequilibrium catalytic processes ........ 186
2. Thermodynamic bases of long-range site–site interactions in proteins and enzymes........ 188
   2.1. General principles........ 188
   2.2. Energy contribution of subunit arrangement........ 190
   2.3. Quaternary constraint energy contribution........ 193
   2.4. Fundamental axioms ........ 198
3. Conformational changes and mutual information of integration in protein lattices ........ 199
   3.1. Conformation change and mutual information of integration of the elementary protein unit ........ 201
   3.2. Ligand binding, conformation changes, and mutual information of integration of protein lattices........ 204
      3.2.1. A simple protein lattice ........ 205
      3.2.2. Mutual information of integration and conformational constraints in the lattice ........ 207
      3.2.3. Integration and emergence in a protein lattice ........ 210
   3.3. Conformational changes in quasi-linear lattices ........ 215
      3.3.1. The basic unit of conformation change ........ 215
      3.3.2. Thermodynamics of spontaneous conformational transitions in a simple quasi-linear protein lattice ........ 216
      3.3.3. Mutual information of integration and conformation changes........ 218

4. Conformational spread and information landscape .................... 223
References .................................................................... 224

*Chapter 10.* Gene networks .................................................. 227

1. An overview of the archetype of gene networks: the bacterial operons ......... 227
   1.1. The operon as a coordinated unit of gene expression ............ 228
   1.2. Repressor and induction ........................................ 230
   1.3. Positive versus negative control ................................ 232
2. The role of positive and negative feedbacks in the expression of gene networks ......... 233
   2.1. Multiple dynamic states and differential activity of a gene ........ 233
   2.2. Formal expression of multiple dynamic states of a gene .......... 234
   2.3. Full-circuits ..................................................... 236
3. Gene networks and the principles of binary logic ....................... 242
4. Engineered gene circuits ................................................ 245
   4.1. The role of feedback loops in gene circuits ........................ 245
   4.2. Periodic oscillations of gene networks ............................ 249
   4.3. The logic of gene networks ....................................... 250
References .................................................................... 252

*Chapter 11.* Stochastic fluctuations and network dynamics ............... 255

1. The physics of intracellular noise ....................................... 256
   1.1. Random walk and master equation ................................ 256
   1.2. Detailed balance ................................................. 258
   1.3. Intracellular noise and the Langevin equation .................... 261
      1.3.1. The Langevin equation of a macromolecule subjected to random collisions ........ 262
      1.3.2. The function $F(t)$ as the expression of the noise-driven properties ......... 264
   1.4. Intracellular noise and the Fokker–Planck equation ............... 266
      1.4.1. The Fokker–Planck equation for one-dimensional motion ......... 266
      1.4.2. Generalized Fokker–Planck equation ........................ 270
2. Control and role of intracellular molecular noise ....................... 270
References .................................................................... 274

Subject Index ................................................................ 275

# Other volumes in the series

Volume 1. *Membrane Structure* (1982)
J.B. Finean and R.H. Michell (Eds.)

Volume 2. *Membrane Transport* (1982)
S.L. Bonting and J.J.H.H.M. de Pont (Eds.)

Volume 3. *Stereochemistry* (1982)
C. Tamm (Ed.)

Volume 4. *Phospholipids* (1982)
J.N. Hawthorne and G.B. Ansell (Eds.)

Volume 5. *Prostaglandins and Related Substances* (1983)
C. Pace-Asciak and E. Granstrom (Eds.)

Volume 6. *The Chemistry of Enzyme Action* (1982)
M.I. Page (Ed.)

Volume 7. *Fatty Acid Metabolism and its Regulation* (1982)
S. Numa (Ed.)

Volume 8. *Separation Methods* (1984)
Z. Deyl (Ed.)

Volume 9. *Bioenergetics* (1985)
L. Ernster (Ed.)

Volume 10. *Glycolipids* (1985)
H. Wiegandt (Ed.)

Volume 11a. *Modern Physical Methods in Biochemistry, Part A* (1985)
A. Neuberger and L.L.M. van Deenen (Eds.)

Volume 11b. *Modern Physical Methods in Biochemistry, Part B* (1988)
A. Neuberger and L.L.M. van Deenen (Eds.)

Volume 12. *Sterols and Bile Acids* (1985)
H. Danielsson and J. Sjovall (Eds.)

Volume 13. *Blood Coagulation* (1986)
R.F.A. Zwaal and H.C. Hemker (Eds.)

Volume 14. *Plasma Lipoproteins* (1987)
A.M. Gotto Jr. (Ed.)

Volume 16. *Hydrolytic Enzymes* (1987)
A. Neuberger and K. Brocklehurst (Eds.)

Volume 17. *Molecular Genetics of Immunoglobulin* (1987)
F. Calabi and M.S. Neuberger (Eds.)

Volume 18a. *Hormones and Their Actions, Part 1* (1988)
B.A. Cooke, R.J.B. King and H.J. van der Molen (Eds.)

Volume 18b. *Hormones and Their Actions, Part 2 – Specific Action of Protein Hormones* (1988)
B.A. Cooke, R.J.B. King and H.J. van der Molen (Eds.)

Volume 19. *Biosynthesis of Tetrapyrroles* (1991)
P.M. Jordan (Ed.)

Volume 20. *Biochemistry of Lipids, Lipoproteins and Membranes* (1991)
D.E. Vance and J. Vance (Eds.) – Please see Vol. 31 – revised edition

Volume 21. *Molecular Aspects of Transfer Proteins* (1992)
J.J. de Pont (Ed.)

Volume 22. *Membrane Biogenesis and Protein Targeting* (1992)
W. Neupert and R. Lill (Eds.)

Volume 23. *Molecular Mechanisms in Bioenergetics* (1992)
L. Ernster (Ed.)

Volume 24. *Neurotransmitter Receptors* (1993)
F. Hucho (Ed.)

Volume 25. *Protein Lipid Interactions* (1993)
A. Watts (Ed.)

Volume 26. *The Biochemistry of Archaea* (1993)
M. Kates, D. Kushner and A. Matheson (Eds.)

Volume 27. *Bacterial Cell Wall* (1994)
J. Ghuysen and R. Hakenbeck (Eds.)

Volume 28. *Free Radical Damage and its Control* (1994)
C. Rice-Evans and R.H. Burdon (Eds.)

Volume 29a. *Glycoproteins* (1995)
J. Montreuil, J.F.G. Vliegenthart and H. Schachter (Eds.)

Volume 29b. *Glycoproteins II* (1997)
J. Montreuil, J.F.G. Vliegenthart and H. Schachter (Eds.)

Volume 30. *Glycoproteins and Disease* (1996)
J. Montreuil, J.F.G. Vliegenthart and H. Schachter (Eds.)

Volume 31. *Biochemistry of Lipids, Lipoproteins and Membranes* (1996)
D.E. Vance and J. Vance (Eds.)

Volume 32. *Computational Methods in Molecular Biology* (1998)
S.L. Salzberg, D.B. Searls and S. Kasif (Eds.)

Volume 33. *Biochemistry and Molecular Biology of Plant Hormones* (1999)
P.J.J. Hooykaas, M.A. Hall and K.R. Libbenga (Eds.)

Volume 34. *Biological Complexity and the Dynamics of Life Processes* (1999)
J. Ricard

Volume 35. *Brain Lipids and Disorders in Biological Psychiatry* (2002)
E.R. Skinner (Ed.)

Volume 36. *Biochemistry of Lipids, Lipoproteins and Membranes* (2003)
D.E. Vance and J. Vance (Eds.)

Volume 37. *Structural and Evolutionary Genomics: Natural Selection in Genome Evolution* (2004)
G. Bernardi

Volume 38. *Gene Transfer and Expression in Mammalian Cells* (2003)
Savvas C. Makrides (Ed.)

Volume 39. *Chromatin Structure and Dynamics: State of the Art* (2004)
J. Zlatanova and S.H. Leuba (Eds.)

Volume 40. *Emergent Collective Properties, Networks and Information in Biology* (2006)
J. Ricard

J. Ricard *Emergent Collective Properties, Networks and Information in Biology*

DOI: 10.1016/S0167-7306(05)40001-0

CHAPTER 1

# Molecular stereospecific recognition and reduction in cell biology

J. Ricard

Classical molecular cell biology is based on the idea that it is legitimate to reduce the properties of a complex biological system to the individual properties of component sub-systems. According to several founding fathers of molecular biology, most biological properties are already present, in potential state, in the structure of some biological macromolecules, in particular DNA, and *revealed* during the building up of the living organism. The aim of this chapter is to show that this belief is often optimistic and based on a number of assumptions that are very rarely, or even never, met in nature in particular the implicit or explicit assumption that biochemical reactions in the cell occur under thermodynamic equilibrium conditions.

*Keywords:* actin filaments, affinity of a diffusion process, ATP–ADP exchange, ATP-driven migration of ions, ATP synthesis, coupled scalar–vectorial processes, dyneins, electrochemical gradient, emergence, fractionation factor, high-level theory, integration, King–Altman rules, kinesins, low-level theory, microtubules, mitotic spindle, Pauling's principle, pre-equilibrium, reduction, steady state, stereospecific recognition, thermodynamic equilibrium, time hierarchy of steps, treadmilling, wavy rate curves.

Since the developments in molecular biology, research in life sciences has often been characterized by analytic and reductionist approaches. It was believed that life processes can be understood by "deconstructing" apparently complicated biological events, or systems, into simpler ones and studying these simple components independently. In this optimistic approach to life processes, living organisms are not considered intrinsically complex. The reductionist approach to biological events is based on the principle of *stereospecific recognition* between macromolecules. Two molecules can "recognize" each other and form a stereospecific complex. This complex can in turn "recognize" another macromolecule thus leading to a larger complex, and so on. In the frame of this reductionist approach, the fundamental biological properties of living systems are already present, at least in a potential state, in different categories of biomolecules, namely nucleic acids and proteins that can be isolated from living cells.

Before discussing whether the idea of reduction can be safely used to study the physical basis of biological processes, it is mandatory to present a coherent definition of reduction, integration, and emergence. In order to be satisfactory, these definitions should meet two requirements: first, they should be expressed in a formal and mathematical way in order to avoid any ambiguity; second, they should conform to common sense for the words reduction, integration, and emergence all have an accepted intuitive meaning.

## *1. The concepts of reduction, integration, and emergence*

Basically, reductionism is the philosophical doctrine that aims at expressing the results and predicates, $T_h$, of a theory from those of another theory, $T_l$, more general and embracing [1–10]. Hence $T_h$ and $T_l$ are called "high-level" and "low-level" theories, respectively. In order to be possible, the reduction of $T_h$ to $T_l$ requires that the set of concepts and predicates, $C_h$, of the "high-level" theory be included in the set of concepts and predicates, $C_l$, of the "low-level" theory [5,8], i.e.,

$$C_h \subset C_l \tag{1}$$

In other words, "high-level" predicates can be reduced to "low-level" predicates. If we aim at understanding biological events in terms of physical predicates, physics will be "low-level" and biology "high-level" theories, respectively.

The term reduction can be given another meaning, however. It can also express the view that the properties of a system can be predicted, at least in part, from a study of the constitutive elements of the system. This statement can be formulated in a loose or in a strict sense. Considered in its loose sense, the term reduction simply means there is an operational advantage in knowing the properties of the parts if we want to know the properties of the whole made up of these parts. There is certainly no logical difficulty in adopting this viewpoint. But the term reduction can also be given a strict, ontological meaning. It then implies it is *sufficient* to know the properties of the parts to know *ipso facto* the properties of the system made up of these parts. This meaning is derived from Descartes' "Règles pour la Direction de l'Esprit" and "Discours de la Méthode" [11,12].

Let us consider a system **XY** made up of two sub-systems **X** and **Y**. Let us assume that it is possible to define a mathematical function $H(X,Y)$ that describes the properties of the system **XY**, or its degrees of freedom. One can also define two other functions, $H(X)$ and $H(Y)$, that describe, as above, the properties, or the degrees of freedom, of **X** and **Y**. One will state that the properties of **XY** can be reduced to properties of **X** and **Y** if [10]

$$H(X, Y) = H(X) + H(Y) \tag{2}$$

If such a reduction applies, this means that **XY** is not a real system but the simple association of **X** and **Y**. The nature of the functions $H(X,Y)$, $H(X)$ and $H(Y)$ will be considered in Chapters 4 and 5. But **XY** can also be a real system that displays some sort of integration of its elements as a coherent "whole". Under this situation one has [10]

$$H(X, Y) < H(X) + H(Y) \tag{3}$$

Then the system **XY** has less potential wealth, or fewer degrees of freedom, than the sum of **X** and **Y**. **XY** is a regular system whose properties cannot be reduced to the properties of **X** and **Y** considered independently. To illustrate this idea, let us consider an ant. It can follow many directions in response to many different signals and possesses many degrees of freedom. Let us now consider the same ant in a line of fellow ants running to their anthill. Its number of degrees of freedom has now decreased to a dramatic extent for our ant is now doing what its fellow ants are doing at precisely the same moment. It has become part of a system and this is precisely a situation similar to that described by Eq. (3). Again, we shall discuss later the significance and the expression of the corresponding $H$ functions.

Now let us assume there exist systems that have more potential wealth, or more degrees of freedom, than their component sub-systems. Then, one will have [10]

$$H(X, Y) > H(X) + H(Y) \tag{4}$$

Hence the system will display properties that are *emergent* relative to those of the component sub-systems, and this system will be considered *complex*. In the frame of this definition, complexity of a system is defined by the emergence of novel and unexpected properties. Coming back to a population of insects, if these insects interact in a rather anarchic manner, i.e., if they alternately fight and cooperate, the insect population will display more degrees of freedom than a population where the insects do not interact.

## 2. *Stereospecific recognition under thermodynamic equilibrium conditions as the logical basis for reduction in biology*

If one believes that ontological reductionism is valid, as most founding fathers of molecular biology thought in the 1960s and 1970s, one has to base this belief upon a firm principle, the principle of stereospecific recognition of biological macromolecules, i.e., DNA, RNA, and proteins [13]. As already outlined, during the building up of living systems macromolecules form more and more elaborate complexes. According to this principle, it is believed that during the formation of these molecular complexes biological properties *that were already present* in a potential state in the macromolecules are *revealed*. To cite Monod, "la construction épigénétique n'est pas une *création*, c'est une *révélation*" [13]. If this idea were correct, it would imply that ontological reductionism is valid, that biological complexity does not exist, and that emergence of novel properties out of a biological system is just an illusion. It is therefore of interest at this point to spend some time discussing the validity of this principle of stereospecific recognition of macromolecules.

The basic idea behind the principle of specific recognition is the belief that the global properties of a system are, in a way, "written in potential state" in the structure and intrinsic properties of the elements of the system. This implies in turn that the intrinsic properties of the isolated elements are independent of their environment. If this were not the case, for instance if the properties of a protein, or of a macromolecular edifice, were qualitatively different depending on its surroundings, one could not consider the overall system solely to reveal the intrinsic properties of its elements, but rather to display emergence of novel properties in response to the interactions existing between the elements and their environment. Let us consider a multimolecular edifice having certain biological properties. Reduction of its global properties to the individual properties of its molecular components implies, at least, two conditions: the existence of a thermodynamic equilibrium between the multimolecular edifice and its molecular components; and the selection of pre-existing conformations when two proteins associate to form a complex. The first condition is required, for it is well known that the properties of a macromolecule are indeed not only dependent upon its "intrinsic" nature but also upon its interactions with the environment. Thus, for instance, as we shall see later, the catalytic properties of an enzyme are usually different depending on whether they are studied under equilibrium, or steady-state conditions. Although it is somewhat difficult to decide what the "intrinsic" properties of an enzyme are, it is logical to consider that they are the properties the enzyme displays when it is in equilibrium with its surroundings. As a matter of fact, if an enzyme reaction departs from an equilibrium state it may *acquire* properties that are *novel* and characteristic of this new state. If, depending on its environment, a macromolecule (the enzyme) can acquire novel properties, it is difficult to understand why a large supramolecular edifice could not. The second condition, which will be studied in detail later, implies that the association of two macromolecules does not *produce* a change in the three-dimensional structure of the reaction partners in order not to alter their properties. These conditions are often believed to be fulfilled. They, nevertheless, need a careful critical discussion.

The first point seems to imply that the cell organelles are either equilibrium structures or, at least, that these structures can be described through equilibrium models. There is no doubt that cell organelles do not exist under thermodynamic equilibrium conditions. They are in fact dissipative structures that require continuous flows of matter and energy. Microtubules, actin filaments, the mitotic spindle, asters are telling examples of this situation [14–21]. Even simple enzyme reactions taking place in the living cell cannot be considered equilibrium processes. Nevertheless it is often considered that equilibrium, or quasi-equilibrium, models allow one to describe nonequilibrium situations. As we shall see later, this belief seems to rely upon the fact that the equilibrium and steady-state treatments of simple one-substrate enzyme reactions lead to indistinguishable rate equations. The second requirement of the stereospecific recognition principle is subtler. It implies that when two proteins associate to form a complex they quite often change their conformation, i.e., their three-dimensional structure. The stereospecific recognition principle requires that the collision of the two proteins stabilizes a *pre-existing* conformation of either partner. This would mean that the conformation of either protein in the complex is already present in the free state. This situation is precisely that which is assumed to take place in the frame of the so-called allosteric model [22]. A different interpretation of the existence of the conformational transition is to assume that it is the collision between the two proteins that *creates* the change in three-dimensional structure. In the context of this induced-fit model, the conformation of the protein within the complex has to be different from that in the free, unbounded state. Some kind of instruction must have been exchanged between the two proteins. The two requirements of the stereospecific recognition principle will be discussed in the following sections of the present chapter.

## 3. *Most biological systems are not in thermodynamic equilibrium conditions*

The aim of this section is to show that most biological systems, even the simplest, depart from thermodynamic equilibrium and cannot usually be described with equilibrium models. We shall attempt to demonstrate this point on different subcellular systems ranging from enzyme reactions to cell organelles.

### 3.1. *Simple enzyme reactions cannot be considered equilibrium systems*

A simple one-substrate, one-product enzyme reaction has initially been looked on as an equilibrium system where an enzyme E binds a substrate S to form an enzyme–substrate complex ES in thermodynamic equilibrium with the free enzyme and the substrate. One has thus

$$E + S \overset{K}{\rightleftarrows} ES$$

where $K$ is the equilibrium constant of the binding process. The enzyme–substrate complex then undergoes catalysis and decomposes to form the product $P$ and regenerates the free enzyme $E$, i.e.,

$$ES \xrightarrow{k} E + P$$

where $k$ is the catalytic rate constant. On the basis of the equilibrium assumption, the equilibrium concentration of the $ES$ complex can be easily calculated and one finds

$$[ES] = \frac{K[E]_0[S]}{1 + K[S]} \tag{5}$$

where $[E]_0$ is the total enzyme concentration (that is the concentration of free and bound enzyme). The enzyme reaction rate, $v$, is assumed to be proportional to the equilibrium concentration of the enzyme–substrate complex [23]. One thus has

$$v = k[ES] = \frac{kK[E]_0[S]}{1 + K[S]} \tag{6}$$

where $k$ is still the catalytic rate constant. Strictly speaking, this classical formulation [24,25] is erroneous, for the enzyme–substrate complex $ES$ cannot decompose in order to regenerate $E$ and give rise to $P$ and, at the same time, be in equilibrium with $E$ and $S$. This formulation, however, can be considered a sensible approximation if the rate constant $k$ is very small relative to the rate constants of substrate binding and release to, and from, the enzyme. The system is then assumed to be in pseudo-equilibrium.

A more realistic description of a one-substrate, one-product enzyme reaction is

$$E \underset{k_{-1}}{\overset{k_1[S]}{\rightleftarrows}} ES \underset{k'}{\overset{k}{\rightleftarrows}} EP \underset{k_{-2}[P]}{\overset{k_2}{\rightleftarrows}} E$$

where the $k$s are still rate constants. Under initial conditions, i.e., at the beginning of the reaction, the concentration of the product $P$ is small enough as to be neglected relative to the substrate concentration. A similar situation is found in the living cell when the product $P$ is taken up as a substrate of the next enzyme reaction. The last step of the above enzyme process is then nearly irreversible and the reaction scheme reduces to

$$E \underset{k_{-1}}{\overset{k_1[S]}{\rightleftarrows}} ES \underset{k'}{\overset{k}{\rightleftarrows}} EP \xrightarrow{k_2} E + P$$

The overall process can be in steady state, far from equilibrium. The corresponding steady state equation given by Briggs and Haldane [25], Peller and Alberty [26] and a number of other authors [27–33] assumes the form

$$v = \frac{k_1 k k_2 [S][E]_0}{k k_2 + k_{-1} k_2 + k_{-1} k' + (k + k' + k_2) k_1 [S]} \tag{7}$$

or

$$v = \frac{\overline{k}\,\overline{K}[S][E]_0}{1 + \overline{K}[S]} \tag{8}$$

Expression (8) is formally identical to expression (6) but $\overline{k}$ and $\overline{K}$ are now different from simple rate and equilibrium constants. Their expression is found to be

$$\begin{aligned} \overline{k} &= \frac{k k_2}{k + k' + k_2} \\ \overline{K} &= \frac{k_1 (k + k' + k_2)}{k k_2 + k_{-1} k_2 + k_{-1} k'} \end{aligned} \tag{9}$$

As Eq. (6) is the mathematical expression of a near-equilibrium model whereas Eq. (8) describes a steady state, nonequilibrium system, one could be tempted to conclude it is perfectly feasible to apply a near-equilibrium formalism to an enzyme network under nonequilibrium conditions. This is precisely what has often been done. In fact, if the reaction process involves two substrates $AX$ and $B$, one can see that the near-equilibrium treatment of a nonequilibrium steady-state model gives rise to spurious results. Under initial nonequilibrium conditions, a two-substrate, two-product enzyme reaction can often be represented as

$$E \underset{k_{-1}}{\overset{k_1[AX]}{\rightleftarrows}} EAX \underset{k_{-2}}{\overset{k_2[B]}{\rightleftarrows}} EAX.B \underset{k'}{\overset{k}{\rightleftarrows}} EA.XB \overset{k_3}{\longrightarrow} \begin{matrix} EA \\ +XB \end{matrix} \overset{k_4}{\longrightarrow} \begin{matrix} E \\ +A \end{matrix}$$

The corresponding steady-state rate equation can be conveniently cast in the following form

$$\frac{v}{[E]_o} = \frac{\alpha[AX][B]}{1 + \beta[AX] + \gamma[B] + \delta[AX][B]} \tag{10}$$

with

$$\begin{aligned} \alpha &= \frac{kk_2k_3}{kk_3 + k_{-2}k_3 + k_{-2}k'} K_1 \\ \beta &= \frac{k_1}{k_{-1}} = K_1 \\ \gamma &= \frac{kk_2k_3}{k_{-1}(kk_3 + k_{-2}k_3 + k_{-2}k')} \\ \delta &= \frac{k_2(k_3k_4 + kk_4 + kk_3 + k'k_4)}{k_4(kk_3 + k_{-2}k_3 + k_{-2}k')} K_1 \end{aligned} \tag{11}$$

The corresponding equilibrium binding equation of the two substrates to the enzyme is

$$\overline{Y} = \frac{K_1K_2[AX][B]}{1 + K_1[AX] + K_1K_2[AX][B]} \tag{12}$$

where $\overline{Y}$ is the fractional saturation of the enzyme by the substrates, $K_1$ and $K_2$ the equilibrium binding constants (i.e., $K_1 = k_1/k_{-1}$, $K_2 = k_2/k_{-2}$). Comparing Eqs. (10) and (12) leads to the conclusion that the steady-state reaction rate is, in general, not proportional to the corresponding substrate-binding isotherm. It is only when the substrate-binding steps are in fast equilibrium that the overall reaction rate becomes proportional to the binding isotherm. As a matter of fact, if the rate constants of substrate release from the enzyme are large relative to the other rate constants, the two binding steps are in fast equilibrium and the corresponding rate equation is

$$\frac{v}{[E]_o} = \frac{\bar{k}K_1K_2[AX][B]}{1 + K_1[AX] + K_1K_2[AX][B]} \tag{13}$$

with

$$\bar{k} = \frac{kk_3k_4}{1 + kk_3 + kk_4 + k'k_4} \tag{14}$$

and the reaction rate is indeed proportional to the substrate-binding function. In most cases it is therefore incorrect to express, as is often done, an enzyme reaction rate with an equilibrium binding model.

### *3.2. Complex enzyme reactions cannot be described by equilibrium models*

Many enzymes are oligomeric, i.e., they are made up of several identical subunits that all bear an active site. These active sites can interact through conformation changes of the corresponding subunits. As already pointed out, the conformation states able to bind a ligand such as a substrate can either pre-exist to the collision between the ligand and the active site of the enzyme, or be induced by the collision itself. Two different models have been put forward in order to explain how subunit interactions alter and modulate ligand binding [22,34]. Both models are based on the equilibrium assumption and, in most cases, fit binding data very well. These models, however, have been extended to chemical reactions catalyzed by multi-sited enzymes on the assumption that there should exist proportionality between binding and rate data. The aim of the present section is to show that this proportionality does not exist, even in the simplest cases. It is therefore usually invalid to fit rate data to equilibrium binding models.

#### *3.2.1. Steady-state rate and induced fit*

Let us consider, as an example, a two-sited dimeric enzyme that binds one substrate only. Moreover, let us assume that the conformation changes of the subunits are induced by the interaction of the enzyme and the substrate (Fig. 1). The substrate-binding isotherm is a very simple 2:2 equation, i.e., the ratio of two polynomials in $[S]^2$. If one considers the "kinetic version" of the same model (Fig. 1) the steady-state rate equation should be, in general, of the 3:3 type. As the number of subunits increases, the degree of the numerator and denominator of the rate and binding equations increases, but the increase is much faster for the rate than for the binding equation. Thus, for a tetramer the binding equation will be 4:4 whereas for the same tetramer, the rate equation will be 10:10. Similarly, for a hexamer, the binding isotherm will be 6:6, but the corresponding reaction rate will be 21:21 [35].

In the case of high-degree reaction rates, one should expect their complexity to be reflected in their rate curve. Although it has been observed that some enzymes exhibit "wavy" or "bumpy" rate curves [35], this is rare and usually not of large amplitude. Most rate curves display a hyperbolic shape, positive or negative cooperativity, or inhibition by excess substrate (Fig. 2). One may therefore wonder whether some constraints between rate constants do not lead to a decrease of

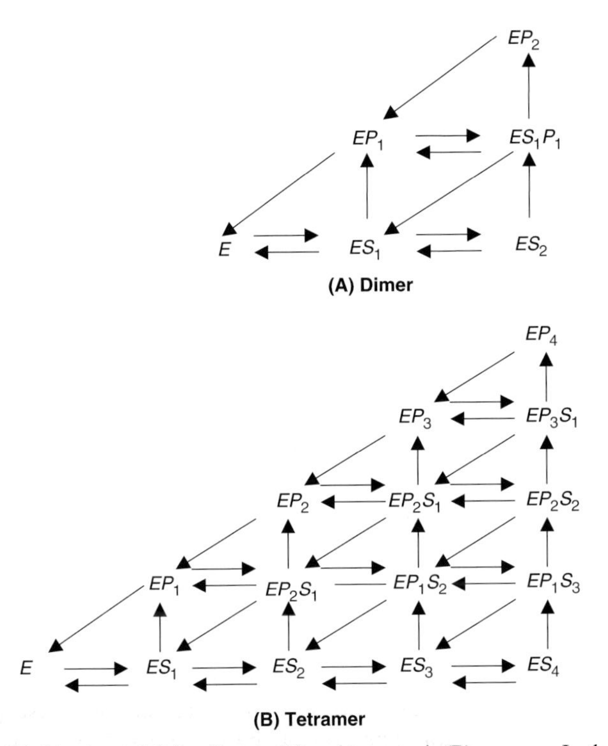


Fig. 1. Simple possible kinetic models for dimeric **(A)** and tetrameric **(B)** enzymes. In these models $S$ is the substrate and $P$ the product. Each of the enzyme process involves substrate binding and release, catalysis and product release.

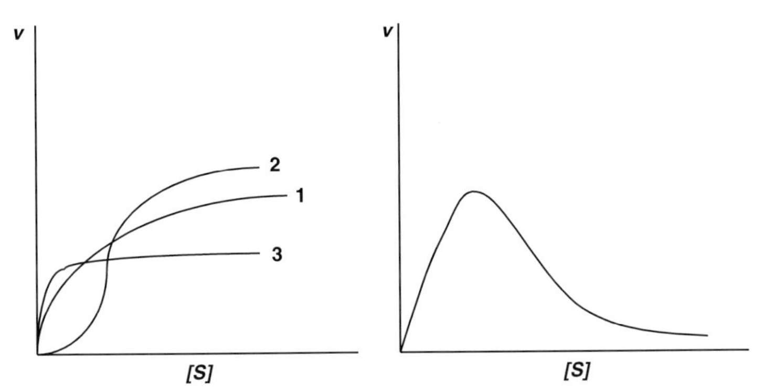


Fig. 2. Typical shapes of the steady-state reaction profile of a multi-sited enzyme. Left: 1 Hyperbolic behavior, 2 Positive cooperativity (sigmoidal behavior), 3 Negative cooperativity. Right: Inhibition by excess substrate. $v$ is the steady state rate, $[S]$ the substrate concentration.

the degree of the rate equation. As "wavy"or "bumpy" curves are clearly not of any functional advantage to the cell, one may speculate that, in the course of evolution, a selective pressure has been exerted on most enzymes so as to favor either of the four types of behavior mentioned above (Fig. 2).

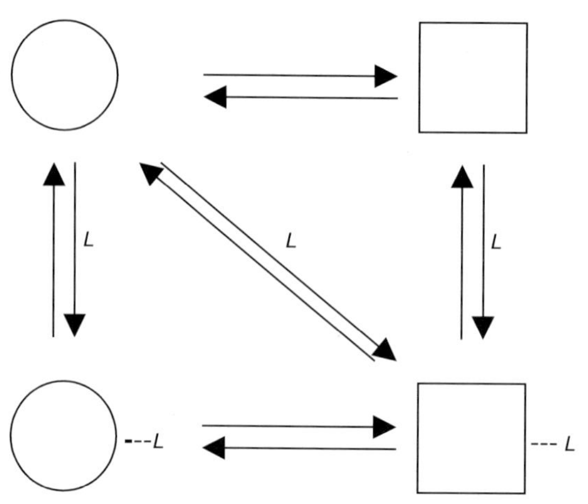


Fig. 3. Pre-equilibrium and induced fit. Pre-equilibrium implies that the enzyme pre-exists under at least two conformation states (the circle and the square). Ligand binding shifts the pre-existing equilibrium. Induced fit implies that the ligand induces the proper conformation of the protein (the square) required for ligand binding. If the pre-equilibrium is strongly shifted towards the circle state the two models become nearly identical.

Let us assume for instance that the substrate and the product induce similar subunit conformations. Then, a number of rate constants assume identical, or very similar values. Hence the resulting rate equation degenerates and becomes of the 2:2 type for a dimer. If one derives the expressions of the binding isotherm ($\overline{\nu} = 2\overline{Y}$) and of the reaction rate ($v/[E]_o$) for the model of Fig. 1, one has

$$\overline{\nu} = 2\overline{Y} = \frac{2K_1[S] + 2K_1K_2[S]^2}{1 + 2K_1[S] + K_1K_2[S]^2} \tag{15}$$

and [36,37]

$$\frac{v}{[E]_o} = \frac{2\overline{k}_1\overline{K}_1[S] + 2\overline{k}_2\overline{K}_1\overline{K}_2[S]^2}{1 + 2\overline{K}_1[S] + \overline{K}_1\overline{K}_2[S]^2} \tag{16}$$

Here, $\overline{\nu}$ is the fractional saturation of the protein computed on the basis of the number of enzyme molecules and $\overline{Y}$ the same fractional saturation estimated on the basis of the number of sites. The parameters that appear in Eqs. (15) and (16) are defined by the following expressions

$$\begin{aligned}
K_1 &= \frac{k_1}{k_{-1}} \quad K_2 = \frac{k_2}{k_{-2}} \\
\overline{K}_1 &= \frac{k_1(k'_1 + k''_1)}{k''_1(k'_1 + k_{-1})} \quad \overline{K}_2 = \frac{k_2(k'_2 + k''_2)}{k''_2(k'_2 + k_{-2})} \\
\overline{k}_1 &= \frac{k'_1k''_1}{k'_1 + k''_1} \quad \overline{k}_2 = \frac{k'_2k''_2}{k'_2 + k''_2}
\end{aligned} \tag{17}$$

Here, $k_1', k_1'', k_2', k_2''$ are rate constants of catalysis and product release [36,37].

Comparison of Eqs. (15) and (16) shows that $v/[E]_o$ is not proportional to $\overline{\nu}$ except if one assumes that $\overline{k}_1 = \overline{k}_2$, $\overline{K}_1 \approx K_1$ and $\overline{K}_2 \approx K_2$. We shall see later this assumption is not acceptable for it is at variance with an important principle of physical chemistry, Pauling's principle [38,39].

*3.2.2. Steady state and pre-equilibrium*

We have considered thus far that subunit conformation changes occurring during substrate binding are induced by the collision of substrate molecules with the active sites of the enzyme. It has been claimed, however, and this is likely to be true in a number of cases, that two, or several, conformation states pre-exist in the absence of any ligand. Substrate binding would thus shift a pre-equilibrium between two, or several, conformation states [22]. A clear-cut distinction between induced fit and pre-equilibrium may seem, however, illusory in many cases. As shown in Fig. 3, if the pre-equilibrium is shifted towards a protein state unable to bind the ligand (exclusive binding) it will be impossible to distinguish that situation from a true induced fit.

In the pre-equilibrium model [22], and its subsequent kinetic version [40], it was assumed that successive ligand binding to different sites did not affect the three-dimensional structure of the protein and that the successive binding constants had all the same value. In the case of a dimeric enzyme which follows the so-called symmetry model with exclusive binding of the ligand to one of the conformation states, the corresponding binding isotherm assumes the form (Fig. 4)

$$\overline{\nu} = 2\overline{Y} = \frac{2K[S](1 + K[S])}{L + (1 + K[S])^2} \tag{18}$$

where $L$ is the equilibrium constant between the two conformations of the free enzyme. It is obvious from Eq. (18) that departure from hyperbolic behavior solely relies upon the value of $L$. If the protein is an enzyme that acts as a catalyst, the relevant steady-state equation is [40]

$$\frac{v}{[E]_o} = \frac{2k'\overline{K}[S](1 + \overline{K}[S])}{L + (1 + \overline{K}[S])^2} \tag{19}$$

where $k'$ and $\overline{K}$ are still the catalytic and apparent binding constant, respectively. The apparent binding constant is defined as

$$\overline{K} = \frac{k_+}{k_- + k'} \tag{20}$$

where $k_+$ and $k_-$ are the rate constants of substrate binding and substrate release, respectively. Although expressions (18) and (19) are very similar, the values of $\overline{\nu}$ and $v/[E]_o$ are not, strictly speaking, proportional, unless $\overline{K} \approx K$.

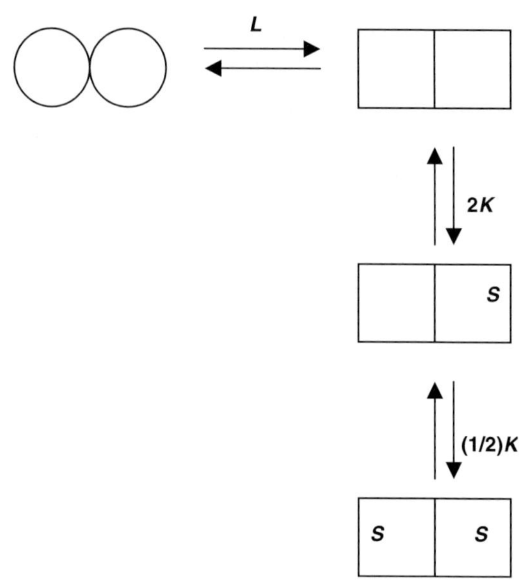


Fig. 4. Symmetry model with exclusive binding for a dimeric enzyme. $L$ is the equilibrium constant of the pre-equilibrium, $S$ the substrate concentration, and $K$ the intrinsic binding constant.

### *3.2.3. Pauling's principle and the constancy of catalytic rate constant along the reaction coordinate*

In order to explain that enzyme catalysis can occur, one has to postulate that the enzyme's active site is "complementary" neither to the substrate nor to the product but rather to the transition state of the chemical reaction, midway between substrate and product. Hence, according to Pauling, the enzyme has a strained conformation when it has bound the substrate and the strain is relieved upon reaching the top of the energy barrier [38,39,41–43]. This principle, which allows many experimental predictions, in particular a high affinity of transition state analogs for the enzymes, offers a simple explanation of the catalytic power of enzymes. If the same idea is applied to oligomeric enzymes, i.e., to enzymes made up of several identical subunits, one has to postulate that quaternary constraints between subunits also have to be relieved at the top of the energy barrier, when the enzyme is bound to the transition state. This idea is hardly compatible with the view that the catalytic rate constant does not vary along the reaction coordinate. In fact, one should expect that the amount of energy released when a substrate molecule is converted into a transition state depends on the number of substrate molecules bound to the enzyme. This would imply in turn that the catalytic rate constant cannot remain unchanged as more substrate molecules are bound to the enzyme. This idea will be discussed at length in a forthcoming chapter. As we shall see later, classical theories of chemical reaction rates [41–43] predict that

$$k_i = \frac{k_B T}{h} \exp(-\Delta G_i^{\neq}/RT) \qquad (21)$$

where $k_B$ and $h$ are the Boltzmann and the Planck constants, $R$ and $T$ the gas constant and the absolute temperature, respectively. $\Delta G_i^{\neq}$ is the so-called free energy of activation of the $i$th step of catalysis. Pauling's principle requires that both intra- and inter-subunit constraints be relieved at the top of the energy barrier. Hence $\Delta G_i^{\neq}$ is the sum of at least two energy contributions, the contribution of the energy of the catalytic act itself and the contribution of quaternary constraints. As we shall see later, it is unrealistic to believe that the latter contribution is independent of the number of substrate molecules bound to the enzyme.

## 4. *Coupled scalar–vectorial processes in the cell occur under nonequilibrium conditions*

Many biological processes taking place in the living cell are more complex than mere enzyme reactions. This is the case for coupled scalar–vectorial processes that associate chemical reactions with the transport of molecules and ions. Transport of ions across biological membranes cannot take place by simple diffusion through the membrane itself for it is mostly made up of lipids and therefore impermeable to ions. Nevertheless, transport can take place through proteins anchored in the membrane. Hence the transport process can imply either facilitated diffusion or an active transfer process. The latter requires coupling between vectorial transport against a concentration gradient and exergonic chemical processes.

### 4.1. *Affinity of a diffusion process*

Let us assume that the available space is separated into two compartments termed cis (′) and trans (″). Let us consider the diffusion of an ion $L$ from the cis to the trans compartment. If the concentrations of $L$ in the two compartments are $[L']$ and $[L'']$, the corresponding electrochemical potentials, $\tilde{\mu}'_L$ and $\tilde{\mu}'_L$, are [21]

$$\begin{aligned} \tilde{\mu}'_L &= \mu_L^0 + RT\ln[L'] + zF\Psi' \\ \tilde{\mu}'_L &= \mu_L^0 + RT\ln[L''] + zF\Psi'' \end{aligned} \tag{22}$$

where $\mu_L^0$ is the standard chemical potential of the ion, $\Psi'$ and $\Psi''$ the electrostatic potentials, $F$ the Faraday constant and $z$ the valence of the ion to which a negative or a positive sign is assigned depending on whether it is an anion or a cation. The affinity of the diffusion process is defined as

$$A_L = -\Delta\tilde{\mu}_L = -(\tilde{\mu}''_L - \tilde{\mu}'_L) \tag{23}$$

if the transport takes place from cis to trans. One can also define the electrostatic potential difference, $\Delta\Psi$, as

$$\Delta\Psi = \Psi'' - \Psi' \tag{24}$$

It then follows that

$$A_L = -\Delta\tilde{\mu}_L = -RT \ln\frac{[L'']}{[L']} - zF\Delta\Psi \tag{25}$$

Then the condition for spontaneous transport from cis to trans to occur is

$$\frac{[L']}{[L'']} = \exp\left(\frac{A_L + zF\Delta\Psi}{RT}\right) > 1 \tag{26}$$

and this condition requires that

$$A_L + zF\Delta\Psi > 0 \tag{27}$$

or

$$-\Delta\tilde{\mu}_L + zF\Delta\Psi > 0 \tag{28}$$

### *4.2. Carriers and scalar–vectorial couplings*

As already referred to in the previous section, ion transport across membranes occurs owing to proteins called carriers. These proteins can allow facilitated diffusion of ligands from one compartment to the other. They also allow exchange of ions between two compartments, coupling between an exergonic chemical reaction and ion transfer against an electrochemical gradient, or alternatively, the use of an electrochemical gradient to allow completion of an endergonic chemical reaction. There exists in cell biology numerous examples of coupled scalar–vectorial processes. For instance, ATP–ADP exchange between mitochondrion and cytosol, extrusion of protons in the inter-membrane space of mitochondrion coupled to the oxidation of NADH, and the synthesis of ATP coupled to proton transport from inter-membrane to the mitochondrial matrix [44].

We are not going to discuss the mechanism of these processes but we shall focus instead on their nonequilibrium character. Let us consider a carrier $X$ that can take two conformations, $X'$ and $X''$, each able to bind ligand $L$ on either side (cis and trans) of the membrane. Let us assume these binding processes fast enough to be considered rapid equilibria relative to the velocity of the conformation changes of the carrier that allows transportation of $L$ from the cis to the trans compartment (Fig. 5). If one assumes that ligand binding and release on both sides of the membrane is fast enough, the concentrations of $X'$ and $X'L'$ on one side, $X''$ and $X''L''$ on the other, can be aggregated as ideal chemical species called $Y'$ and $Y''$, i.e.,

$$\begin{aligned} Y' &= [X'] + [X'L'] \\ Y'' &= [X''] + [X''] \end{aligned} \tag{29}$$

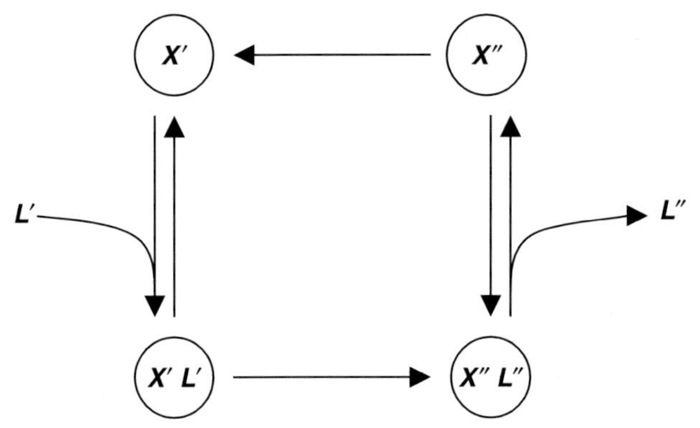


Fig. 5. Simple nonequilibrium kinetic model that allows to explain transport of a ligand from cis (′) to trans (″) compartment. $L'$ and $L''$ represent the ligand in compartments cis and trans, respectively. $X'$ and $X''$ are the carrier in these compartments.

The aggregation process based on a time hierarchy of ligand binding and release relative to carrier conformation change allows considerable simplification of the kinetic model. One can now define the fractionation factors as

$$f'_1 = \frac{1}{1 + K'_1[L']} \qquad f''_1 = \frac{1}{1 + K''_1[L'']}$$
$$f'_2 = \frac{K'_1[L']}{1 + K'_1[L']} \qquad f''_2 = \frac{K''_1[L'']}{1 + K''_1[L'']} \tag{30}$$

where $K'_1$ and $K''_1$ are the ligand binding constants on either side of the membrane. The application of King–Altman–Hill [46–49] rules to the model system of Fig. 5 yields the relevant steady-state equation of the net flow, i.e.,

$$J = \frac{k_{-2} f''_2 k_1 f'_1 \left( \frac{k_2 f'_2 k_{-1} f''_1}{k_{-2} f''_2 k_1 f'_1} - 1 \right)}{k_2 f'_2 + k_{-2} f''_2 + k_{-1} f''_1 + k_1 f'_1} \tag{31}$$

Under this form it becomes evident that the condition required for a net flow oriented cis–trans is

$$\frac{k_2 f'_2 k_{-1} f''_1}{k_{-2} f''_2 k_1 f'_1} > 1 \tag{32}$$

Making use of expressions (30) this relationship is equivalent to

$$\frac{K_2 K'_1}{K_1 K''_1} \frac{[L']}{[L'']} > 1 \tag{33}$$

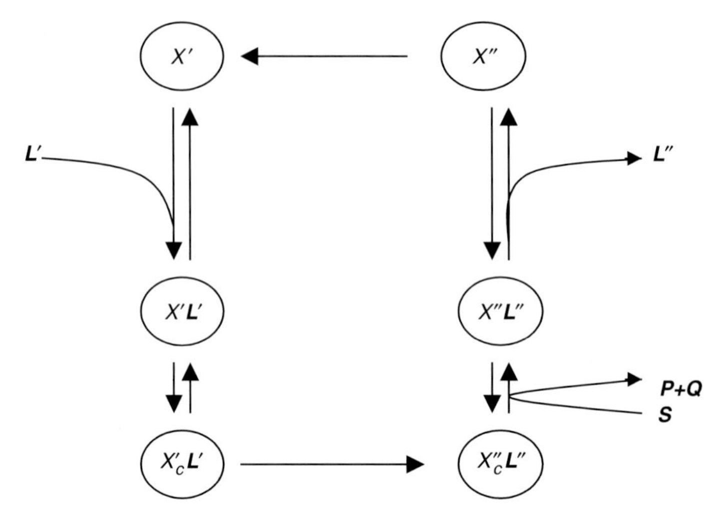


Fig. 6. Active transport of a ligand coupled to ATP hydrolysis. ATP hydrolysis ($S \rightarrow P + Q$) drives the active transport of ligand $L$ from cis to trans compartment.

Moreover thermodynamics imposes that

$$K_2 K_1' = K_1 K_1'' \tag{34}$$

Hence expression (33) is equivalent to

$$\frac{[L']}{[L'']} = \exp\left(-\frac{\Delta\tilde{\mu}_L - zF\Delta\Psi}{RT}\right) > 1 \tag{35}$$

As expected, this relationship is identical to Eq. (26).

An exergonic chemical reaction such as the hydrolysis of ATP can drive the migration of ions against an electrochemical gradient. The kinetic model that describes this situation is shown in Fig. 6. It implies that conformation changes of the carrier and ion migration are driven by the hydrolysis of ATP. One can derive from this model the following fractionation factors

$$\begin{aligned} f_1' &= \frac{1}{1 + K_1'[L'] + K_1'K_2'[L']} \qquad & f_1'' &= \frac{1}{1 + K_1''[L''] + K_1''K_2''[L'']} \\ f_3' &= \frac{K_1'K_2'[L']}{1 + K_1'[L'] + K_1'K_2'[L']} \qquad & f_3'' &= \frac{K_1''K_2''[L'']}{1 + K_1''[L''] + K_1''K_2''[L'']} \end{aligned} \tag{36}$$

It is therefore possible to write the kinetic model in a compact form and to derive the relevant steady-state rate of the net flow which can be written as

$$J = \frac{k_1 f_1' k_{-3} f_3'' \left(\frac{k_{-1} f_1'' k_3 f_3'}{k_1 f_1' k_{-3} f_3''} - 1\right)}{k_{-1} f_1'' + k_1 f_1' + k_{-3} f_3'' + k_3 f_3} \tag{37}$$

From this expression it is obvious that the net flow will be oriented cis–trans if

$$\frac{k_{-1}f_1''k_3f_3'}{k_1f_1'k_{-3}f_3''} > 1 \tag{38}$$

which is equivalent to

$$\frac{K_3K_2'K_1'}{K_1K_1''K_2''}\frac{[L']}{[L'']}\frac{[S]}{[P][Q]} > 1 \tag{39}$$

where $S$, $P$, and $Q$ are ATP, ADP, and phosphate, respectively. As thermodynamics requires that

$$K_1'K_2'K_3 = K_1K_1''K_2'' \tag{40}$$

expression (39) reduces to

$$\frac{[L']}{[L'']}\frac{[S]}{[P][Q]} > 1 \tag{41}$$

Let us consider the hydrolysis of ATP into ADP and phosphate, i.e.,

$$S \overset{\Delta G^0}{\rightleftarrows} P + Q$$

the equilibrium is shifted towards the right and the standard free energy $\Delta G^0$ is negative. The affinity of ATP hydrolysis is then

$$A_S = -\left(\Delta G^0 + RT\ln\frac{[P][Q]}{[S]}\right) \tag{42}$$

and the ratio $[P][Q]/[S]$ can be expressed as

$$\frac{[P][Q]}{[S]} = \exp\left(-\frac{\Delta G^0 + A_S}{RT}\right) \tag{43}$$

Similarly the reciprocal of this ratio assumes the form

$$\frac{[S]}{[P][Q]} = \exp\left(\frac{\Delta G^0 + A_S}{RT}\right) \tag{44}$$

and expression (41) can be rewritten as

$$\frac{[L']}{[L'']}\frac{[S]}{[P][Q]} = \exp\left\{\frac{-(\Delta\tilde{\mu}_L - zF\Delta\Psi) + \Delta G^0 + A_S}{RT}\right\} > 1 \tag{45}$$

In order to be fulfilled, this expression requires that

$$-(\Delta\tilde{\mu}_L - zF\Delta\Psi) + \Delta G^0 + A_S > 0 \tag{46}$$

which is equivalent to

$$zF\Delta\Psi - \Delta\tilde{\mu}_L + RT\ln\frac{[S]}{[P][Q]} > 0 \tag{47}$$

Ion transport against an electrochemical gradient implies that

$$zF\Delta\Psi - \Delta\tilde{\mu}_L < 0 \tag{48}$$

and expression (47) will be fulfilled if $RT\,\mathrm{ln}\{[S]/[P][Q]\}$ is positive and large enough. This situation can be expected to occur only if the ATP concentration is large relative to ADP and phosphate concentrations. This implies that ATP hydrolysis is shifted relative to its thermodynamic equilibrium.

The same reasoning can also be used to explain the synthesis of ATP from ADP and phosphate but then expression (41) has to be replaced by

$$\frac{[L']}{[L'']}\frac{[P][Q]}{[S]} > 1 \tag{49}$$

This expression implies that

$$zF\Delta\Psi - \Delta\tilde{\mu}_L - (\Delta G^0 + A_S) > 0 \tag{50}$$

which is equivalent to

$$zF\Delta\Psi - \Delta\tilde{\mu}_L + RT\ln\frac{[P][Q]}{[S]} > 0 \tag{51}$$

In the present case

$$zF\Delta\Psi - \Delta\tilde{\mu}_L > 0 \tag{52}$$

hence relation (51) will be fulfilled even if the ratio $[P][Q]/[S]$ is smaller than one. It therefore appears that the electrochemical gradient drives ATP synthesis well beyond its equilibrium concentration. These theoretical considerations offer a sensible explanation to the fact that in the living cell, and in particular in mitochondria, the ATP concentration is much larger than what would have been expected on the basis of a simple thermodynamic equilibrium between ATP and ADP plus phosphate. Again, there is no doubt that the important processes of energy conversion in the living cell take place in the living cell *only* because at least some important chemical reactions significantly depart from thermodynamic equilibrium.

## 5. *Actin filaments and microtubules are nonequilibrium structures*

Actin filaments and microtubules are dynamic structures that depart from thermodynamic equilibrium. Both are supramolecular edifices made up of specific proteins. A microtubule is a hollow tube consisting of 13 protofilaments composed of alternating $\alpha$ and $\beta$ tubulin. Moreover microtubules are not permanent entities of the cell. They grow and shrink through polymerization of tubulin and depolymerization of microtubules. As a matter of fact, microtubules are supramolecular structures that form the transient mitotic spindle. Actin filaments are made up of globular actin (G-actin), a protein consisting of a single polypeptide chain. G-actin polymerizes as a left-handed helix called the actin filament.

Of particular interest for the present chapter is the mechanism of polymerization-depolymerization of microtubules and actin filaments. Broadly speaking, the mechanism is very similar for the two supramolecular edifices. The two ends of a microtubule are not equivalent. One end, called "plus", grows thanks to polymerization whereas the other end, called "minus", depolymerizes. This process is controlled by guanosine triphosphate (GTP) hydrolysis and exchange. $\alpha$ tubulin binds GTP which can then be neither hydrolyzed nor exchanged. $\beta$ tubulin can also bind GTP but is soon hydrolyzed to guanosine diphosphate (GDP) which is then exchanged for another GTP molecule present in the medium. After GTP has been hydrolyzed to GDP the microtubule tends to disassemble. A somewhat similar situation is observed for actin filaments, but in this case ATP replaces GTP. Of particular interest, is the fact that both actin filaments and microtubules can be steady state polymers, i.e., they grow at one end and disassemble at the other end at the same rate, in such a way that their length remains constant and that newly incorporated monomers move along the polymer from the "minus" to the "plus" end. This phenomenon is called treadmilling [49,50].

The kinetics and thermodynamics of treadmilling can be discussed in physico-chemical terms [49,50]. At both ends of an actin filament two events are taking place. The first one is the association of a monomer of actin bearing an ATP molecule, designated $A_S$ (not to be confused with the affinity of a chemical reaction), with the $\alpha$ end of an actin filament. During the polymerization process ATP ($S$) is hydrolyzed and both ADP ($P$) and phosphate ($Q$) remain bound to the actin monomer

$$(\alpha) - A_P - A_P + A_S \underset{\alpha_{-1}}{\overset{\alpha_1}{\rightleftarrows}} (\alpha) - A_P - A_P - A_{PQ}$$

The same process takes place, but with a different rate, at the $\beta$ end of the actin filament, i.e.,

$$(\beta) - A_P - A_P + A_S \underset{\beta_{-1}}{\overset{\beta_1}{\rightleftarrows}} (\beta) - A_P - A_P - A_{PQ}$$

The rate constants $\alpha_1$ and $\alpha_{-1}$, $\beta_1$ and $\beta_{-1}$ are of course different but, if the actin filament is in steady state, i.e., if its length is constant, the ratios of the rate constants should be equal

$$\frac{\alpha_1}{\alpha_{-1}} = \frac{\beta_1}{\beta_{-1}} \tag{53}$$

The second type of event that occurs at the two ends of the polymer is the dissociation of the last actin monomer bearing ADP and phosphate ($A_{PQ}$) and the simultaneous dissociation of these two ligands from the monomer, namely

$$(\alpha) - A_P - A_P - A_{PQ} \rightleftarrows A + P + Q + (\alpha) - A_P - A_P$$

$$(\beta) - A_P - A_P - A_{PQ} \rightleftarrows A + P + Q + (\beta) - A_P - A_P$$

The dissociation process at the two ends of the actin filament allows the binding of ATP to the free actin monomer, i.e.,

$$A + S \rightleftarrows A_S$$

Hence the overall process at either end of a filament is

$$(\alpha) - A_P - A_P - A_{PQ} + S \underset{\alpha_{-2}}{\overset{\alpha_2}{\rightleftarrows}} (\alpha) - A_P - A_P + P + Q + A_S$$

or

$$(\beta) - A_P - A_P - A_{PQ} + S \underset{\beta_{-2}}{\overset{\beta_2}{\rightleftarrows}} (\beta) - A_P - A_P + P + Q + A_S$$

Again $\alpha_2 \neq \beta_2$ and $\alpha_{-2} \neq \beta_{-2}$ but if the actin filament is in steady state

$$\frac{\alpha_2}{\alpha_{-2}} = \frac{\beta_2}{\beta_{-2}} \tag{54}$$

If $c_{AS}$ is the concentration of the monomer $A_S$, the dynamics of the polymerization process can be described by the following equations

$$\begin{aligned} \frac{dn_\alpha}{dt} &= (\alpha_1 + \alpha_{-2})c_{AS} - (\alpha_{-1} + \alpha_2) \\ \frac{dn_\beta}{dt} &= (\beta_1 + \beta_{-2})c_{AS} - (\beta_{-1} + \beta_2) \end{aligned} \tag{55}$$

where $n_\alpha$ and $n_\beta$ are the number of monomers added to the $\alpha$ end and $\beta$ end of the actin filament. The rate of motion of a monomer along a polymer, i.e., the treadmill flow $\tilde{J}$ is therefore

$$\tilde{J} = \frac{dn_\alpha}{dt} = -\frac{dn_\beta}{dt} \tag{56}$$

If $\bar{c}_{AS}$ is the steady-state concentration of the monomer $A_S$ one has

$$(\alpha_1 + \alpha_{-2})\bar{c}_{AS} - (\alpha_{-1} + \alpha_2) = -(\beta_1 + \beta_{-2})\bar{c}_{AS} + (\beta_{-1} + \beta_2) \tag{57}$$

or

$$\bar{c}_{AS} = \frac{\alpha_2 + \beta_2 + \alpha_{-1} + \beta_{-1}}{\alpha_1 + \beta_1 + \alpha_{-2} + \beta_{-2}} \tag{58}$$

The expression for the treadmill flow can now be expressed as

$$\tilde{J} = \frac{\alpha_1\beta_{-1} - \alpha_{-1}\beta_1 + \alpha_{-2}\beta_2 - \alpha_2\beta_{-2} + \alpha_1\beta_2 - \alpha_2\beta_1 + \alpha_{-2}\beta_{-1} - \alpha_{-1}\beta_{-2}}{\alpha_1 + \beta_1 + \alpha_{-2} + \beta_{-2}} \tag{59}$$

From Eqs. (53) and (54) one has

$$\alpha_1\beta_{-1} = \alpha_{-1}\beta_1 \qquad \alpha_2\beta_{-2} = \alpha_{-2}\beta_2 \qquad \frac{\alpha_1\alpha_2}{\alpha_{-1}\alpha_{-2}} = \frac{\beta_1\beta_2}{\beta_{-1}\beta_{-2}} \tag{60}$$

and Eq. (59) reduces to

$$\tilde{J} = \frac{(\alpha_1\beta_2 - \alpha_2\beta_1)\left(1 - \frac{\alpha_{-1}\alpha_{-2}}{\alpha_1\alpha_2}\right)}{\alpha_1 + \beta_1 + \alpha_{-2} + \beta_{-2}} \tag{61}$$

Combining reactions

$$(\alpha) - A_P - A_P + A_S \underset{\alpha_{-1}}{\overset{\alpha_1}{\rightleftarrows}} (\alpha) - A_P - A_P - A_{PQ}$$

and

$$(\alpha) - A_P - A_P - A_{PQ} + S \underset{\alpha_{-2}}{\overset{\alpha_2}{\rightleftarrows}} (\alpha) - A_P - A_P + P + Q + A_S$$

already referred to leads to ATP dephosphorylation, namely

$$S \underset{\alpha_{-1}\alpha_{-2}}{\overset{\alpha_1\alpha_2}{\rightleftarrows}} P + Q$$

and the same reasoning would apply to the $\beta$ end of the actin filament. What these results show is that it is hydrolysis of ATP that drives the treadmill flow. Moreover this process takes place away from thermodynamic equilibrium. Equation (61) also shows that if $\alpha_1 = \beta_1$ and $\alpha_2 = \beta_2$ the polymer is not polarized and therefore no treadmill can ever occur.

## *6. The mitotic spindle is a dissipative structure*

Probably one of the best illustration of the idea that supramolecular edifices in the living cell are not in thermodynamic equilibrium is offered by the mitotic spindle. This transient supramolecular system is present during cell division and associates microtubules and chromosomes. Association of the two partners occurs thanks to kinetochores, specific regions of the chromosomes. The building up of the mitotic spindle basically relies on the activity of two types of motor proteins, namely dyneins and kinesins. Motor proteins can bind to supramolecular edifices such as microtubules and slide along them. Dyneins slide towards the minus end whereas kinesins move in the direction of the plus end. Motion of motor proteins along a fibrous structure requires consumption of ATP. In fact, motor proteins have a globular head that binds and hydrolyzes ATP thus providing the energy required for the motor to move. These simple properties are sufficient to explain auto-organization of the mitotic spindle (Fig. 7). Kinesins are bound to both a chromosome and a microtubule whereas dyneins are bound to (at least) two microtubules, so one can easily understand that as soon as dyneins slide towards the minus end and kinesins move in the opposite direction of the microtubules, a spindle will spontaneously form under the influence of these forces. But this process requires energy, so much so that mitotic spindles can be considered

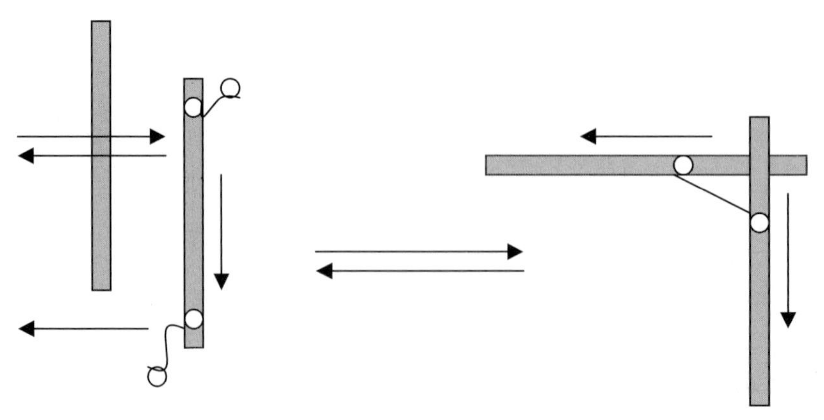

Fig. 7. Schematic representation of the building up of a mitotic spindle. Left: Kinesins reversibly bind to microtubules. Right: Kinesins slide along two microtubules thus leading to the building up of a spindle. Adapted from Ref. [51].

dissipative structures. Since microtubules can disassemble, the mitotic spindle cannot be a permanent entity of the cell. The formation of the mitotic spindle has been observed *in vitro* and modeled [51].

## 7. *Interactions with the environment, nonequilibrium, and emergence in biological systems*

As already outlined in this chapter, the application of analytic and reductionist approaches to biological phenomena relies on the view that the global, collective properties of a system are already present, in a potential state, in the structure and the function of at least some of the elements of the system. There is little doubt that most studies aimed at deciphering the structure of large genomes were performed with the belief that most structural and functional properties of living organisms were encoded in these genomes. One has to realize that a dynamic state, for instance a steady state, of a very simple system implies that this system interacts with its surroundings. Hence the properties that can be observed and studied under these conditions do not refer solely to this system considered in isolation, but also to the interactions it displays with its environment. It is therefore impossible to hold the view that all the global properties of a large system made up of many simple systems are encoded in the structure of the component sub-systems. Hence one is led to the idea that indeed novel properties should *emerge* out of the interactions of the component sub-systems.

These general ideas have received support from the experimental data referred to in this chapter. Even systems as simple as enzymes that catalyze chemical reactions display properties that are qualitatively different depending on whether the reactions are in equilibrium or in steady state. It is therefore impossible to speak of the *intrinsic* properties of the enzymes without taking account of the interactions they have with their environment. The fact that there is in general no simple relationship between the substrate binding isotherm and the steady-state equation of an enzyme illustrates this point.

In the same vein, ATP synthesis in mitochondria can be conceived of and explained *only* because there is a coupling between ATP-synthase, the enzyme responsible for ATP synthesis, and the electrochemical potential [52]. Hence ATP synthesis *emerges* out of this coupling. The activity of ATP-synthase alone could in no way have explained ATP synthesis. It is the merit of Mitchell [52], to have shown that it is precisely the interaction between two different physico-chemical events that generates this *novel* remarkable property.

The dynamics of actin filaments and microtubules offers another striking example of a property that can in no way be understood if one does not take into account the interactions between actin filaments (or microtubules) and their surroundings. The process of treadmilling, i.e., the "motion" of actin molecules along an actin filament can be explained in physical terms only through the interaction of this supramolecular edifice with ATP consumption. It is evident that the property of treadmilling is "written" neither in the structure of actin filaments nor in the

50. Hill, T.L. (1980) Bioenergetic aspects and polymer length distribution in steady state head-to-tail polymerization of actin and microtubules. Proc. Natl. Acad. Sci. USA 77, 4803–4807.
51. Surrey, Th., Nedelec, F., Leibler, S., and Karsenti, E. (2001) Physical properties determining self-organization of motors and microtubules. Science 292, 1167–1171.
52. Mitchell, P. (1961) Coupling of phosphorylation to electron and hydrogen transfer by a chemi-osmotic type of mechanism. Nature 191, 144–148.
53. Kauffman, S.A. (1993) The Origins of Order. Oxford University Press, Oxford.
54. Kauffman, S.A. (1996) At Home in the Universe. Penguin Books, London.
55. Bak, P. (1996) How Nature Works. MIT Press, Cambridge.
56. Goldbeter, A. (1996) Biochemical Oscillations and Cellular Rhythms. The Molecular Bases of Periodic and Chaotic Behaviour. Cambridge University Press, Cambridge.
57. Callagher, R. and Appenzeller, T. (1999) Beyond reductionism. Science 284, 79.
58. Goldenfeld, N. and Kadanoff, L.P. (1999) Simple lessons from complexity. Science 284, 87–89.
59. Whitesides, G.R. and Ismagilov, R.F. (1999) Complexity in chemistry. Science 284, 89–92.
60. Parish, J.K. and Edelstein-Keshet, L. (1999) Complexity, patterns and evolutionary trade offs in animal aggregation. Science 284, 99–101.
61. Arthur, W.B. (1999) Complexity and the economy. Science 284, 107–109.
62. Albert, R. and Barabasi, A.L. (2002) Statistical mechanics of complex networks. Rev. Mod. Phys. 74, 47–97.
63. Atwell, L.H., Hopfield, J.J., Leibler, S., and Murray, A.W. (1999) From molecular to modular cell biology. Nature 402 Suppl, C47–C52.

J. Ricard *Emergent Collective Properties, Networks and Information in Biology*

DOI: 10.1016/S0167-7306(05)40002-2

CHAPTER 2

# Mathematical prelude: elementary set and probability theory

J. Ricard

The study of networks, information, organization, and emergence in biochemical systems requires some knowledge in mathematics. The present chapter offers a brief presentation of set algebra and probability theory. It emphasizes the study of classical distribution functions and discusses the properties of moments and cumulants of probability distributions.

*Keywords:* binomial distribution, bivariate moments, characteristic function, cumulants, covariance, distribution function, image, Laplace–Gauss distribution, mapping, mapping into, mapping onto, Markov process, mean of the binomial distribution, mean of the Laplace–Gauss distribution, mean of the Poisson distribution, moments, monovariate moments, one-to-one mapping, Poisson distribution, probability, relations and graphs, sets, Stieltjes integrals, variance of the binomial distribution, variance of the Laplace–Gauss distribution, variance of the Poisson distribution.

Most chapters of this book will be devoted to networks, information, organization and emergence. This study requires some preliminary knowledge of set and probability theory. The aim of this chapter is to offer the bases of these theories.

## 1. *Set theory*

### 1.1. *Definition of sets*

A set is a collection of objects called elements of the set [1,2]. Usually, one uses capital letters to denote sets and small letters to designate the elements of the set. One expresses the fact that an element $a$ belongs to a set $A$ by

$$a \in A \tag{1}$$

If $a$ is not an element of $A$, one writes

$$a \notin A \tag{2}$$

A set, $X$, is denoted by its elements $x \in X$ that have a certain property $P(x)$. If $Y$ is the set of all the $x$'s, one writes

$$Y = \{x; x \in X, P(x)\} \tag{3}$$

Thus for instance if $x$ belongs to the set of natural numbers $N$, to the set of integers $Z$ and to the set of positive integers and zero $Z^+$ one has

$$\begin{aligned} Y &= \{x; x \in N\} \\ Y &= \{x; x \in Z\} \\ Y &= \{x; x \in Z^+\} \end{aligned} \tag{4}$$

If a set, $A$, only contains elements belonging to another set $X$ one has

$$A \subset X \tag{5}$$

and $A$ is defined as a subset of $X$. Any set can be defined as a subset of another set and *in fine* of a universal set $U$. Thus for instance, the set of rational numbers is a subset of the set of real numbers. A set that contains no element is called an empty set $\emptyset$. Two sets $X$ and $Y$ are defined as equal if, and only if,

$$X \subset Y \quad \text{and} \quad Y \subset X \tag{6}$$

Equality is then expressed by

$$X = Y \tag{7}$$

This means that every element of $X$ is an element of $Y$, and conversely.

### *1.2. Operations on sets*

A first operation that can be performed on sets is their union. The union of two sets $A$ and $B$ is the set of all the elements $x$ that belong either to set $A$ or to set $B$, or to both. This is written as

$$A \cup B = \{x; x \in A \text{ or } x \in B\} \tag{8}$$

A second operation is the intersection of $A$ and $B$. It is the set of all the elements $x$ that belong to both $A$ and $B$. The intersection of $A$ and $B$ is then expressed as

$$A \cap B = \{x; x \in A \text{ and } x \in B\} \tag{9}$$

The elements $x$ that belong to the universal set $U$ but which do not belong to the set $A$ is the complement, $A^c$, of $A$. Hence one has

$$A^c = \{x; x \in U, x \notin A\} \tag{10}$$

Relationships (10), (11) and (12) are illustrated in Fig. 1.

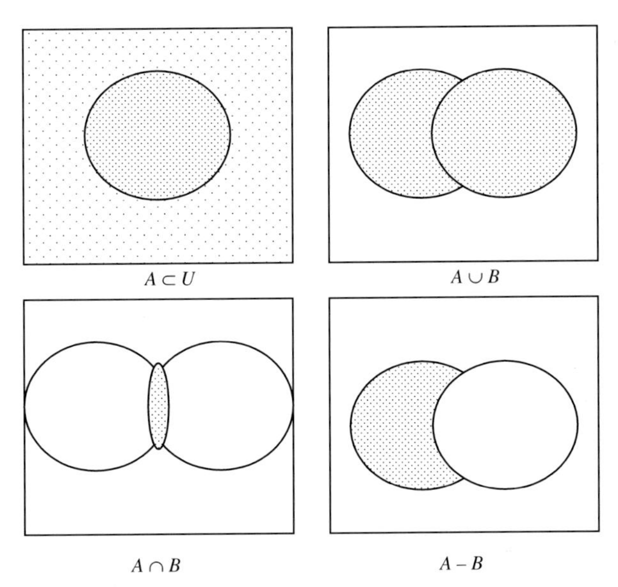


Fig. 1. Operations on sets – Top left: The set $A$ (dark) is a subset of the universal set $U$. Top right: The union (dark) of the two sets $A$ and $B$. Bottom left: The intersection (dark) of the two sets $A$ and $B$. Bottom right: The difference (dark) between the two sets $A$ and $B$.

The difference between two sets $A$ and $B$ is defined as the set of the elements $x$ that belong to $A$ but not to $B$ i.e.,

$$A - B = \{x; x \in A, x \notin B\} \tag{11}$$

This relationship is illustrated in Fig. 1. The union of sets follows the following rules [2]

$$\begin{aligned} A \cup (B \cup C) &= (A \cup B) \cup C && \text{(associativity)} \\ A \cup B &= B \cup A && \text{(commutativity)} \\ A \cup (B \cap C) &= (A \cup B) \cap (A \cup C) && \text{(distributivity)} \\ A \cup \emptyset &= A \\ A \cup A^c &= U \end{aligned} \tag{12}$$

These relationships are valid regardless the nature of the universal set $U$. Somewhat similar relationships hold for the intersection of sets i.e.,

$$\begin{aligned} A \cap (B \cap C) &= (A \cap B) \cap C && \text{(associativity)} \\ A \cap B &= B \cap A && \text{(commutativity)} \\ A \cap (B \cup C) &= (A \cap B) \cup (A \cap C) && \text{(distributivity)} \\ A \cap U &= A \\ A \cap A^c &= \emptyset \end{aligned} \tag{13}$$

The difference of sets has the following properties

$$\begin{aligned} &A - B = \emptyset \quad \text{if, and only if,} \quad A \subset B \\ &A - B = A \quad \text{if, and only if,} \quad A \cap B = \emptyset \\ &A - (B \cup C) = (A - B) \cap (A - C) \\ &A - (B \cap C) = (A - B) \cup (A - C) \end{aligned} \tag{14}$$

The cartesian product of two sets $X$ and $Y$ is the set of all the ordered pairs $(xy)$ with $x \in X$ and $y \in Y$.

### *1.3. Relations and graphs*

In general, a relation $R$ is defined as a definite subset of the cartesian product of two sets $X$ and $Y$ i.e.,

$$R \subset X \times Y \tag{15}$$

In this relation, $X$ is called the domain and $Y$ the range. The subset $R$ is made up of pairs $(x_i y_j)$ that are taken in a definite order (except if $x_i = y_j$) in such a way that

$$x_i R y_j \neq y_j R x_i \tag{16}$$

The letter $R$ expresses some kind of rule that defines the relation between $x_i$ and $y_j$. It is possible to express this relation by a graph. If for instance

$$R = i + j \leq 4 \quad (i, j \in N) \tag{17}$$

the points corresponding to the elements of the two sets ($x \in X$ and $y \in Y$) are called the nodes of the graph and connected as shown in Fig. 2.

In its usual meaning, the term graph refers to the elements of the same set. In this perspective, a graph $G$ is defined as a subset of the cartesian product of set $S$ by itself, i.e.,

$$G \subset S \times S \tag{18}$$

then the graph will be defined by the relation $x_i R x_j$. As previously, $R$ defines a rule that associates $x_i$ and $x_j$. For instance, if

$$R = i < j \quad (i, j \in N) \tag{19}$$

and if

$$S = \{1, 2, 3, 4\} \tag{20}$$

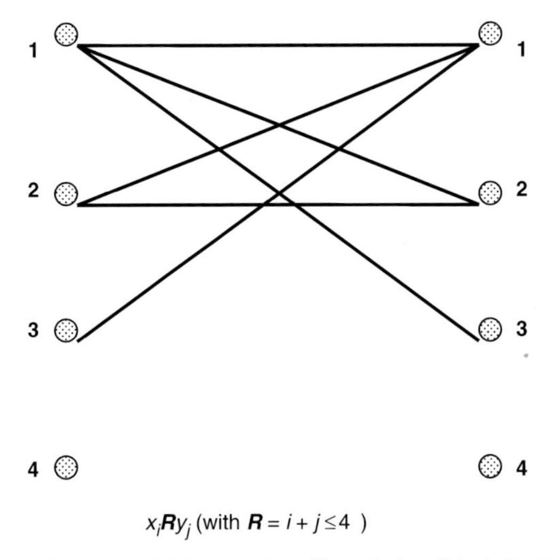


Fig. 2. A relation between discrete variables $x_i$ and $y_j$. The relation $\boldsymbol{R}$ is defined by $i+j \leq 4$.

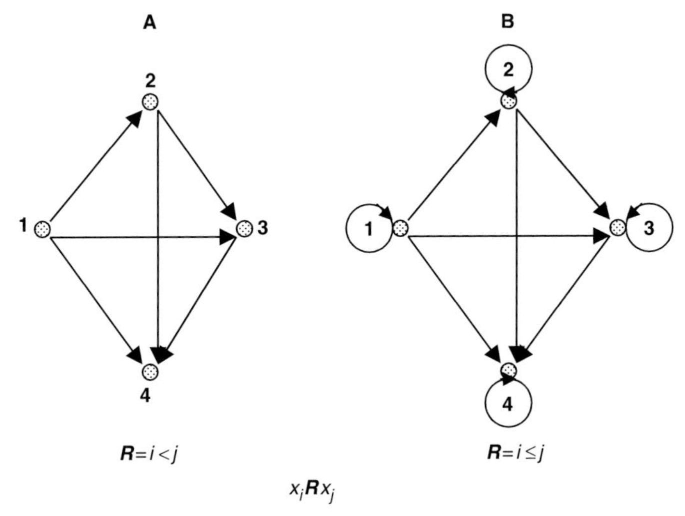


Fig. 3. Two examples of graphs. The topology of the graphs is based on either relation $\boldsymbol{R}=i<j$(A) or relation $\boldsymbol{R}=i \leq j$(B).

The graph $G$ will be made up of the pairs

$$G = \{1, 2;\ 1, 3;\ 1, 4;\ 2, 3;\ 2, 4;\ 3, 4\} \tag{21}$$

and each pair is a vertex of the graph which is shown in Fig. 3A. Relation (21) can be extended to the case where $x_i = x_j$, then

$$R = i \leq j \quad (i, j \in N) \tag{22}$$

and the graph is then defined by

$$G = \{1, 1; 1, 2; 1, 3; 1, 4; 2, 2; 2, 3; 2, 4; 3, 3; 3, 4; 4, 4\} \tag{23}$$

and is shown in Fig. 3B.

### *1.4. Mapping*

Mapping is a special type of relation that associates *all the elements*, $x$, of a set $X$, called the domain, to the elements of another set $Y$ in such a way that no element $x$ can be connected to several elements $y$. This operation is designated by $f: X \to Y$. As with any relation, a mapping is a subset of the cartesian product of set $X$ by set $Y$ that associates with each element $x \in X$ *a unique image* $y \in Y$ through the relation $fx = y$. But some elements $y$ may remain unconnected. More specifically, the image of $X$ by $f$ is defined as

$$fX = \{y \in Y; y = fx, x \in X\} \tag{24}$$

The ordered pairs $(xy)$ as defined by the relation $fx = y$ can be considered either a function or a graph. The range of the function consists of the elements of $Y$ involved in the mapping $f: X \to Y$. If the range is a subset of $Y$ then $X$ is said to be mapped *into* $Y$ (Fig. 4). But the range can also be the entire set $Y$, and $X$ will be mapped *onto* $Y$ (Fig. 4). Last, there may exist an one-to-one correspondence of the elements of $X$ and $Y$ (Fig. 4).

This is related to the question of whether inverse mapping is meaningful. In fact, this means that one wishes to map the image onto the original set. As the function $f$ is a subset of the cartesian product $X \times Y$, one should expect the inverse function $f^{-1}$ to be a subset of the product $Y \times X$. If $f$ is one-to-one, then each element of $X$ corresponds to an element of $Y$. Hence the inverse function $f^{-1}$ requires that each element of $Y$ corresponds to an element of $X$. This can be set as

$$\begin{aligned} f &= \{(x, y) \in X \times Y; (y, x) \in f^{-1}\} \\ f^{-1} &= \{(y, x) \in Y \times X; (x, y) \in f\} \end{aligned} \tag{25}$$

This condition is the only one that allows inverse mapping. We shall see later that the problem of mapping is of crucial importance for understanding the logic of communication processes.

## *2. Probabilities*

### *2.1. Axiomatic definition of probability and fundamental theorems*

Let us consider a set, $\Omega$, composed of elements of equal likelihood. This set is called the probability sample space [3–5]. Let us consider in this space disjoint subsets

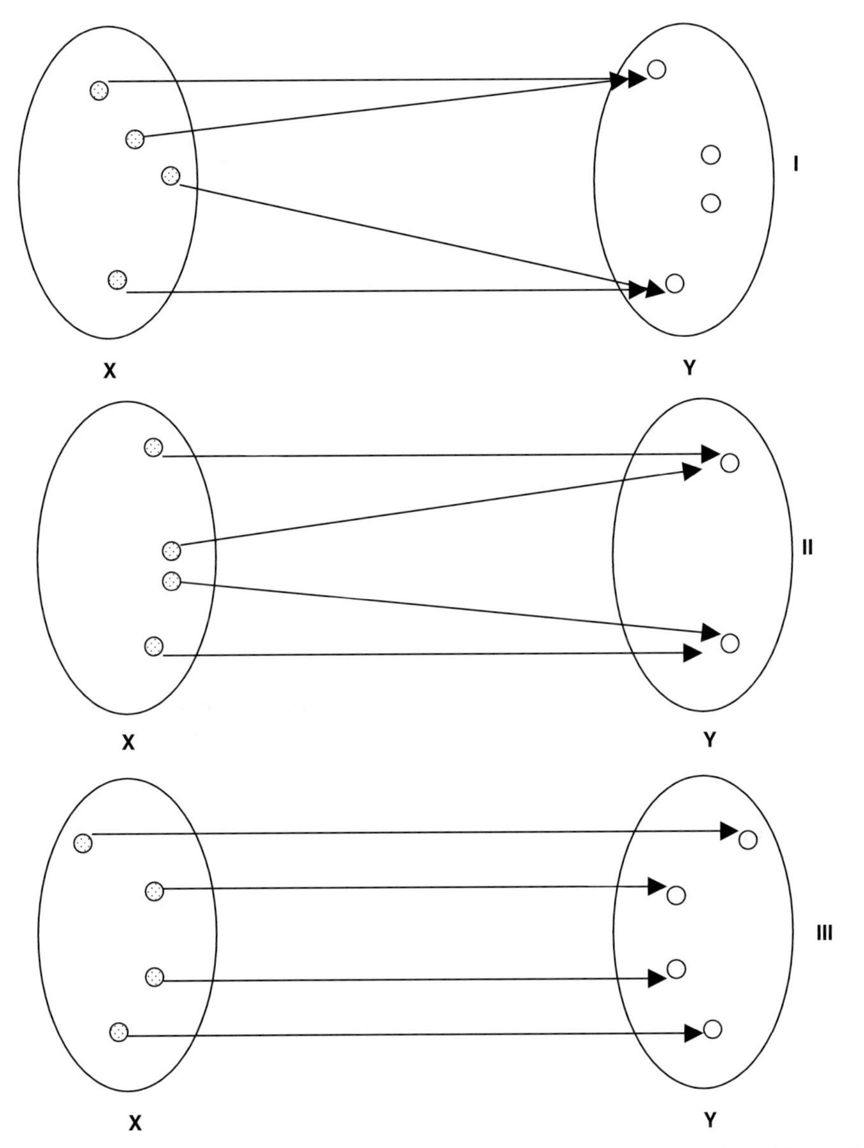


Fig. 4. Mapping – **I** Mapping **X** into **Y**. **II** Mapping **X** onto **Y**. **III** One-to-one mapping of **X** onto **Y**.

$A, B, \ldots$ i.e., subsets such that

$$A \cap B = \emptyset \tag{26}$$

In this discrete system, probabilities are numbers that characterize subsets $A, B, \ldots$ as well as the set $\Omega$ itself. Let us call these numbers $p(A), p(B), \ldots$ and $p(\Omega)$. The probabilities should meet the three following axioms [3–5].

For every subset $A, B, \ldots$ one should have

$$0 \leq p(A) \leq 1,\ 0 \leq p(B) \leq 1,\ \ldots \tag{27}$$

The probability of the union of the disjoint subsets should be equal to the sum of the probabilities of the subsets i.e.,

$$p(A \cup B \cup \ldots) = p(A) + p(B) + \cdots \tag{28}$$

The probability of the set $\Omega$ is one, i.e.,

$$p(\Omega) = 1 \tag{29}$$

The only expression of a probability that meets these axioms is (for subset $A$ for instance)

$$p(A) = \frac{n(A)}{n(\Omega)} \tag{30}$$

where $n(A)$ is the number of elements of the subset A and $n(\Omega)$ that of the probability sample space. $n(A)$ and $n(\Omega)$ are also called cardinal of $A$, $|A|$, and cardinal of $\Omega$, $|\Omega|$. A similar expression indeed holds for $p(B)$. Now let us consider two subsets, $A$ and $B$, of the set $\Omega$ (Fig. 5). Let us now assume that $A$ and $B$ overlap. Then, one has

$$\begin{aligned} n(A) &= n(A \cap B) + n(A \cap B^c) \\ n(B) &= n(A \cap B) + n(A^c \cap B) \end{aligned} \tag{31}$$

and

$$n(\Omega) = n(A \cap B) + n(A \cap B^c) + n(A^c \cap B) + n(A^c \cap B^c) \tag{32}$$

where $n(A)$, $n$(B), and $n(\Omega)$ are still the number of elements of $A$, $B$ and $\Omega$. Likewise, $n(A \cap B)$, $n(A \cap B^c)$, $n(A^c \cap B)$, and $n(A^c \cap B^c)$ are the numbers of elements of the

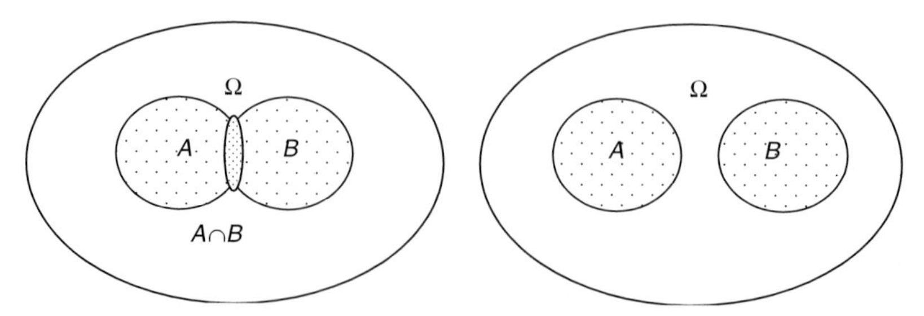


Fig. 5. Probabilities and sets. Left: The two subsets $A$ and $B$ overlap. Right: The two subsets $A$ and $B$ do not overlap. See text.

corresponding subsets. One has

$$p(A \cup B) = \frac{n(A \cap B) + n(A \cap B^c)}{n(\Omega)} + \frac{n(A \cap B) + n(A^c \cap B)}{n(\Omega)} - \frac{n(A \cap B)}{n(\Omega)} \tag{33}$$

which is equivalent to

$$p(A \cup B) = p(A) + p(B) - p(A \cap B) \tag{34}$$

It is then evident that if subsets $A$ and $B$ are disjoint, relationship (34) reduces to axiom (28).

Let us now consider the probability of occurrence of $B$ given that $A$ is known. This probability, which is called *conditional probability,* is written $p(B/A)$. It can be derived from Eq. (31). One has

$$p(B/A) = \frac{n(A \cap B)}{n(A \cap B) + n(A \cap B^c)} \tag{35}$$

Similarly, one can derive the expression of $p(A/B)$, i.e., the conditional probability of $A$ given that $B$ is known, from Eq. (31)

$$p(A/B) = \frac{n(A \cap B)}{n(A \cap B) + n(A^c \cap B)} \tag{36}$$

Dividing numerators and denominators of Eqs. (35) and (36) by $n(\Omega)$ yields

$$p(B/A) = \frac{\dfrac{n(A \cap B)}{n(\Omega)}}{\dfrac{n(A \cap B) + n(A \cap B^c)}{n(\Omega)}} = \frac{p(A \cap B)}{p(A)} \tag{37}$$

and

$$p(A/B) = \frac{\dfrac{n(A \cap B)}{n(\Omega)}}{\dfrac{n(A \cap B) + n(A^c \cap B)}{n(\Omega)}} = \frac{p(A \cap B)}{p(B)} \tag{38}$$

Hence one has

$$p(A \cap B) = p(A)p(B/A) = p(B)p(A/B) \tag{39}$$

which is the celebrated Bayes relationship that associates joint and conditional probabilities. The joint probability, $p(A \cap B)$, is often written $p(AB)$.

So far the concept of probability has been defined for discrete systems. But a probability can also be given for a continuous function, $f(x)$, defined from $-\infty$ to $+\infty$. Let us assume that $f(x)$ is a probability density whose values are function

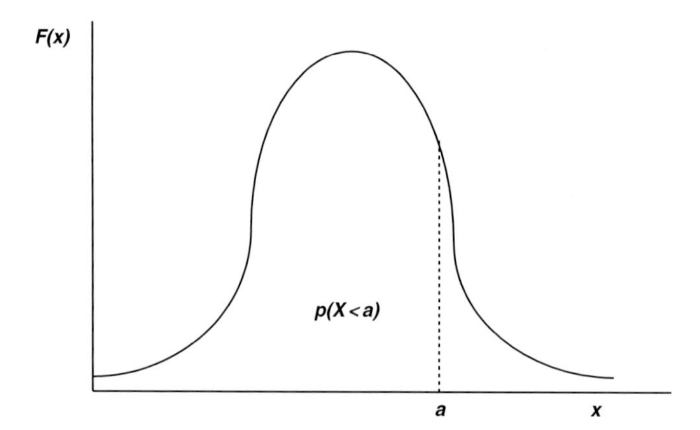


Fig. 6. Probabilities and continuous distribution functions. The probability $p(X < a) = F(a)$ is equal to the area comprised between the curve and the abscissa $x = a$. See text.

of $x$. The probability that a random variable $X$ be included in the interval defined by $x$ and $x + dx$ is

$$p(x < X < x + dx) = p(x)dx \tag{40}$$

And the probability that $X$ be smaller than $x$ is

$$p(X < x) = \int_{-\infty}^{x} p(x)dx = F(x) \tag{41}$$

The function $F(x)$ is called a *distribution function*. Its properties are important and will be briefly discussed in the next section. It follows from the definition of probability that

$$\int_{-\infty}^{+\infty} p(x)dx = 1 \tag{42}$$

and any probability value is equal to a part of the area comprised between the curve and the abscissa (Fig. 6).

## *2.2. Properties of the distribution function and the Stieltjes integral*

The distribution function $F(x)$ displays three properties:

it does not decrease with $x$;
it approaches zero when $x$ approaches $-\infty$ and one when $x$ approaches $+\infty$;

the distribution function of a discrete random variable is continuous on the left and discontinuous on the right.

It is this last property which is particularly important for it allows to express with the same mathematical formalism the parameters (the so-called moments) of a probability function whether this function is continuous or discontinuous. This property also serves as a basis for the definition of the Stieltjes integral [6].

Let us consider a discontinuous distribution of probability

$$F(x_i) = \sum_{i=-\infty}^{r} p(x_i) \tag{43}$$

or a continuous distribution of probability

$$F(x) = \int_{-\infty}^{x} p(x)dx \tag{44}$$

Thanks to the concept of the Stieltjes integral, it is possible to express both relationships (43) and (44) with the same mathematical formalism. It is therefore useful to recall what the Stieltjes integral is. Let us consider a continuous function $\Psi(x)$. It is well known that the definite integral of this function in the interval $(a, b)$ is

$$\int_{a}^{b} \Psi(x)dx = \mathrm{Lim} \sum_{i} \Psi(\xi_i)(x_{i+1} - x_i) \tag{45}$$

The limit involved in this expression is reached when the increment of $x_i$ approaches zero, i.e., $\Delta x_i \rightarrow 0$, moreover the discrete variable $\xi_i$ is such that $x_{i+1} < \xi_i < x_i$. Now let us consider a second function $F(x)$, which can be either continuous or discontinuous, and two values $F(x_{i+1})$ and $F(x_i)$ of this function. The Stieltjes integral, $I_s$, of $\Psi(x)$, defined with respect to $F(x)$, can be expressed as

$$I_s = \int_{a}^{b} \Psi(x)dF = \mathrm{Lim} \sum_{i} \Psi(\xi_i)\{F(x_{i+1}) - F(x_i)\} \tag{46}$$

The Stieltjes integral $I_s$ reduces to the ordinary integral

$$\int_{a}^{b} \Psi(x)p(x)dx \tag{47}$$

if $F(x)$ is continuous. In the interval $(-\infty, x)$ and when $\Psi(x) = 1$ Eq. (47) becomes identical to expression (44).

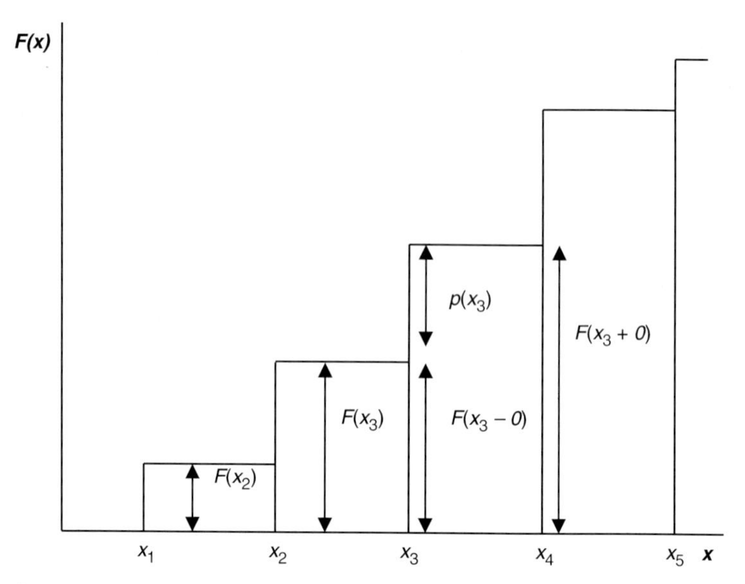


Fig. 7. The distribution function $F(x)$ of a discrete variable is continuous on the left and discontinuous on the right. The step function illustrates the fact that $F(x-0) = F(x)$ and $F(x+0) = F(x) + p(x)$. See text.

Let us now assume that $F(x)$ is a step-function as shown in Fig. 7. Simple inspection of this figure shows that on the left of a discontinuity, and close to it, $F(x)$ is continuous. Thus for instance

$$p(X < x_3) = F(x_3) = F(x_3 - 0) \tag{48}$$

and on the right of the same discontinuity, and close to it, $F(x)$ is discontinuous for

$$F(x_3 + 0) = F(x_3) + p(x_3) = F(x_4) \tag{49}$$

More generally speaking,

$$\begin{aligned} F(x-0) &= F(x) \\ F(x+0) &= F(x) + p(x) \end{aligned} \tag{50}$$

The first relation (50) expresses the continuity of $F(x)$ on the left and the second shows the discontinuity on the right. Hence the difference $F(x_{i+1}) - F(x_i)$ which appears in Eq. (46) can be expressed as

$$F(x_{i+1}) - F(x_i) = F(x_i + 0) - F(x_i) = dF(x_i) = p(x_i) \tag{51}$$

Similarly one has

$$F(x_i) - F(x_i - 0) = dF(x_i) = 0 \tag{52}$$

This implies that the Stieltjes integral for $F(x)$ discontinuous can be written as

$$I_s = \int_a^b \Psi(x)dF = \sum_i \Psi(x_i)p(x_i) \tag{53}$$

As previously, it follows that, for the interval $(-\infty, x)$, this expression reduces to Eq. (43) if $\Psi(x_i) = 1$. Hence one can expect that parameters of a distribution function should be expressed in terms of expressions involving Stieltjes integrals whether or not the function $F(x)$ is continuous.

Owing to the fact that the function $F(x)$ can be continuous or discontinuous, the expressions of probabilities that can be derived from this function depend on its continuous or discontinuous character. If $F(x)$ is continuous, we know that

$$p(a < X < b) = F(b) - F(a) \tag{54}$$

but if $F(x)$ is a step-function, as shown in Fig. 7, the situation is more complex. Thus for instance one can have for $a < b$

$$p(X < b) = p(X \leq a \quad \text{or} \quad a < X < b) \tag{55}$$

The theorem on the union of probabilities implies that

$$p(X < b) = p(X \leq a) + p(a < X < b) \tag{56}$$

and

$$F(b) = F(a + 0) + p(a < X < b) \tag{57}$$

Hence it follows that

$$p(a < X < b) = F(b) - F(a + 0) \tag{58}$$

which is to be compared to expression (54) that is valid for a continuous function. One can demonstrate in the same way that

$$\begin{aligned} p(a \leq X < b) &= F(b) - F(a) \\ p(a < X \leq b) &= F(b + 0) - F(a + 0) \\ p(a \leq X \leq b) &= F(b + 0) - F(a) \end{aligned} \tag{59}$$

These results are illustrated in Fig. 8.

## 3. *Probability distributions*

### 3.1. *Binomial distribution*

Most probability distributions derive from the so-called binomial distribution [6–10]. If one repeats many times an experiment with two possible outcomes called success

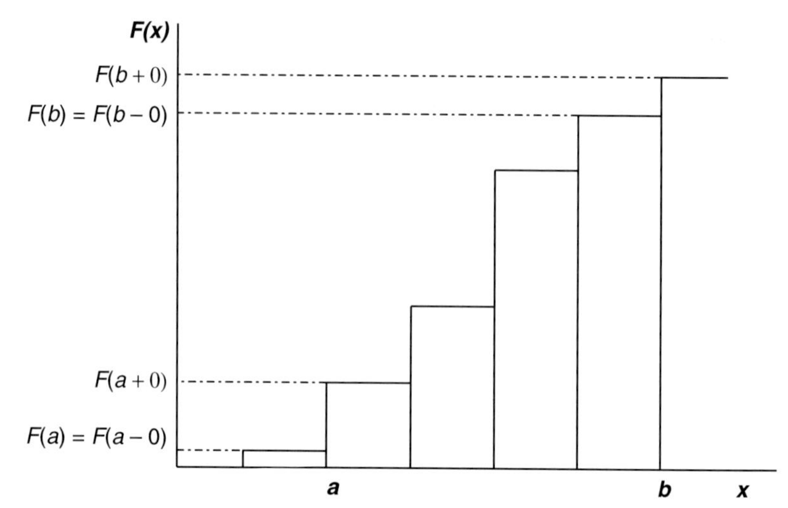


Fig. 8. Relations between probabilities and discrete distribution functions. This figure illustrates the following, equalities: $p(a \leq X < b) = F(b) - F(a)$; $p(a < X \leq b) = F(b+0) - F(a+0)$; $p(a \leq X \leq b) = F(b+0) - F(a)$. See text.

and failure, the probability of success is denoted $p$ and that of failure $q$. Indeed one should have $p+q=1$. The probability of having $k$ successes out of $n$ trials is denoted as $p(k)$ and is equal to the general term of the development of $(p+q)^n$ i.e.,

$$p(k) = \binom{n}{k} p^k q^{n-k} \tag{60}$$

Hence the corresponding probability distribution is discontinuous. Moreover the distribution approaches symmetry only for relatively large values of $k$.

Like any probability distribution, binomial distribution is characterized by its mean and variance. We shall see in the next section that both of them are moments of this distribution. The mean and variance are defined as

$$\begin{aligned} \langle k \rangle &= \sum_k p(k)k \\ \operatorname{var}(k) &= \sum_k p(k)k^2 - \langle k \rangle^2 \end{aligned} \tag{61}$$

The expression $p(k)k$ is

$$p(k)k = \frac{n!}{(k-1)!(n-k)!} p^k q^{n-k} \tag{62}$$

Setting

$$k = k' + 1 \quad \text{and} \quad n = n' + 1 \tag{63}$$

it follows that

$$n - k = n' - k' \tag{64}$$

and

$$p(k)k = \frac{(n'+1)!}{k'!(n'-k')!} p^{k'+1} q^{n'-k'} \tag{65}$$

As

$$(n'+1)! = n'!n \tag{66}$$

and

$$p^{k'+1} = p^{k'} p \tag{67}$$

expression (65) can be rewritten as

$$p(k)k = np \frac{n'!}{k'!(n'-k')!} p^{k'} q^{n'-k'} \tag{68}$$

But one can see that

$$\frac{n'!}{k'!(n'-k')!} p^{k'} q^{n'-k'} = p(k') \tag{69}$$

and therefore Eq. (68) can be rewritten as

$$p(k)k = np \times p(k') \tag{70}$$

Hence one has

$$\langle k \rangle = \sum_{k'} np \times p(k') = np \sum_{k'} p(k') \tag{71}$$

Since

$$\sum_{k'} p(k') = 1 \tag{72}$$

one finally obtains

$$\langle k \rangle = np \tag{73}$$

The expression of the variance of the binomial distribution can be derived by the same reasoning. One has (Eq. 62)

$$p(k)k^2 = p(k)k \times k = np \times p(k')k \tag{74}$$

and

$$np \sum_k p(k') \times k = np \sum_{k'} p(k')(k'+1) \tag{75}$$

Moreover, one can write

$$np \sum_{k'} p(k')(k'+1) = np\left\{\sum_{k'} p(k')k' + \sum_{k'} p(k')\right\} \tag{76}$$

and

$$np\left\{\sum_{k'} p(k')k' + \sum_{k'} p(k')\right\} = np(\langle k'\rangle + 1) \tag{77}$$

Making use of the second Eq. of (61), one finds

$$\text{var}(k) = np\,(\langle k'\rangle + 1) - \langle k\rangle^2 \tag{78}$$

As $\langle k'\rangle = n'p$, Eq. (78) becomes

$$\text{var}(k) = np(n'p+1) - n^2p^2 = np\{1 - (n-n')p\} \tag{79}$$

Moreover $n - n' = 1$, therefore Eq. (79) reads

$$\text{var}(k) = np(1-p) = npq \tag{80}$$

and defines the variance of the binomial distribution.

### *3.2. The Poisson distribution*

The Poisson distribution is a special case of the binomial distribution obtained when $n \to \infty$ and $p \to 0$ in such a way that the product $np$ remains constant and equal to $\mu$. Under these conditions, the binomial distribution can be expressed as

$$p(k) = \frac{n(n-1)(n-2)\cdots(n-k+1)}{k!}\left(\frac{\mu}{n}\right)^k\left(1-\frac{\mu}{n}\right)^n\left(1-\frac{\mu}{n}\right)^{-k} \tag{81}$$

which can be rearranged to

$$p(k) = \frac{\mu^k}{k!}\left(1-\frac{\mu}{n}\right)^n \left\{\frac{1}{n}(n-1)\frac{1}{n}(n-2)\cdots\frac{1}{n}(n-k+1)\right\}\left(1-\frac{\mu}{n}\right)^{-k} \tag{82}$$

and to

$$p(k) = \frac{\mu^k}{k!}\left(1-\frac{\mu}{n}\right)^n \left\{\left(1-\frac{1}{n}\right)\left(1-\frac{2}{n}\right)\cdots\left(1-\frac{k-1}{n}\right)\right\}\left(1-\frac{\mu}{n}\right)^{-k} \tag{83}$$

As $n$ becomes very large, one has

$$\left(1-\frac{\mu}{n}\right)^n \approx e^{-\mu} \tag{84}$$

and

$$\left(1-\frac{1}{n}\right)\left(1-\frac{2}{n}\right)\cdots\left(1-\frac{k-1}{n}\right)\left(1-\frac{\mu}{n}\right)^{-k} \approx 1 \tag{85}$$

Hence the equation of the Poisson distribution is

$$p(k) = \frac{\mu^k}{k!}e^{-\mu} \tag{86}$$

This distribution is indeed discontinuous and becomes nearly symmetrical when the value of $\mu$ increases.

We have previously shown that for a binomial distribution

$$\langle k \rangle = np \tag{87}$$

and

$$\mathrm{var}(k) = npq = np(1-p) \tag{88}$$

As any Poisson distribution is a special case of a binomial distribution with $p$ very small and $n$ very large in such a way that the product $np$ has a definite value, $\mu$, and that $1-p \approx 1$, it follows that the mean and the variance of a Poisson distribution are

$$\begin{aligned} \langle k \rangle &= \mu \\ \mathrm{var}(k) &= \mu \end{aligned} \tag{89}$$

The mean and the variance are thus equal.

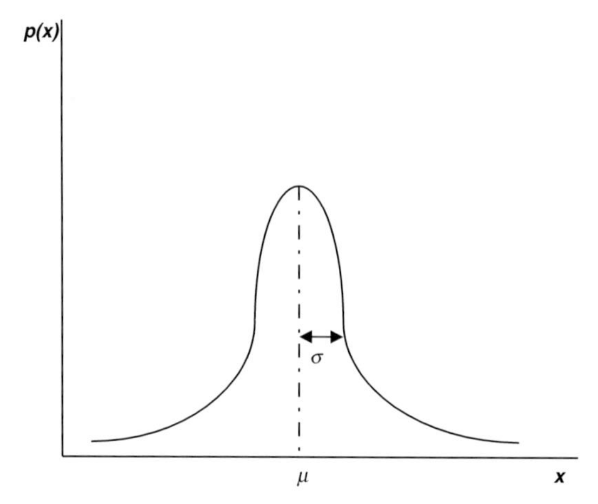


Fig. 9. Mean and standard deviation of the Laplace–Gauss distribution. $\mu$ is the mean, $\sigma$ the standard deviation.

### *3.3. The Laplace–Gauss distribution*

The Laplace–Gauss distribution is so widespread in nature that it is often called the *normal law*. The Laplace–Gauss distribution can be derived from the binomial distribution when $n \to \infty$ and both $p$ and $q \neq 0$. This derivation can be effected thanks to Moivre's theorem that will not be presented here. It is a continuous bell-shaped distribution (Fig. 9) that follows the equation

$$p(x) = \frac{1}{\sqrt{2\pi\sigma^2}} e^{-(x-\mu)^2/2\sigma^2} = y \tag{90}$$

where $\mu$ is the mean and $\sigma^2$ the variance of $x$. Both are moments that will be discussed in the next section. The corresponding curve of the distribution is symmetrical with a maximum, called mode, at $\mu$. The distance that separates the symmetry axis from either of the inflection points of the curve is the square root, $\sigma$, of the variance. It is termed the standard deviation of the distribution.

A much simpler expression of the normal law can be given through a simple transformation of variable. Let us set

$$X = \frac{x - \mu}{\sigma} \tag{91}$$

and

$$Y = \sigma y \tag{92}$$

Then the expression of the normal probability distribution is

$$Y = p(X) = \frac{1}{\sqrt{2\pi}} e^{-X^2/2} \tag{93}$$

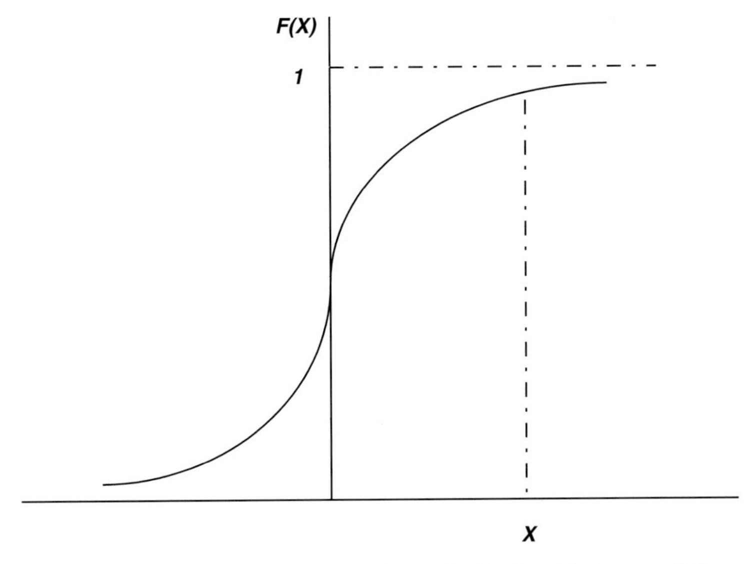


Fig. 10. Probabilities and the reduced Laplace–Gauss distribution. The curve $F(X)$ represents the variation of the probability that the reduced variable be smaller than a given value.

and is called the reduced normal probability distribution. It is in fact a normal distribution with a mean equal to zero and a variance equal to one. As already pointed out, the probability that $X$ be comprised between $a$ and $b$ is

$$p(a<X<b) = \frac{1}{\sqrt{2\pi}} \int_a^b e^{-X^2/2} dX \tag{94}$$

Similarly, the probability that $X$ be smaller than a given value is

$$F(X) = \frac{1}{\sqrt{2\pi}} \int_{-\infty}^{X} e^{-X^2/2} dX \tag{95}$$

and a plot of $F(X)$ *versus* $X$ yields a sigmoidal curve that approaches one as $X \to \infty$ (Fig. 10).

## 4. *Moments and cumulants*

### 4.1. *Moments*

Moments can, in a way, be conceived as a mathematical description of a discontinuous, or continuous, probability function. If the term variate is used, as generally done, to refer to a random variable, moments are either mono- or multivariate [6–10].

*4.1.1. Monovariate moments*

If we have a discontinuous linear or nonlinear function, $\lambda(x_i)$, of a random variable and the probability of occurrence, $p(x_i)$, of the $x_i$ values, the general expression of monovariate moments of degree $r$ is

$$\mu_r = \sum_i p(x_i)\lambda(x_i) \tag{96}$$

If now the probability function is a continuous probability density $p(x)$ and if the corresponding linear, or nonlinear function, of the random variable is $\lambda(x) = (x - a)^r$, where $a$ and $r$ are two constants, the monovariate moments of degree $r$ will be defined as

$$\mu_r = \int_{-\infty}^{+\infty} \lambda(x)p(x)dx \tag{97}$$

which implies that $\lambda(x)$ is defined between $-\infty$ and $+\infty$. As referred to previously, both expressions (96) and (97) can be rewritten in condensed form as

$$\mu_r = \int_{-\infty}^{+\infty} \lambda(x)dF \tag{98}$$

The Stieltjes integral reduces to (97) if $F(x)$ is continuous for then,

$$dF = p(x)dx \tag{99}$$

and to (96) if $F(x)$ is discontinuous for then, one has

$$dF = F(x_i + 0) - F(x_i) = p(x_i) \tag{100}$$

and

$$dF = F(x_i) - F(x_i - 0) = 0 \tag{101}$$

The simplest monovariate moment, called moment of degree one, is obtained when $\lambda(x) = x - a$ ($a$ is a constant). Then expression (98) becomes

$$\mu_1 = \int_{-\infty}^{+\infty} (x - a)dF \tag{102}$$

The monovariate moment of degree one is the mean defined from an arbitrary origin of $x$, $a$. If $a = 0$ then $\lambda(x) = x$ and the mean, $\mu_1'$, is defined as

$$\mu_1' = \int_{-\infty}^{+\infty} x dF \tag{103}$$

The general expression of monovariate moments is

$$\mu_r = \int_{-\infty}^{+\infty} (x - a)^r dF \tag{104}$$

A particularly interesting situation is obtained when $r = 2$ and $a$ is equal to the mean $m$. Then

$$\mu_2 = \sigma^2 = \int_{-\infty}^{+\infty} (x - m)^2 dF \tag{105}$$

and this expression defines the variance of the distribution. If the variable is counted from its mean, the previous equation becomes

$$\mu_2' = \int_{-\infty}^{+\infty} x^2 dF \tag{106}$$

and one can easily show that there exists a simple relationship between $\mu_2$ and $\mu_2'$. One has

$$\mu_2 = \int_{-\infty}^{+\infty} x^2 dF - 2m \int_{-\infty}^{+\infty} x dF + m^2 \int_{-\infty}^{+\infty} dF \tag{107}$$

As the Stieltjes integral in the interval $(-\infty, +\infty)$ is equal to one, expression (107) is equivalent to

$$\mu_2 = \mu_2' - 2m^2 + m^2 = \mu_2' - m^2 \tag{108}$$

which is a very convenient equation to calculate the numerical value of the variance.

If we consider the general expression

$$\mu_{r,a} = \int_{-\infty}^{+\infty} (x - a)^r dF \tag{109}$$

where the values of $x$ are measured from an arbitrary value, $a$, and

$$\mu_{r,m} = \int_{\infty}^{+\infty} (x-m)^r dF \tag{110}$$

where the $x$ are measured from the mean, $m$. One can derive a relationship between (109) and (110). One has

$$\mu_{r,a} = \int_{-\infty}^{+\infty} (x-a)^r dF = \int_{-\infty}^{+\infty} \{(x-m)+(m-a)\}^r dF \tag{111}$$

Setting $D = m - a$ this expression becomes

$$\mu_{r,a} = \int_{-\infty}^{+\infty} \{(x-m)+D\}^r dF \tag{112}$$

which can be re-expressed as

$$\mu_{r,a} = \int_{-\infty}^{+\infty} \left\{(x-m)^r + \binom{r}{1}(x-m)^{r-1}D + \cdots + \binom{r}{p}(x-m)^{r-p}D^p + \cdots + D^r\right\} dF \tag{113}$$

and one finally obtains

$$\mu_{r,a} = \mu_{r,m} + r\mu_{r-1,m}D + \cdots + \binom{r}{p}\mu_{r-p,m}D^p + \cdots + D^r \tag{114}$$

*4.1.2. Bivariate moments*

The concept of moment can easily be extended to several variables in such a way one can define multivariate, and in particular, bivariate moments. The general expression of a bivariate moment is

$$\mu_{r,s} = \int_{-\infty}^{+\infty}\int_{-\infty}^{+\infty} \lambda(x,y) dF \tag{115}$$

This integral should be taken in the Stieltjes sense in such a way that this definition covers both continuous and discontinuous probability distributions. Probably the simplest bivariate moment is obtained when

$$\lambda(x,y) = x + y \tag{116}$$

Then one has

$$\mu_{1,1}(x+y) = \int_{-\infty}^{+\infty}\int_{-\infty}^{+\infty} (x+y)dF \tag{117}$$

If the two variables $x$ and $y$ are independent

$$F(x,y) = F(x)F(y) \tag{118}$$

and

$$dF = dF_x dF_y \tag{119}$$

Hence it follows that

$$\int_{-\infty}^{+\infty}\int_{-\infty}^{+\infty} (x+y)dF = \int_{-\infty}^{+\infty}\int_{-\infty}^{+\infty} xdF + \int_{-\infty}^{+\infty}\int_{-\infty}^{+\infty} ydF \tag{120}$$

and taking account of expression (118), Eq. (120) becomes

$$\int_{-\infty}^{+\infty}\int_{-\infty}^{+\infty} (x+y)dF = \int_{-\infty}^{+\infty} xdF_x \int_{-\infty}^{+\infty} dF_y + \int_{-\infty}^{+\infty} dF_x \int_{-\infty}^{+\infty} ydF_y \tag{121}$$

Moreover one has

$$\int_{-\infty}^{+\infty} dF_x = \int_{-\infty}^{+\infty} dF_y = 1 \tag{122}$$

Equation (121) becomes

$$\mu_{1,1}(x+y) = \mu_1(x) + \mu_1(y) \tag{123}$$

with

$$\begin{aligned} \mu_1(x) &= \int_{-\infty}^{+\infty} xdF_x \\ \mu_1(y) &= \int_{-\infty}^{+\infty} ydF_y \end{aligned} \tag{124}$$

Hence the bivariate moment of the sum of two independent random variables is equal to the sum of the monovariate moments of these variables.

In the same spirit if $\lambda(x, y) = xy$

$$\int_{-\infty}^{+\infty}\int_{-\infty}^{+\infty} xy dF = \int_{-\infty}^{+\infty} x dF_x \int_{-\infty}^{+\infty} y dF_y \tag{125}$$

Therefore as $x$ and $y \rightarrow +\infty$ one has

$$\mu_{1,1}(xy) = \mu_1(x)\mu_1(y) \tag{126}$$

The moment of the product of two independent variables is equal to the product of the monovariate moments of these variables. It is worth stressing again that as the integration is to be taken in the Stieltjes sense, these results are valid when both random variables are discontinuous.

In fact the general explicit expression of a bivariate moment is

$$\mu_{r,s}(x, y) = \int_{-\infty}^{+\infty}\int_{-\infty}^{+\infty} (x-a)^r (y-b)^s dF \tag{127}$$

The most frequently used bivariate moment is the so-called covariance

$$\mu_{1,1} = \int_{-\infty}^{+\infty}\int_{-\infty}^{+\infty} (x - m_x)(y - m_y) dF \tag{128}$$

where $m_x$ and $m_y$ are the means of $x$ and $y$, respectively. One can also define the expression

$$\mu'_{1,1}(x, y) = \int_{-\infty}^{+\infty}\int_{-\infty}^{+\infty} xy dF \tag{129}$$

and try to find out a relationship between $\mu_{1,1}(x, y)$ and $\mu'_{1,1}(x, y)$. In Eqs. (128) and (129), $x$ and $y$ are not necessarily independent. As $dF$ represents in fact a joint probability, the theorem on conditional probabilities allows to write

$$dF = dF_x dF_{y/x} = dF_y dF_{x/y} \tag{130}$$

Hence one has

$$\mu_{1,1}(x, y) = \mu'_{1,1}(x, y) - m_x m_y \tag{131}$$

or written in more familiar form

$$\mathrm{cov}(x, y) = \langle xy \rangle - \langle x \rangle \langle y \rangle \tag{132}$$

### *4.2. Cumulants and characteristic functions*

There are two characteristic functions. The first one denoted $\Phi(t)$ is the mean of the function $\lambda(x) = e^{itx}$ where $i$ is, as usual, the symbol $\sqrt{-1}$. This means that

$$\Phi(t) = \int_{-\infty}^{+\infty} e^{itx} dF = \mu(e^{itx}) \tag{133}$$

As

$$\exp(itx) = \sum_{k=0}^{+\infty} \frac{(it)^k}{k!} x^k \tag{134}$$

it follows that

$$\Phi(t) = \int_{-\infty}^{+\infty} \sum_{k=0}^{+\infty} \frac{(it)^k}{k!} x^k dF = \sum_{k=0}^{+\infty} \frac{(it)^k}{k!} \int_{-\infty}^{+\infty} x^k dF \tag{135}$$

which is equivalent to

$$\Phi(t) = \sum_{k=0}^{+\infty} \frac{(it)^k}{k!} \mu'_k = 1 + \mu'_1(it) + \mu'_2 \frac{(it)^2}{2!} + \cdots + \mu'_r \frac{(it)^r}{r!} + \cdots \tag{136}$$

Hence the first characteristic function can be expressed in terms of the moments.

The second characteristic function, $\Psi(t)$, is defined as

$$\Psi(t) = \ln \Phi(t) \tag{137}$$

Hence it follows that

$$1 + \mu'_1(it) + \mu'_2 \frac{(it)^2}{2!} + \cdots + \mu'_r \frac{(it)^r}{r!} + \cdots = \exp\left\{ \kappa_1(it) + \kappa_2 \frac{(it)^2}{2!} + \cdots + \kappa_r \frac{(it)^r}{r!} + \cdots \right\} \tag{138}$$

In this expression, the kappas are the so-called cumulants. In fact, both the moments and the cumulants can be used to define a probability distribution. Probably the simplest example of the definition of a probability distribution through its moments or cumulants is the Laplace–Gauss distribution. The two characteristic equations of the normal law are

$$\Phi(t) = \exp(it\mu_1') \exp\left(-\frac{\mu_2' t^2}{2}\right) \tag{139}$$

and

$$\Psi(t) = it\mu_1' - \frac{\mu_2' t^2}{2} \tag{140}$$

where, as usual, $\mu_1'$ and $\mu_2'$ are the mean and the variance, respectively. From these equations, it follows that above degree two, both moments and cumulants are equal to zero. Hence, any distribution that meets these requirements can be considered Laplace-Gaussian.

Another interesting property of characteristic functions is their composition rules. If $z = x + y$ and if $\Phi_z(t), \Phi_x(t)$ and $\Phi_y(t)$ are the corresponding characteristic functions, one has

$$\Phi_z(t) = \mu\{\exp(itz)\} = \mu\{\exp[it(x+y)]\} = \mu\{\exp(itx)\exp(ity)\} \tag{141}$$

Hence it follows that

$$\Phi_z(t) = \Phi_x(t)\Phi_y(t) \tag{142}$$

which implies that

$$\Psi_z(t) = \Psi_x(t) + \Psi_y(t) \tag{143}$$

Therefore it appears that the first characteristic function of the sum of two random variables is the product of their corresponding characteristic functions whereas the second characteristic is the sum of the characteristics of these functions.

## 5. *Markov processes*

Let us assume a system $S$ that can exist under a number of states $S_1, S_2, \ldots, S_n$ [11,12]. Let us consider, at time $t$, one of these states denoted $S_i(t)$. This state is generated from different other states already present at time $t-1$. Let us call these states $S_j(t-1)$. Moreover, we assume that between times $t-1$ and $t$ one state only, for instance $S_4(t-1)$, is converted into state $S_i(t)$ (Fig. 11). The probability that the state $S_i(t)$ is formed from a specific state $S_j(t-1)$ is a joint probability denoted $p\{S_i(t), S_j(t-1)\}$. Hence this probability is equal to

$$p\{S_i(t), S_j(t-1)\} = p\{S_j(t-1)\} \times p\{S_i(t)/S_j(t-1)\} \tag{144}$$

The conditional probability that appears in this expression is the probability of change of a definite state $S_j$ (let us say state $S_4$) into $S_i$ in the unit time interval. Hence it can be considered a transition probability $\tau_{ji}$ (Fig. 11). One thus has

$$p\{S_i(t), S_j(t-1)\} = p\{S_j(t-1)\}\tau_{ji} \tag{145}$$

If this relationship is fulfilled, the process is defined as a Markov process, and again this requires that between times $t-1$ and $t$, only one state $S_j(t-1)$ is transformed to $S_i(t)$. Even though this model postulates that only one state, $S_j(t-1)$, can be converted to $S_i(t)$, many states (up to $n$) have the ability to undergo this conversion. Hence the probability of occurrence of state $S_i(t)$ is

$$p\{S_i(t)\} = \sum_{j=1}^{n} p\{S_i(t), S_j(t-1)\} = \sum_{j=1}^{n} p\{S_j(t-1)\}\tau_{ji} \tag{146}$$

The transition probabilities can be put in matrix form as

$$\mathrm{T} = \begin{bmatrix} \tau_{11}\ldots.\tau_{1i}\ldots.\tau_{1n} \\ \ldots\ldots\ldots\ldots\ldots\ldots \\ \tau_{j1}\ldots.\tau_{ji}\ldots.\tau_{jn} \\ \ldots\ldots\ldots\ldots\ldots\ldots \\ \tau_{n1}\ldots.\tau_{ni}\ldots.\tau_{nn} \end{bmatrix} \tag{147}$$

The elements of this matrix are the probabilities that the system be in state $S_j$ at time $t-1$ and in state $S_i$ at time $t$. Hence one has

$$\sum_j \tau_{ji} = 1 \tag{148}$$

and the **T** matrix is in fact a stochastic matrix.

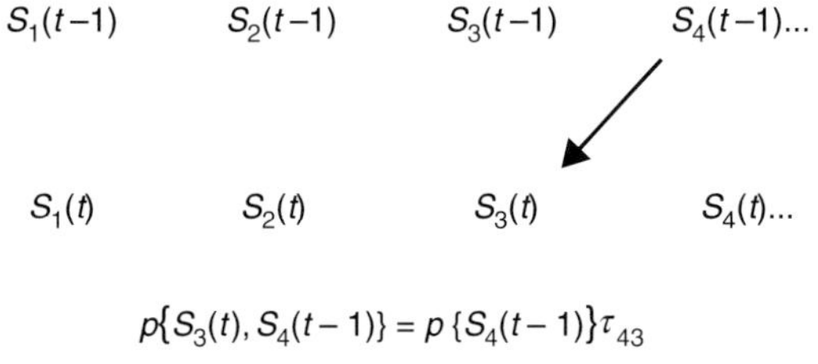

$S_1(t-1)$ $S_2(t-1)$ $S_3(t-1)$ $S_4(t-1)\ldots$

$S_1(t)$ $S_2(t)$ $S_3(t)$ $S_4(t)\ldots$

$$p\{S_3(t), S_4(t-1)\} = p\{S_4(t-1)\}\tau_{43}$$

Fig. 11. Simple example of a Markov process wherein the state $S_4$ (at time $t-1$) is being converted into the state $S_3$ (at time $t$) (see text).

One can define, at time $t$, a vector, the state vector $\mathbf{p\{S(t)\}}$, that collects the probabilities that the system is in states $1, 2, \ldots, n$, i.e.,

$$\mathbf{p}\{\mathbf{S(t)}\} = [p\{S_1(t)\} \ldots p\{S_i(t)\} \ldots p\{S_n(t)\}] \quad (149)$$

Once we know a state vector at time $t$ one can determine a state vector at any other time, for instance at time $t-1$. One has

$$\mathbf{p}\{\mathbf{S(t-1)}\} = \mathbf{p}\{\mathbf{S(t)}\}\mathbf{T}^{-1} \quad (150)$$

In a way, the internal logic of a Markov process is written in the transition matrix $\mathbf{T}$.

## *6. Mathematics as a tool for studying the principles that govern network organization, information, and emergence*

As it is apparent in the title of this book, most of the following chapters will be devoted to the principles that define the organization of simple networks. It is probably useful, at the end of this chapter, to convince the reader that the above mathematical topics can be, and should be, used in order to understand the physical bases of organization, information, and emergence of simple model networks. As we shall see, a network is a system that can only be studied in probabilistic terms. The modern concept of probability itself requires a preliminary knowledge of set theory. The very idea of communication and information, which are now commonplace in life sciences, are in fact rooted in the concept of mapping which is, as we realize now, a fundamental problem of set theory. The mathematical formulation of information theories, which can be used to describe in quantitative terms both communication and organization processes, is based on the mathematical definition and properties of monovariate and bivariate moments. As we shall see in the last chapter of this book, molecular noise certainly plays an important role in many biological phenomena. As a matter of fact, it can be used by the cell to generate unexpected properties. Hence, among different mathematical models, Markov processes represent an interesting way to approach the physical study of intracellular noise.

## *References*


1. Krivine, J.L. (1998) Théorie des Ensembles. Cassini, Paris.
2. Clark, A. (1971) Elements of Abstract Algebra. Dover Publications, Inc., New York.
3. Kolmogogorov, A.N. (1956) Foundations of the Probability Theory. Chelsea Publishing Company, New York.
4. Feller, W. (1950) Probability Theory and its Applications. John Wiley and Sons, Inc., New York.
5. Parzen, E. (1960) Modern Probability Theory and its Applications. John Wiley and Sons, Inc., New York.

6. Kendall, M.G. (1948) The Advanced Theory of Statistics. Vol. 1 Charles Griffin and Company Limited, London.
7. Kendall, M.G. (1950) The Advanced Theory of Statistics. Vol. 2 Charles Griffin and Company Limited, London.
8. Cramer, H. (1954) Mathematical Methods of Statistics. Princeton University Press.
9. Risser, R. and Traynard, C.E. (1957) Les Principes de la Statistique Mathématique. Vol. 1 Gauthier-Villars, Paris.
10. Risser, R. and Traynard, C.E. (1958) Les Principes de la Statistique Mathématique. Vol. 2 Gauthier-Villars, Paris.
11. Hofstetter, E.M. (1964) Random processes. In: Margenau, H. and Murphy G. (eds.) The Mathematics of Physics and Chemistry. Vol. 2. Van Nostrand. Reinhold Company, New York.
12. Kemeny, J., Snell, J.L., and Knapp, A.W. (1976) Denumerable Markov Chains. Springer-Verlag, New York.

J. Ricard *Emergent Collective Properties, Networks and Information in Biology*

DOI: 10.1016/S0167-7306(05)40003-4

CHAPTER 3

# Biological networks

J. Ricard

Elements that interact constitute a network whose nodes, or vertices, are the elements in interaction and the edges, or links, the interactions between the elements. In this perspective, any biological system can be considered a network. This chapter offers a general view of the science of networks as applied to biological systems. The following points are of particular interest: random graphs, percolation as a model for the emergence of organization in a network, small-world and scale-free networks with a particular emphasis on metabolic networks, and attack tolerance of networks.

*Keywords:* attack tolerance of networks, binary matrix of a graph, Cayley tree (Bethe lattice), clustering coefficient, cluster size distribution, connectedness, diameter of a network, evolution of random graphs, graphs, metabolic networks, networks, network clustering coefficient, path-length of a network, percolation, percolation probability, percolation threshold, phase transition in random graphs, power law, random networks, scale-free networks, small-world networks.

Elements that interact constitute a network, or a graph, whose nodes, or vertices, are the elements in interaction, and the edges, or links, the interactions between the elements. A network can be defined independently of the physical nature of the nodes and of the edges. Indeed networks are present everywhere in the world and the very concept of network transgresses the boundaries between various scientific fields. Nearly every kind of collective property, or behavior, can be described by a network. Thus, in biology for instance, an enzyme reaction can be viewed as a network whose nodes are the enzyme states occurring during the reaction and where edges are the various processes of conversion of these enzyme states. In biology, networks are present at all levels, molecular (enzyme and metabolic networks), cellular (genetic networks), multicellular (neural networks), organismic (predator–prey networks), and populational (social networks) levels [1,2]. Owing to its generality and importance, the science of networks is expanding rapidly, crossing the traditional boundaries between disciplines spanning statistical physics, chemistry, molecular, cellular, organismic and population biology, sociology, economics, and other social sciences. The study of networks has benefited from at least two favorable situations: the development and extensive use of statistical mechanics in many different fields; the increased computing power that has allowed scientists both to investigate networks containing millions of nodes and to computerize the data, leading in turn to databases pertaining to many different networks.

Although it is probably impossible to present, even briefly, the main results that have been obtained on networks, the present chapter will attempt to summarize some of the most striking results concerning biological networks.

## *1. The concept of network*

As outlined above, a network, whatever its physical nature, can be represented by a graph. A graph is a set of points, called nodes, connected by edges according to a certain topology. From a mathematical viewpoint, a graph $G$ is a subset of the cartesian product (see Chapter 2) of the set $S$ by itself, i.e.,

$$G \subset S \times S \tag{1}$$

If $x_i$ and $x_j$ ($\forall x_i, x_j \in S$) are the nodes of the graph we associate the nodes by a given relation $x_i R x_j$ (see Chapter 2). For instance if the set $S$ is

$$S = \{1, 2, 3, 4\} \tag{2}$$

and if the relation $x_i R x_j$ means $x_i$ smaller than $x_j$ the graph $G$ will be defined as

$$G = \{(12), (13), (14), (23), (24), (34)\} \tag{3}$$

and is shown in Fig. 1.

If we write the elements of the cartesian product $S \times S$ in a table

| $S \times S$ | 1 | 2 | 3 | 4 |
|---|---|---|---|---|
| 1 | 11 | 12 | 13 | 14 |
| 2 | 21 | 22 | 23 | 24 |
| 3 | 31 | 32 | 33 | 34 |
| 4 | 41 | 42 | 43 | 44 |

the graph can be described by a binary matrix

$$M = \begin{bmatrix} 0 & 1 & 1 & 1 \\ 0 & 0 & 1 & 1 \\ 0 & 0 & 0 & 1 \\ 0 & 0 & 0 & 0 \end{bmatrix} \tag{4}$$

Roughly speaking, there exist three different types of networks. Regular networks are graphs whose nodes are connected according to a fixed rule. Contrary to this situation, random networks are graphs whose nodes are randomly connected. Somewhere between these two extreme cases one can also find networks that are fuzzy organized, i.e., they are neither random nor regular (Fig. 2). In fact most,

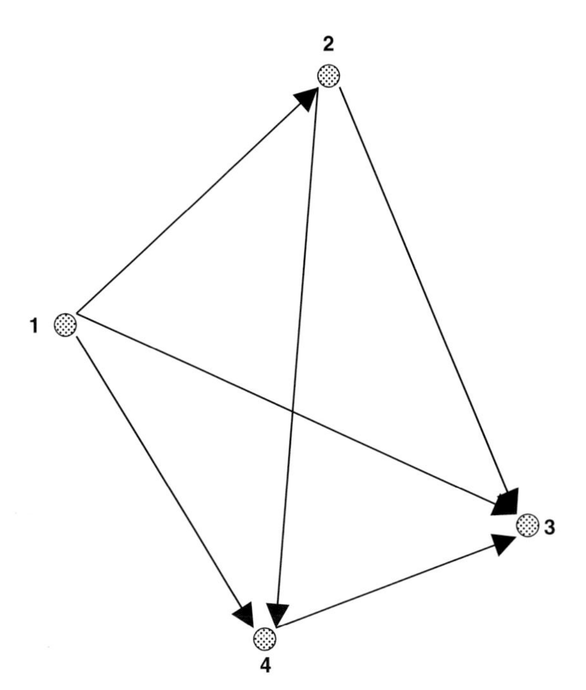


Fig. 1. Construction of a graph out of a set. The graph is constructed from the set $S=\{1,2,3,4\}$ by making use of the relation $R=x_i<x_j$.

perhaps all, large networks of the real world are fuzzy organized. A particularly interesting type of fuzzy organized network is called small world. In plain terms the very idea of small world means that, in spite of their large size, there is always in this type of graph a short path between two nodes. Empirical studies performed in social psychology have shown that most large social networks display what is called the "six degrees of separation" [3]. These types of networks are typical small worlds.

In spite of the diversity of networks that have been studied so far, for instance World Wide Web, Internet, film actor collaboration network, science collaboration network, cellular and protein networks [1,4–8] etc. ... it is striking that some general underlying principle can be uncovered in these complex and large networks suggesting there may exist a true science of complexity with its specific laws.

## 2. *Random networks*

It has often been tempting to consider, as a first approximation, that real, large networks are in fact random graphs. As we shall see later this is not the case. Nevertheless there is no doubt that the study of random graphs is important in order to understand the behavior of large, real networks. Random graphs have been studied, from a theoretical viewpoint, by Erdos and Renyi [9] as well as

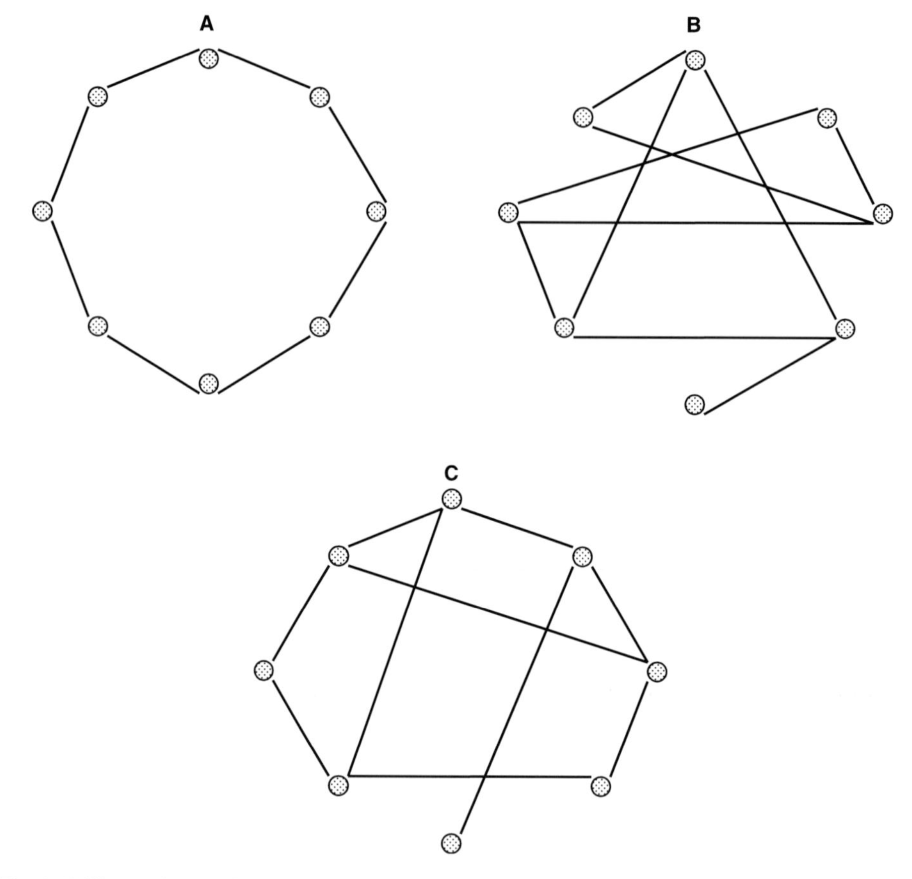


Fig. 2. Different classes of graphs. A: Regular; B: Random; C: Fuzzy-organized.

Bollobas [10]. If a graph possesses $N$ nodes the largest possible number of edges one can expect is

$$n(\max) = \frac{N(N-1)}{2} \tag{5}$$

If the graph possesses a number of edges, $n$, smaller than the maximum possible number, i.e.,

$$n < n(\max) \tag{6}$$

the number of graphs, $n(G)$, one can construct with $N$ nodes and $n$ edges is

$$n(G) = \binom{n(\max)}{n} \tag{7}$$

Hence the probability of obtaining one of these graphs, $p(G)$, is derived from the general term of the binomial distribution, i.e.,

$$p(G) = p^n(1-p)^{[N(N-1)/2]-n} \tag{8}$$

where $p$ is the probability of connecting the nodes. Given this probability $p$, the mean number of edges of the graphs, $\langle n \rangle$, is indeed the mean of the binomial distribution, i.e.,

$$\langle n \rangle = p\frac{N(N-1)}{2} \tag{9}$$

An interesting, but intuitively evident, implication of the expression (7) is that as the number of edges $n$ of the random graph increases and approaches the maximum number $n(\max)$, the number of graphs diminishes for the graphs tend to coalesce to form a giant network. Finally, when $n \to n(\max)$, $p \to 1$ for then all the nodes are connected. An interesting result obtained by Erdos and Renyi is that a giant network with novel properties appears suddenly. We shall come back to this problem in the next section devoted to percolation, for the growth of a random graph which is called evolution by mathematicians, is formally similar to percolation.

In order to understand intuitively what is occurring during random graph evolution let us consider the following toy problem [11,12]. Let us assume we have on a table a large number of buttons and pieces of thread. We randomly connect two buttons with a piece of thread and put this pair down on the same table. Again we randomly pick up two buttons, connect them and put the pair down on the table. We repeat the same procedure many times. If there is a large number of buttons on the table there is a high probability that, at the beginning of the process of connecting buttons, one connect buttons that have not been connected. However as we go on connecting buttons, the probability of connecting previously connected buttons increases. The interesting, and surprising, result, however, is that the size of the cluster increases suddenly. As the ratio of the number of pieces of thread to the number of buttons reaches the value 0.5 there is some kind of *phase transition* and a large number of buttons suddenly appear connected as a giant cluster (Fig. 3). This toy model has been used by Kauffman [11,12] as an illustration of the possible coalescence of enzyme reactions that form an integrated metabolic network during the processes involved in the origin of living systems.

During the evolution process leading to a giant cluster it is evident that subgraphs are transiently formed. A graph $F$ can be considered a subgraph of $G$ if all the nodes and the edges of $F$ also belong to $G$. Simple examples of subgraphs are trees and cycles, i.e., poorly connected structures. A tree of order $k$ has $k$ nodes and $k-1$ edges (Fig. 4). A cycle of order $k$ is a closed structure having $k$ nodes and $k$ edges. More complex strongly connected subgraphs, can also be formed (Fig. 4).

Indeed the degree of connection varies in a random graph from node to node. A probability, $p(k)$, expresses the idea that a randomly chosen node is connected to other nodes by $k$ edges. The connection of a node to others is called the node degree.

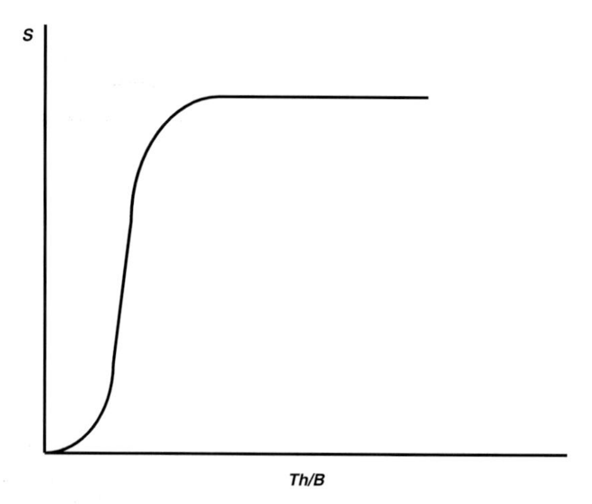


Fig. 3. Emergence of a giant cluster. Buttons are randomly connected by pieces of thread. A giant cluster collecting most buttons appears suddenly when the ratio of the number of pieces of thread over the number of buttons reaches the value 0.5. *S* is the size of the cluster, *Th* the number of pieces of thread, and *B* the number of buttons. Adapted from Ref. [11] and [12].

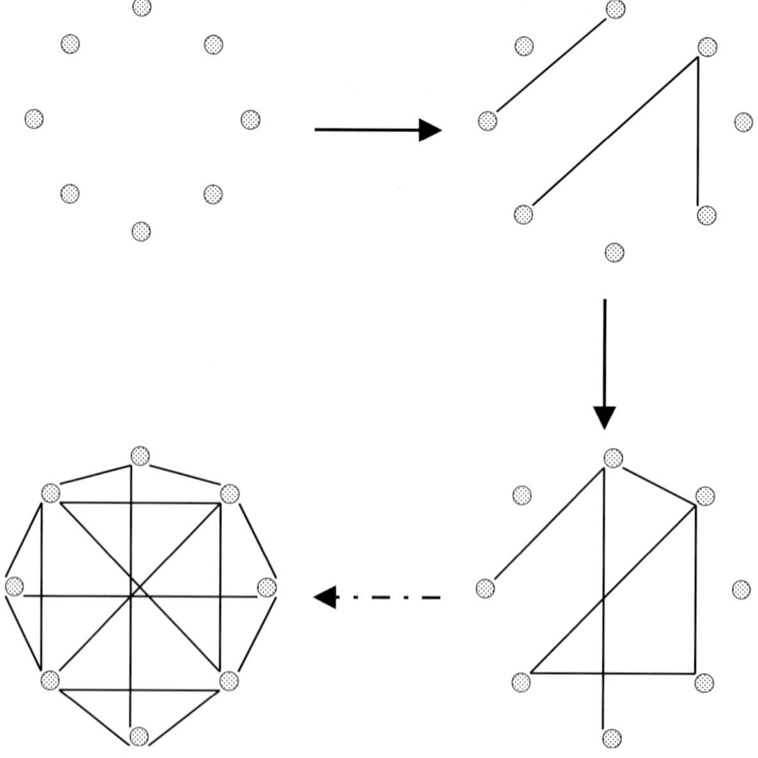

Fig. 4. Formation of trees and cycles during the process of node connection. When connecting disconnected nodes, trees and cycles usually appear. In this figure, a tree (top right) and a cycle (bottom right) are transiently formed.

In a random network, a great majority of nodes have the same node degree equal to the mean node degree $\langle k \rangle$ of the probability distribution. If a random graph has a connection probability, $p$, the degree, $k$, of nodes follows the binomial distribution

$$p(k) = \binom{N-1}{k} p^k (1-p)^{N-1-k} \tag{10}$$

and for large values of $N$ this expression becomes very close to a Poisson distribution

$$p(k) = \frac{\langle k \rangle^k}{k!} e^{-\langle k \rangle} \tag{11}$$

whose mean is indeed $\langle k \rangle$.

Two given nodes that constitute a pair are separated by a minimum number of steps, or edges. The diameter $\langle d \rangle$ is defined as the mean of the minimum number of steps that separate the nodes of all the possible couples. Clearly the diameter of a random graph increases with the number of nodes. All random graphs with the same number of nodes $N$ and the same connection probability $p$ have the same diameter. One can easily see that, in random graphs, the number of nodes $N$ is approximately equal to

$$N \approx \langle k \rangle^{\langle d \rangle} \tag{12}$$

Hence the diameter $\langle d \rangle$ is approximately equal to

$$\langle d \rangle \approx \frac{\ln N}{\ln \langle k \rangle} \tag{13}$$

With this definition of diameter in mind, it appears that poorly connected random graphs with unconnected nodes can have an infinite diameter for one needs an infinite number of steps to travel from a node to another one if they are unconnected. One can also define a clustering coefficient $C$ as

$$C = \frac{\langle k \rangle}{N} \tag{14}$$

## 3. *Percolation as a model for the emergence of organization in a network*

One of the interesting results presented in the previous section is that, in the course of evolution of a random graph, a giant cluster can form only if the probability of occurrence of the edges becomes larger than a critical probability [13–17]. This is precisely what is occurring during percolation. It is therefore of interest to present a brief overview of the physical processes of percolation in order to understand how a giant network can emerge.

Let us consider a two-dimensional lattice of sites. Let us then randomly occupy these sites with a probability $p$. This means that the ratio of the number of occupied sites over the total number of sites is equal to $p$. Some of the occupied sites are isolated whereas others belong to clusters. In these clusters, let us connect the adjacent sites in the two cardinal directions. This means there can be no nearest neighbor (Fig. 5). The process of forming bonds between sites is called site percolation. One could also consider a process of bond percolation that would be equivalent to the previous one (Fig. 5).

If the value of the probability $p$ is smaller than a critical value $p_c$, small clusters are formed and coexist with isolated free and occupied sites. However, as soon as the probability $p$ is larger than the critical probability $p_c$ a giant cluster is formed that connects the top with the bottom, or the left with right of the lattice. If one does not care about the boundaries of the lattice, the giant cluster can be considered an infinite cluster. The appearance of the infinite cluster is usually associated with the emergence of novel properties of the system which has undergone a geometric phase transition. If we plot the probability of occurrence, $p(C_\infty)$, of the infinite cluster, $C_\infty$, as a function of the probability, $p$, of occupation of sites, one has the type of curve shown in Fig. 6.

There is a quantity which is of particular interest for the study of percolation, the probability that a site belongs to clusters of a given size, $s$. This probability, $p(C_s)$, is the ratio of the number of sites belonging to clusters of size $s$ over the total number

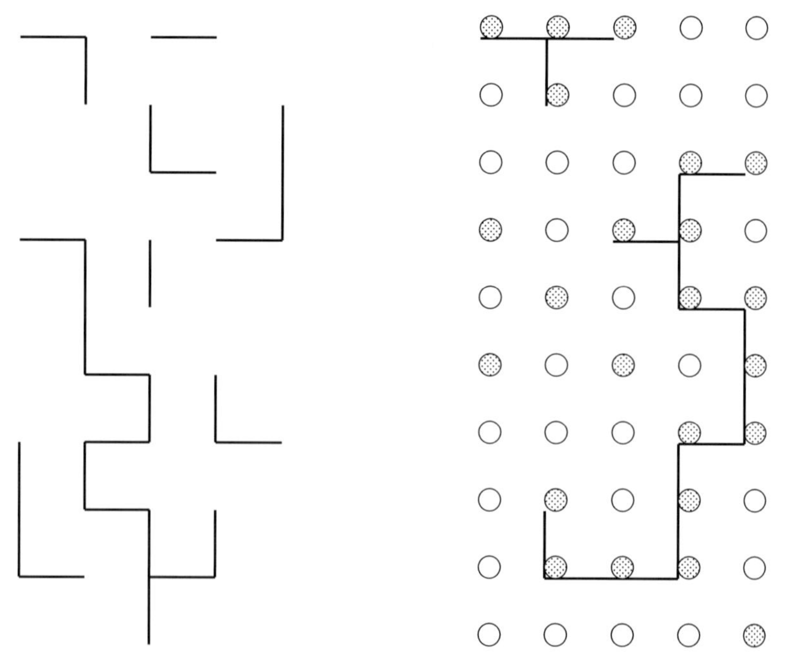

Fig. 5. Site and bond percolation. Left: Bond percolation. Right: Site percolation. In either case it is assumed that a giant cluster is formed as well as small clusters. Isolated sites and bonds are also present.

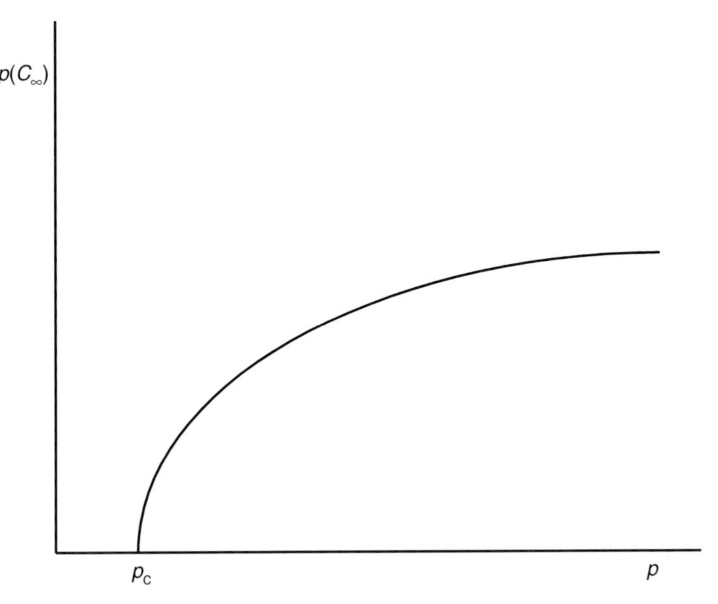


Fig. 6. Probability of emergence of a giant cluster as a function of the probability of site occupation. A giant cluster can be formed, with a probability of occurrence $p(C_\infty)$, as soon as the probability of site occupation $p$ becomes larger than the critical probability $p_c$. Adapted from Ref. [17].

of sites. With this definition in mind, it appears that $\sum_{s<\infty} p(C_s)$ is the probability that a site belongs to a cluster of *any* size provided that $s<\infty$. If we consider a site, it is either free, with probability $1-p$, it can also belong to a cluster of any size $s$ (provided that $s<\infty$), with probability $\sum_s p(C_s)$, or it can be occupied, together with other sites, by the infinite cluster. The last probability is equal to the product of $p$ by $p(C_\infty)$, i.e., the product of the probability, $p$, that the site is occupied times the probability, $p(C_\infty)$, that the infinite cluster exists. For a given site, the sum of all these probabilities should be equal to one. Hence one has

$$1 - p + \sum_s p(C_s) + p \times p(C_\infty) = 1 \tag{15}$$

From this equation one can derive the expression of the percolation probability, i.e.,

$$p(C_\infty) = 1 - \frac{\sum_s p(C_s)}{p} \tag{16}$$

Now let us consider two occupied sites that are a distance $r$ apart. We call connectedness, $g_p(r)$, the probability that the two sites belong to the same cluster when the lattice is filled with probability $p$ (Fig. 7). If the distance $r$ that separates the two sites is very small it is highly probable that the two sites belong to the same cluster. Hence if $r=0$ one has $g_p(r)=1$. Conversely if the distance $r$ is large it is

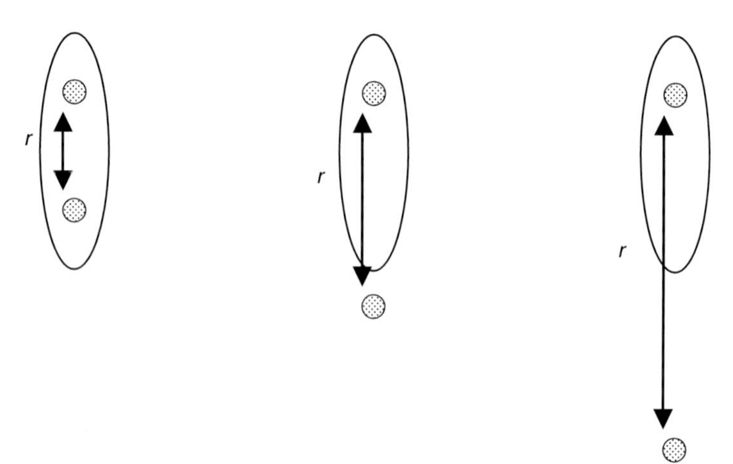


Fig. 7. Connectedness of two sites filled with a given probability. If two occupied sites are a distance $r$ apart and if $r$ is small, it is highly probable that they belong to the same cluster. However, as the distance $r$ increases, it becomes more and more likely that the sites belong to different clusters. Hence the connectedness of the two sites decreases with the distance $r$. The cluster is schematically depicted by an ellipse.

highly probable that the two sites do not belong to the same cluster. Hence as $r \to \infty$ one should have $g_p(r) \to 0$. It therefore appears that the degree of connectedness, $g_p(r)$ as a function of $r$ is very likely to be exponential. Moreover the decay of $g_p(r)$ as a function of $r$ should also depend on the probability used to fill the lattice. If this probability is small the size of the clusters is small as well and the decline of $g_p(r)$ with the distance should be steep. Conversely, if $p$ is much larger the size of the clusters has increased and the decline of $g_p(r)$ as a function of $r$ is no longer steep anymore. Hence we should define another variable, $\xi(p)$, whose value depends on the probability $p$ used to fill the lattice. $\xi(p)$ is equal to zero for $p=0$ and goes to infinity as $p$ approaches the critical probability $p_c$. $\xi(p)$ can be considered a correlation length. Hence one should have

$$g_p(r) \approx \exp\left\{-\frac{r}{\xi(p)}\right\} \tag{17}$$

If a site is connected to many others, and if we sum over all the sites in interaction with the initial one, we shall obtain the mean cluster size, $\langle s \rangle_p$, for a given probability of filling the lattice. Hence one has

$$\sum_r g_p(r) = \langle s \rangle_p \tag{18}$$

The internal logic of the percolation process can be best understood if this process takes place in one dimension only, which is indeed an unrealistic situation. Let us consider one-dimensional ideal clusters containing $s$ occupied sites for a given probability, $p$, of filling the lattice. As each cluster is limited at both ends by an empty site of probability $1-p$ and contains $s$ occupied sites, the probability that a site is an end site of the cluster is

$$p(C_s) = (1-p)^2 p^s \tag{19}$$

From this equation it appears that the condition for obtaining an infinite cluster is $p = p_c = 1$. As there cannot exist any probability such that $p > 1$, no phase transition is ever to be expected in the case of a one-dimensional percolation system. It is nevertheless interesting to study this ideal situation in order to understand the logic of the percolation process. It is indeed of interest to perform this analysis for a probability $p$ close to $p_c$.

As $p \to 1$, one can expand $\ln p$ in series

$$\ln p = p - 1 - \frac{(p-1)^2}{2} + \frac{(p-1)^3}{3} - \cdots \tag{20}$$

As $p < 1$, this series is of necessity convergent. Moreover if $p$ is close to one just the first term of the expansion can be kept. The expression of the probability $p$ can then be written as

$$p = e^{\ln p} \tag{21}$$

and taking advantage of Eq. (20) this is equivalent to

$$p = \mathrm{Lim}\ e^{p-1} = \mathrm{Lim}\ e^{p-p_c} \tag{22}$$

Here the limit implies that $p \to 1$ or $p \to p_c$. Hence Eq. (19) can be rewritten as

$$p(C_s) \approx (1 - e^{p-p_c})^2 e^{(p-p_c)s} \tag{23}$$

If $p$ is close enough to $p_c$

$$e^{p-p_c} \approx 1 + p - p_c \tag{24}$$

and

$$(1 - e^{p-p_c})^2 \approx (p - p_c)^2 \tag{25}$$

Hence Eq. (23) becomes

$$p(C_s) \approx (p - p_c)^2 e^{(p-p_c)s} \tag{26}$$

and as the size, $s$, of the cluster increases, the probability of occurrence of the end site of the cluster declines exponentially for $p < p_c$.

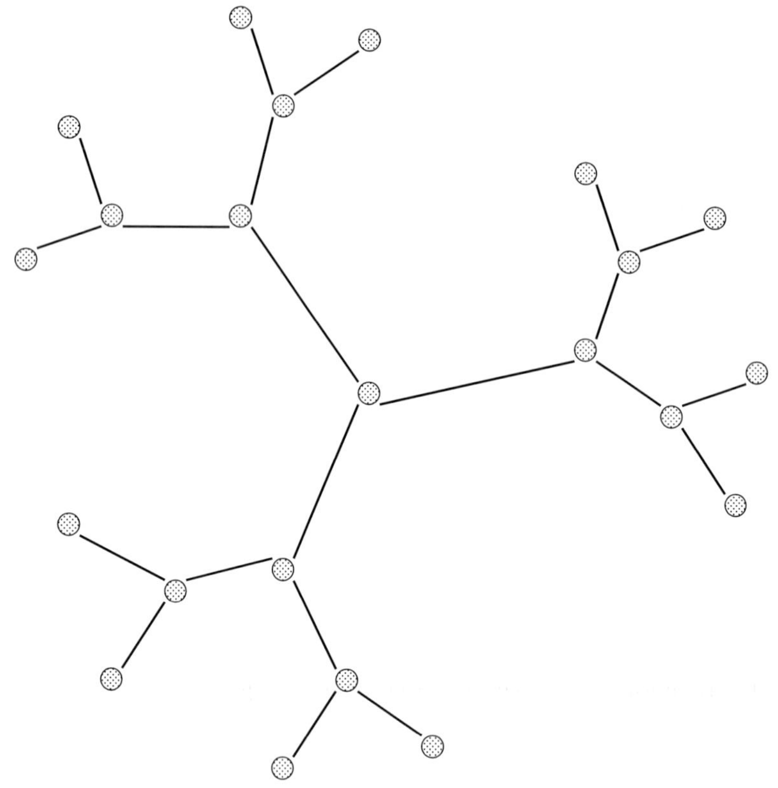

Fig. 8. A Cayley tree.

A very interesting system for studying percolation theory is the so-called Cayley tree, or Bethe lattice (Fig. 8). A Cayley tree is a graph in which every node, except those located at the periphery of the tree, has $z$ neighbors [18,19]. Hence $z$ edges originate from each node, with the exception of the nodes located at the periphery of the tree, which have only one edge each. It therefore appears that the structure of the Cayley tree is such that the nodes are located on concentric shells. We know that the surface $S$ and volume $V$ of a regular d-dimensional object obeys the scaling relationship

$$S \propto V^{1-1/d} \tag{27}$$

Hence, for large $d$ the number of nodes on the surface becomes proportional to the total number of nodes.

A Cayley tree is an interesting structure for it is a model of both a graph and a percolation system whose edges are nonmetric. This situation is characteristic of graphs but does not occur for site or bond percolation systems. The shells bearing the nodes are a distance $l$ apart. Hence going from shell to shell means that the number of sites, or of nodes, is each time multiplied by $z-1$. One can easily see that the number of sites, $N(\lambda)$, located at the nonmetric distance $\lambda$ of the initial, or central, site is

$$N(\lambda) = z(z-1)^{\lambda-1} \tag{28}$$

In the case of Fig. 8, for a Cayley tree having $z=3$ we have $N(\lambda_1)=3$, $N(\lambda_2)=6$, and $N(\lambda_3)=12$. The number of occupied sites is $\lambda$ (because there is an occupied site at the origin of $\lambda$). If, as done previously, we derive the expression of the correlation function $g_p(\lambda)$ as a function of the distance $\lambda$ between sites, this function will be the product of $N(\lambda)$ with the probability $p^\lambda$. Hence one has

$$g_p(\lambda) = z(z-1)^{\lambda-1}p^\lambda \quad (\forall\lambda \le z) \tag{29}$$

If $z$ is large enough

$$g_p(\lambda) \approx (z-1)^\lambda p^\lambda = [(z-1)p]^\lambda \tag{30}$$

It then follows that

$$g_p(\lambda) \approx \exp\{\ln[(z-1)p]^\lambda\} = \exp\{\lambda \ln(z-1)p\} \tag{31}$$

Hence the function $g_p(\lambda)$ will diverge or converge exponentially depending on whether $(z-1)p>1$, or $(z-1)p<1$, respectively. The critical value of the probability, $p_c$, which can be defined as the percolation threshold is thus

$$p_c = \frac{1}{z-1} \tag{32}$$

It is worth stressing that percolation theory, whatever the complexity of the system, obeys general scaling relationships near the percolation threshold. The cluster size distribution, $n_s(p)$, can be written as

$$n_s(p) \approx s^{-\tau}\Phi_-\left(|p-p_c|^{1/\sigma}s\right) \quad (p \le p_c) \tag{33}$$

and

$$n_s(p) \approx s^{-\tau}\Phi_+\left(|p-p_c|^{1/\sigma}s\right) \quad (p \ge p_c) \tag{34}$$

where $\tau$ and $\sigma$ are critical exponents, $\Phi_-(|p-p_c|^{1/\sigma}s)$ and $\Phi_+(|p-p_c|^{1/\sigma}s)$ are decreasing and increasing functions of $s$ that involve the cutoff $|p-p_c|^{1/\sigma}$. As expected, Eq. (26) is consistent with general expression (33) above. Equation (26) can be rewritten as

$$n_s(p) = s^{-2}[(p-p_c)s]^2 e^{(p-p_c)s} \tag{35}$$

Here $\tau = 2$ and $\sigma = 1$, moreover the scaling function assumes the form

$$\Phi_{-}\left(|p - p_{c}|^{1/\sigma} s\right) = [(p - p_{c})s]^{2} e^{(p - p_{c})} \tag{36}$$

Thus as $p < p_c$ the scaling function $\Phi_{-}$ declines as $s$ increases and so does $n_s(p)$ which approaches a power law with $\tau = 2$.

## 4. *Small-world and scale-free networks*

For years the study of real networks has been done assuming, either explicitly or implicitly, they were random graphs. We have previously mentioned that for such graphs the probability distribution that the nodes be linked to $k$ edges varies according to a Poisson law. This means in practice that most nodes have the same connectivity equal to $\langle k \rangle$ or, put in other words, that not very many nodes are highly or poorly connected. Moreover, as the number of nodes of a random network increases, one should expect its diameter to increase accordingly. The probability distribution of node connectivity and the variation of network diameter as a function of the number of nodes are both features that can be studied experimentally. We shall examine first, in the present section, the validity of the second prediction and leave the first one for the moment. This study will be performed in the next section.

### 4.1. *Metabolic networks are fundamentally different from random graphs*

Sequencing the genome of different living organisms has allowed the characterization of most of their genes and the corresponding enzymes as well. As this wealth of information can be found in databases [20–22], it is then possible, in most cases, to construct the metabolic networks spanning the phylogenetic tree from mycoplasms and bacteria up to eukaryotes. In these networks, each node is a metabolite and the edge connecting two metabolites is the enzyme chemical reaction allowing the conversion of one reagent into another. When this analysis was performed it was found that although the number of metabolites increases together with the number of edges of the network, the corresponding diameter remains remarkably constant for the 43 different organisms studied [23]. This result is certainly not fortuitous and can be understood only if one assumes that the connectivity between certain nodes of the metabolic networks increases together with the number of nodes (Fig. 9).

As the edges of metabolic networks describe dynamic unidirectional, or approximately unidirectional, processes every node can be associated with incoming and outgoing links. If one plots the average number per node of incoming, $\langle l_{\mathrm{i}} \rangle$, or outgoing, $\langle l_{\mathrm{o}} \rangle$, links as a function of the number of nodes, one obtains, for the 43 organisms studied, an increasing curve (Fig. 9). These results unambiguously show that node connectivity increases with the number of nodes. One can also demonstrate that the degree of connection varies dramatically from node to node. Thus if one randomly removes nodes from a network mimicking the metabolism of

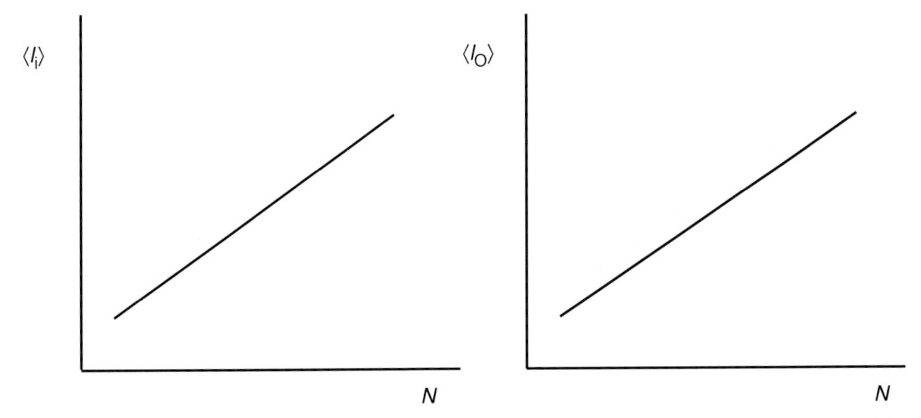


Fig. 9. Variation of node connectivity as a function of the number of nodes. Node connectivity is expressed by the mean number per node of incoming, $\langle l_i \rangle$, and outgoing, $\langle l_o \rangle$, links. The two curves show that connectivity increases with the number of nodes. Adapted from Ref. [23].

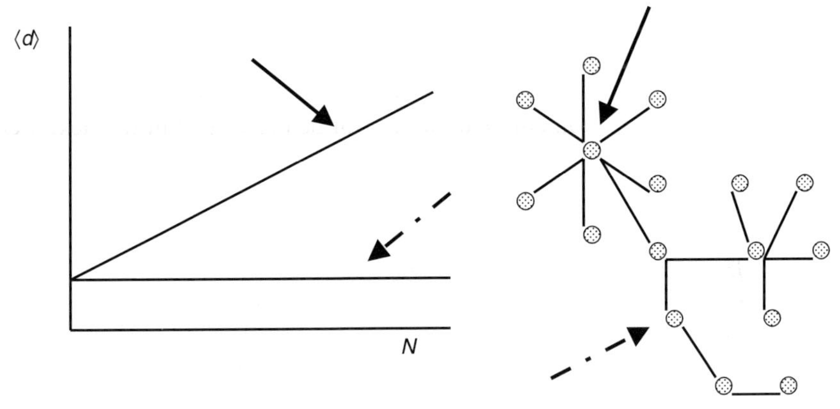


Fig. 10. Variation of network diameter as a function of the number of nodes. Left: Up to 8% of poorly connected nodes are removed (dotted arrow) and the network diameter is still independent of the number of nodes. Some highly connected nodes are removed and the network diameter becomes dependent on the number of nodes (full arrow). Right: A poorly connected node is removed (dotted arrow). A highly connected node is removed (full arrow). Adapted from Ref. [23].

*E. coli* up to 8% of the nodes, the network diameter does not vary. If, however, some *specific* nodes are removed, depending on the nature of these nodes, the diameter of the network can increase (Fig. 10). This implies that the nodes which have been removed were highly connected whereas the others were not. Moreover, in the 43 organisms studied by Jeong *et al.* [23], only 4% of the metabolites are common to all these organisms and it is precisely the nodes pertaining to these metabolites that are the most highly connected.

Hence it seems there exists an evolutionary pressure that tends to keep the mean diameter of the metabolic network constant whatever the number of metabolites involved in the network. Moreover this evolutionary pressure tends to generate a quite irregular degree of connection of the nodes, some nodes being highly connected whereas others are not. This implies that metabolic networks, and probably most biological networks, are either small-world, or scale-free, or both [23,24]. Hence there exists both short paths between distant nodes and irregular degree of connection between them. One can therefore wonder about the nature of the functional advantages brought about by this situation. There are at least two functional advantages. The first one is evident. It implies that whatever the complexity of the network there always exist some paths that allow relatively rapid mutual conversion of metabolites. Thus most metabolites in *E. coli* are separated by only three steps. The second functional advantage is less evident. Removing metabolites (or nodes) from the network is equivalent to mutations that inactivate enzymes, and therefore result in suppressing metabolic reactions. The results presented in Fig. 10 show that, if the process of removing metabolites occurs randomly, the diameter of the network remains unchanged. As mutations, which result in removing metabolites, take place randomly one can conclude that metabolic networks are, to a large extent, attack tolerant which is no doubt an important functional advantage.

## 4.2. *Properties of small-world networks*

Small-world networks have a structure midway between strict organization and total randomness. One can construct a small-world network from a regular graph by randomly removing edges with a certain probability $p$. If $p = 0$ this means that no rewiring has been effected and the network is regular (Fig. 11). If $p=1$ the process of rewiring the edges is complete and the graph is random (Fig. 11). If $0 < p < 1$ some edges connected to their nodes have been rewired thus creating shortcuts between nodes and therefore the small-world feature. The procedure followed by Watts and Strogatz [8] to obtain a small-world graph is simple. In a regular ring lattice they choose a node and the edge that connect this node to its nearest neighbor in a clockwise direction. The edge is randomly reconnected with probability $p$ to another node and the process is carried on in the clockwise direction. In order to avoid disconnection of the nodes one should have the condition $N \gg k \gg \ln N \gg 1$ where, as usual, $N$ is the number of nodes and $k$ their degree of connection.

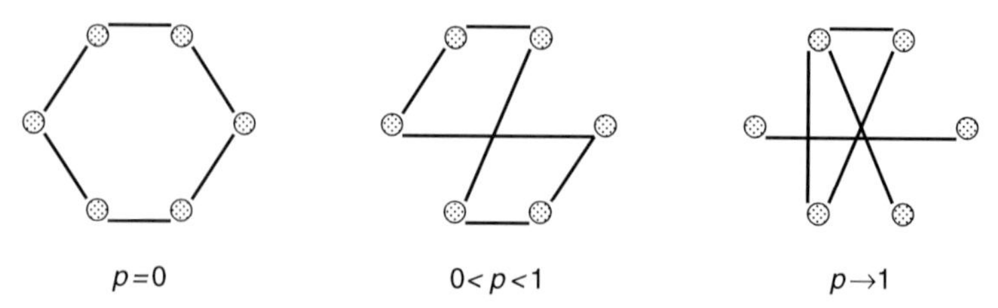


Fig. 11. Generation of a small-world from a regular network. Starting from a regular network ($p=1$) the nodes are randomly rewired with a certain probability $p$. The small-world feature appears when $0 < p < 1$. When $p=0$ the network is random. Adapted from Ref. [8].

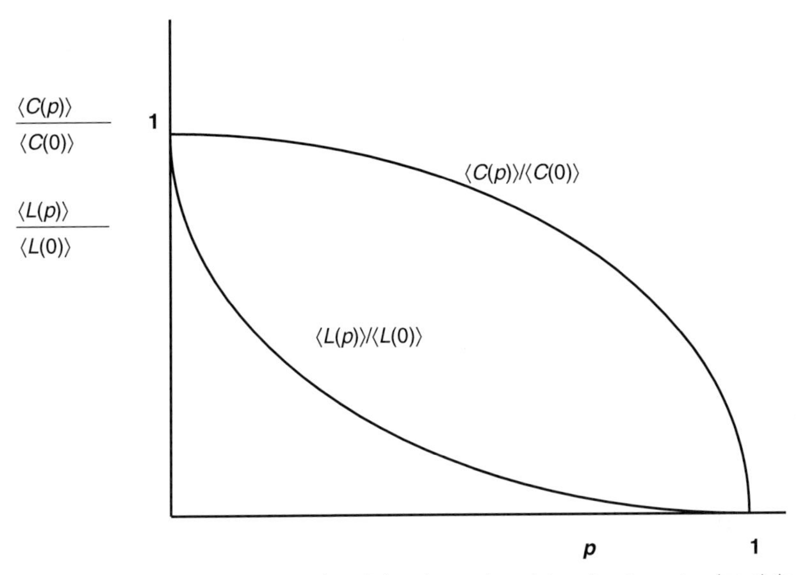


Fig. 12. Effect of rewiring on the ratios $\langle L(p)\rangle/\langle L(0)\rangle$ and $\langle C(p)\rangle/\langle C(0)\rangle$. The ratios $\langle L(p)\rangle/\langle L(0)\rangle$ and $\langle C(p)\rangle/\langle C(0)\rangle$ are plotted as a function of the probability of rewiring, $p$. For a significant increase of $p$ values the ratio $\langle C(p)\rangle/\langle C(0)\rangle$ does not vary significantly whereas the ratio $\langle L(p)\rangle/\langle L(0)\rangle$ tends to bottom out. It is in this range of $p$ values that the network displays the small-world feature. Adapted from Ref. [8].

Two parameters are important to characterize small-world networks: the path-length $\langle L\rangle$ and the clustering coefficient $\langle C\rangle$. $\langle L\rangle$ is the smallest possible number of steps between two nodes averaged over all possible node pairs. It is therefore a collective property of the network whose value will of necessity decrease as the probability of rewiring increases (Fig. 12). The clustering coefficient $\langle C\rangle$ is defined as follows. Let us consider a node $i$ surrounded by other nodes that form a *clique*. If there are $N_i$ nodes in the clique, the maximum number of edges one can expect is equal to $N_i(N_i - 1)/2$. If $n_i$ is the number of edges that are actually connected to node $i$, the ratio

$$C_i = \frac{2n_i}{N_i(N_i - 1)} \tag{37}$$

is defined as the clustering coefficient of this node $i$. The mean clustering coefficient is the average of all the $C_i$ values. This means clustering coefficient is in fact a local property averaged over different regions of the network. The mean value of both $L$ and $C$ are functions of the rewiring probability $p$. They can therefore be written $\langle L(p)\rangle$ and $\langle C(p)\rangle$, respectively. When $p = 0$ the mean values of $L$ and $C$, $\langle L(0)\rangle$ and $\langle C(0)\rangle$, are specific features of regular, organized networks. One can plot the ratios $\langle L(p)\rangle/\langle L(0)\rangle$ and $\langle C(p)\rangle/\langle C(0)\rangle$ as a function of $p$ (Fig. 12). What this figure shows is a sharp decline of $\langle L(p)\rangle$ as $p$ increases whereas $\langle C(p)\rangle$ does not vary significantly under these conditions. For small $p$ values each shortcut connects nodes that are usually far apart. Under these conditions, the lattice is a highly clustered large world.

It is only for large $p$ values that $\langle C(p) \rangle$ drops significantly. Hence there exists a broad range of probability $p$ for which the network is highly clustered whereas there exist shortcuts between nodes. This is the typical situation for a small-world network.

### 4.3. *Scale-free networks*

We have seen previously that, for random graphs, the probability distribution follows the Poisson law. This implies that above the average connection value $\langle k \rangle$, the probability $p(k)$ decays exponentially. In fact this is not what occurs in the real world. Most real networks have a node distribution that decays according to a power law [23,25,26]

$$p(k) = k^{-\gamma} \tag{38}$$

This means that a very large number of nodes are poorly connected and that few of them are highly connected (Figs. 13 and 14).

The highly connected nodes are often called "hubs". Actually this topology (Fig. 13) describes networks that are not in a static state, but are growing and display preferential attachment of new free nodes to highly connected vertices of the network. In fact Barabasi and Albert [25] have proposed that the probability $p(k_i)$ that a new node be connected to node $i$ of a network depends on the degree, $k_i$, of node $i$, i.e.,

$$p(k_i) = \frac{k_i}{\sum_j k_j} \tag{39}$$

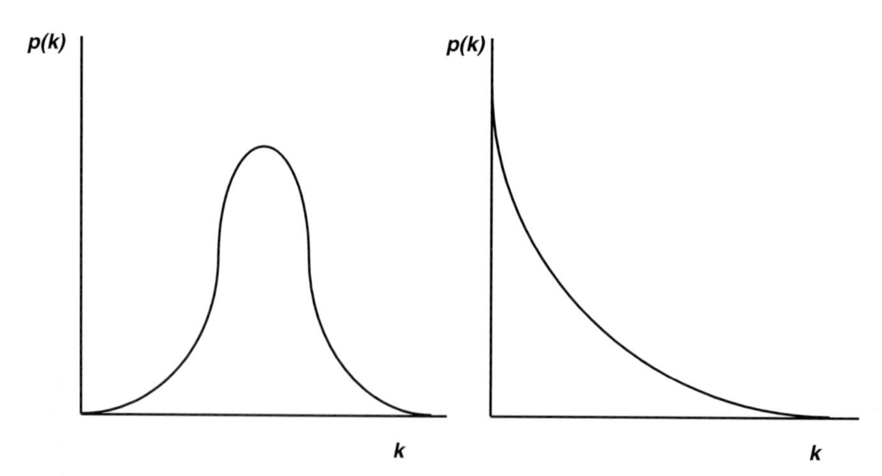


Fig. 13. Probability distribution of node connection for random and scale-free networks. Left: Node connection is distributed according to a Poisson law. Right: Node connection is distributed according to a declining power-law.

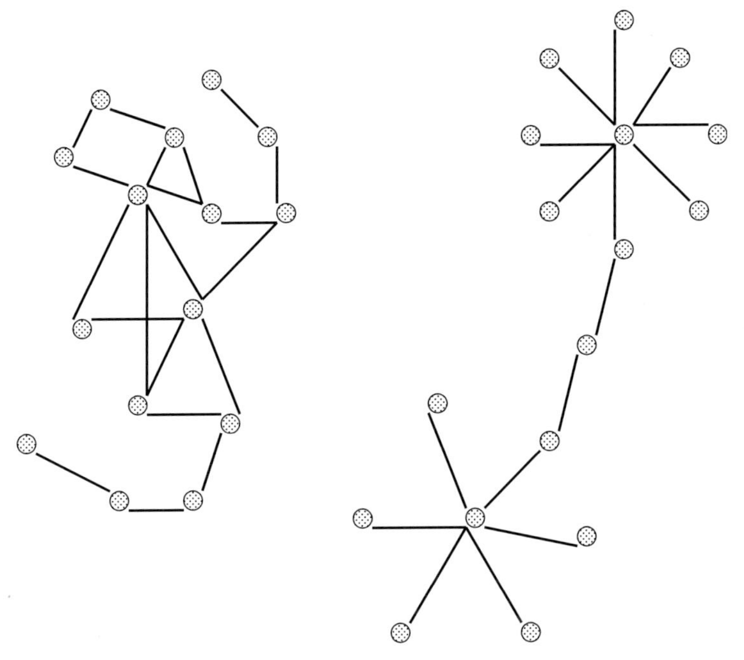

Fig. 14. In random networks, a majority of nodes possess an average degree of connection and in scale-free graphs few nodes are highly connected whereas a large number of nodes possess a poor degree of connection. Left: A random network. Right: A scale-free graph with two nodes highly connected (hubs).

where the $k_j(\forall j \in N)$ are the degrees of the other nodes. It is evident from this expression that preferential attachment to node $i$ will be more likely if the degree of connection, $k_i$, of node $i$ is large.

According to the Barabasi–Albert model [25] the mean path length $\langle L \rangle$ increases with the number of nodes $N$, but this increase is smaller than that observed for a random network (Fig. 15, left). Moreover the mean clustering coefficient $\langle C \rangle$ of a scale-free network declines with the number of nodes but this decline is smaller than the corresponding decrease observed for a random network (Fig. 15, right). As we shall see in the next section, scale-free networks possess a remarkable property: they are tolerant to many different kinds of random attacks, for instance to genetic mutation that randomly inactivate some enzymes thus leading to node removal.

## 5. *Attack tolerance of networks*

Most biological networks display a remarkable tolerance to external and internal perturbations. An important question is therefore to know the origin of the robustness displayed by metabolic, genetic, and other biological networks. It is striking to note that this robustness is not limited to networks present in living systems. For example, although some of their components quite often malfunction, complex communication networks are usually robust. Local failures rarely generate

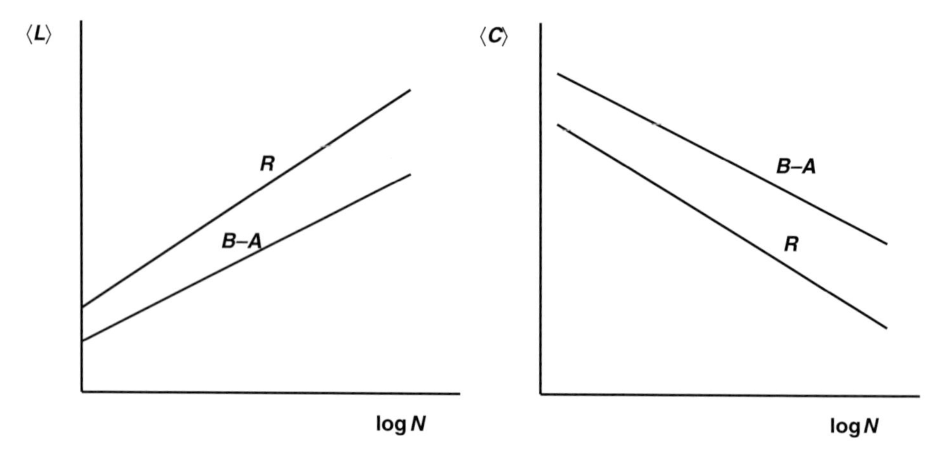


Fig. 15. Variation of the mean path length $\langle L \rangle$ and of the mean clustering coefficient $\langle C \rangle$ as a function of the number of nodes for random and scale-free graphs. Left: Variation of the mean path length $\langle L \rangle$. Both random graphs ($R$) and Barabasi–Albert scale-free graphs ($B$–$A$) are dependent on the number of nodes, but scale-free networks are less sensitive to a variation of the number of nodes. Right: Variation of the mean clustering coefficient $\langle C \rangle$. Both random ($R$) and scale-free graphs ($B$–$A$) are dependent on the number of nodes, but the latter are less sensitive to an increase of the number of nodes. Adapted from Ref. [1].

a global breakdown of network functioning. One can therefore wonder whether robustness of networks can be assigned to their topology. It is possible to test this idea by removing, either randomly or intentionally, nodes or edges of a network and seeing whether or not a majority of nodes remain connected. One will notice that such a situation is the converse of a percolation process. Put in other words, this means there should exist a critical probability, $p_c(N)$, which is both node dependent and involved in the process of node disconnection. One should then expect that if the probability of node removal $p$ is smaller than $p_c(N)$ the network remains, to a large extent, connected. Conversely, if $p > p_c(N)$ it tends to dismantle. Moreover one can notice that node and edge removal are not strictly equivalent. As the removal of a node implies that the corresponding edges cannot function, it is easy to see that node removal generates more damage than edge removal. Upon removal of a fixed number of nodes the network breaks down into giant and smaller clusters and in isolated nodes as well. Under these conditions the distance between some nodes may dramatically increase (Fig. 16). If one removes an equal number of edges much less damage is observed in the network (Fig. 16).

One can express quantitatively [26–29] the effect of node removal, by plotting the variation of size of a giant cluster obtained after node removal as a function of the number of nodes removed. Node removal can be effected randomly or preferentially on the most connected nodes. The effect of node removal can also be followed by plotting the variation of the mean path length $\langle L \rangle$ as a function of the fraction of the nodes removed [26–29]. The results are shown in Fig. 17.

In the case of a random network, the size of the giant cluster declines more sharply for preferential than for random node removal. In the case of preferential

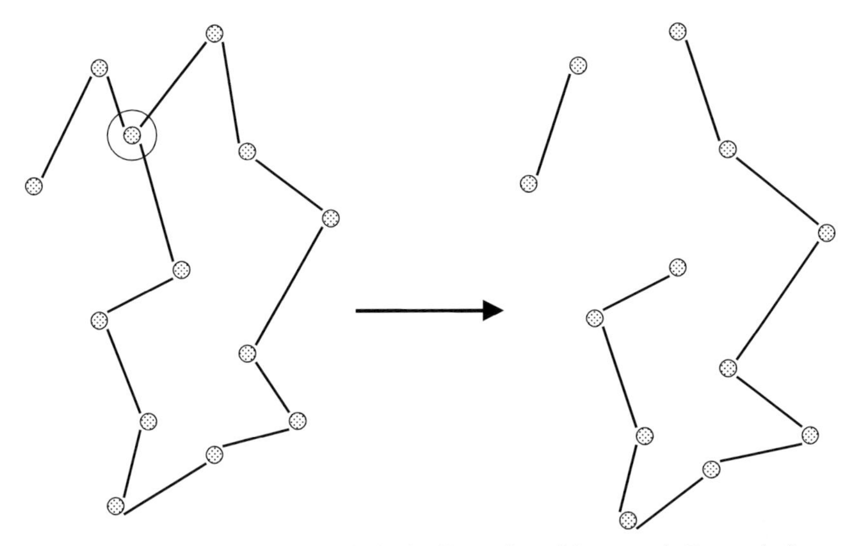

Fig. 16. Removal of a connected node results in the dismantling of the network. Removal of most of the other nodes would not lead to the breaking down of the network. Adapted from Ref. [1].

node removal, the giant cluster breaks down completely for a fraction of nodes removed equal to $f_c = 0.3$. For a random removal, in contrast, the network dismantles only if the fraction of the nodes removed is much higher ($f_c = 0.75$). Similarly, the mean path length $\langle L \rangle$ increases for both random and preferential node removal, and peaks at the critical fraction of nodes removed. Above this critical value it falls off dramatically. This is indeed due to the breakdown of the giant cluster. The situation is qualitatively similar, but quantitatively different, for scale-free graphs. After preferential removal of nodes, the size of the giant cluster falls off more sharply ($f_c = 0.8$) than for a random graph. Alternatively, the giant cluster is more tolerant to random node removal ($f_c = 0.8$) than its random counterpart. As previously, the mean path length $\langle L \rangle$ increases and peaks at the corresponding $f_c$ value. However, in this case, the variation of mean path length is smooth and of weak amplitude (Fig. 17).

The processes that have been described are equivalent to an inverse percolation transition. Below $f_c$, the giant cluster exists and it disappears above this value. From a functional viewpoint, scale-free graphs are more robust than simple random graphs for the attack on the nodes is very likely to be random. This is precisely the case for genetic mutations. If, however these attacks are directed to the most connected nodes the whole system will swiftly dismantle.

## 6. *Towards a general science of networks*

The properties of networks, which have been presented in this chapter, do not appear to be specific to biological systems. It therefore seems legitimate to think that any

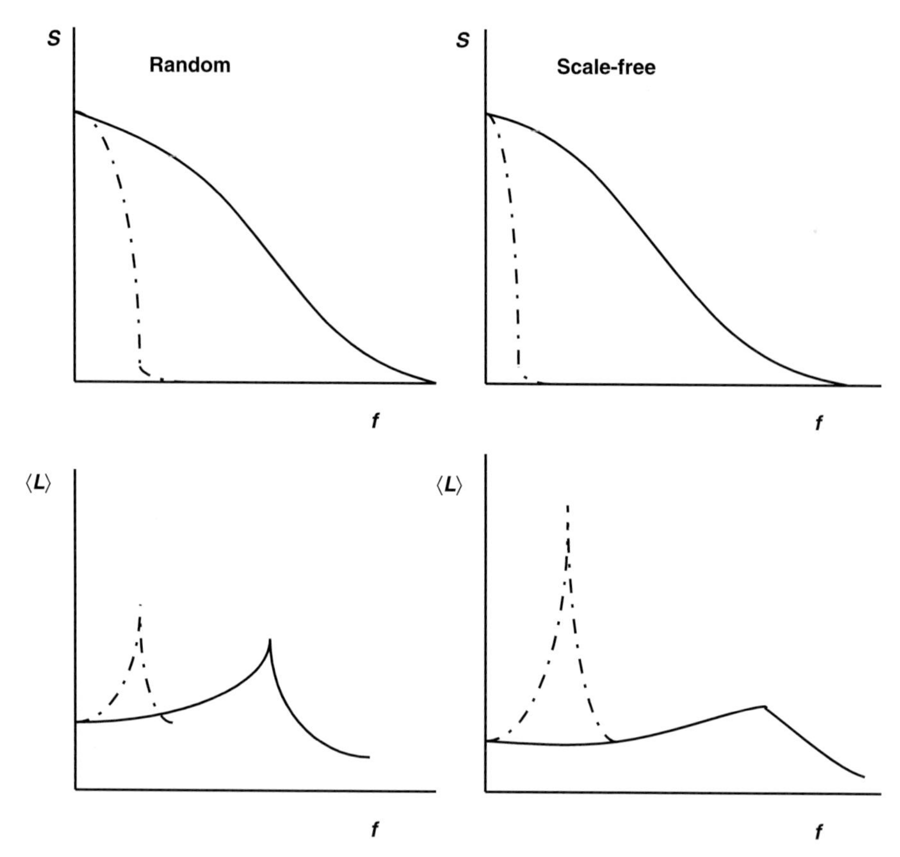


Fig. 17. Effect of random and preferential node removal on the organization of random and scale-free graphs. Left: Random networks. Effect of random node removal (full lines) and preferential removal of highly connected nodes (dotted lines) on size ($S$) and mean path length ($\langle L\rangle$) of random networks. Both effects are expressed as a function of the fraction of nodes removed ($f$). Right: Scale-free networks. The effect of random (full lines) and preferential (dotted line) node removal are analyzed on size ($S$) and mean path length ($\langle L\rangle$) of scale-free networks. See text. Adapted from Ref. [27].

network, whatever its physical nature, shares many properties in common with all other networks. Thus for instance the World Wide Web is a scale-free network [1,29]. The same conclusions seem to apply to Internet [1,7], the film actor collaboration network [8,30], the science collaboration graph [4–6], the telephone call network [31], the science citation network [4–6], and many others [32–50]. Most of these graphs, although not all of them [51], seem to be either small-world or scale-free, or both. The scale-free property implies that during evolution of the network the highly connected nodes tend to establish new links with additional nodes. This situation is sometimes called the "Mathieu effect". A well-known example of this situation is the web of air connections in the United States where only seven vertices in about forty are highly connected leading to the scale-free character of the network. Opposite to this situation is the graph of motorway connections in the USA. The network is random and its node connections, as expected, follow a Poisson distribution.

The small-world and scale-free features of most real-world networks strongly suggest they represent a functional advantage for these graphs. This is precisely the case. Thanks to the small-world character, the nodes are usually much closer than one would have expected for a regular or random graph. An empirical rule in society networks called the six degrees of separation rule [3] states that, at most, six steps separate two nodes even in very large networks. Such a rule implies the existence of a shortcut in the path that connects two nodes and therefore means that the graph is small-world. The scale-free feature also represents a functional advantage for a network. As we have already seen in these graphs, a few nodes are highly connected whereas the vast majority of other vertices are not. This topology is strongly tolerant to internal, or external, random perturbations. In the case of biological networks the perturbations could, for instance, be unfavorable mutations that would result in the disconnection of nodes from the rest of the network. More generally speaking, attack tolerance of a network means that the disconnection of some nodes does not lead to a breakdown of the whole network. If the attack on the nodes occurs randomly it is highly probable that the most connected nodes will not be affected since these nodes represent a small fraction of the graph vertices. The network will then be robust and its global, or collective, properties will not be altered even though some of its local characters are dramatically changed. If, however, the attack is not random but targeted to the most connected vertices the whole system will swiftly collapse.

Another important feature of networks is the occurrence of abrupt changes in their topology similar to percolation processes, or to phase transitions. This means that a complex network can suddenly appear out of disconnected sub-networks or can break down into isolated nodes and sub-networks. In a way, the coalescence of isolated processes as a coherent network can be considered a process of emergence. We shall come back to this matter later in this book.

Perhaps one of the main interests of the science of networks is to bring together and unify scientific fields that until then were considered distinct or even independent, namely statistical mechanics, cell biochemistry and biology, evolutionary biology, ecology, sociology etc.... No doubt the science of networks represents a great leap forward in the attempt to unify sciences.

## *References*


1. Albert, R. and Barabasi, A.L. (2002) Statistical mechanics of complex networks. Rev. Mod. Phys. 74, 47–97.
2. Barabasi, A.L. (2002) Linked: The New Science of Networks. Perseus Publishing, New York.
3. Milgram, S. (1967) The small world problem. Psychol. Today 2, 60–67.
4. Newman, M.E.J. (2001) The structure of scientific collaboration networks. Proc. Natl. Acad. Sci. USA 98, 404–409.
5. Redner, S. (1998) How popular is your paper? An empirical study of the citation distribution. Eur. J. Phys. B 4, 131–134.
6. Seglen, P.O. (1997) The skewness of science. J. Am. Soc. Inform. Sci. 43, 628–638.
7. Faloustos, M., Faloustos, P., and Faloustos, C. (1999) On power-law relationships of the internet topology. Comp. Comm. Rev. 29, 251–262.

8. Watts, D.J. and Strogatz, S.H. (1998) Collective dynamics of "small world" networks. Nature 393, 440–442.
9. Erdos, P. and Renyi, A. (1960) On the evolution of random graphs. Publi. Math. Inst. Hung. Acad. Sci. 5, 17–61.
10. Bollobas, B. (1985) Random Graphs. Academic Press, London.
11. Kauffman, S.A. (1995) At Home in the Universe. Penguin Books, London.
12. Kauffman, S.A. (1993) The Origins of Order. Oxford University Press, New York.
13. Stauffer, D. and Aharony, A. (1992) Introduction to Percolation Theory. Taylor and Francis, London.
14. Bunde, A. and Havlin, S. (eds.) (1994) Fractals in Science. Springer-Verlag, Berlin.
15. Bunde, A. and Havlin, S. (eds.) (1996) Fractals and Disordered Systems. Springer-Verlag, Berlin.
16. Grimmett, G. (1999) Percolation. Springer-Verlag, Berlin.
17. Adami, C. (1998) Introduction to Artificial Life. Springer-Verlag, Berlin.
18. Ma, S.K. (1976) Modern Theory of Critical Phenomena. Benjamin, Reading.
19. Stanley, H.E. (1971) Introduction to Phase Transitions and Critical Phenomena. Oxford University Press, New York.
20. Karp, P.D., Kummenacker, M., Paley, S., and Wagg, J. (1999) Integrated pathway-genome databases and their role in drug discovery. Trends Biothchnol. 17, 275–281.
21. Kanehisa, M. and Goto, S. (2000) KEGG: Kyoto encyclopedia of genes and genomes. Nucleic Acid Res. 28, 27–30.
22. Overbeck, R. *et al.* (2000) WITT: integrated system for high-throughput genome sequence analysis and metabolic reconstruction. Nucleic Acid Res. 28, 123–125.
23. Jeong, H., Tombor, B., Albert, R., Oltvai, Z.N., and Barabasi, A.L. (2000) The large scale organization of metabolic networks. Nature 407, 651–654.
24. Fell, D.E. and Wagner, A. (2000) The small world of metabolism. Nature Biotech. 18, 1121–1122.
25. Barabasi, A.L. and Albert, R. (1999) Emergence of scaling in random networks. Science 286, 509–512.
26. Barabasi, A.L. and Albert, R. (1999) Mean-field theory for scale-free networks. Physica A 272, 173–197.
27. Albert, R., Jeong, H., and Barabasi, A.L. (2000) Error and attack tolerance of complex networks. Nature 406, 378–382.
28. Albert, R. and Barabasi, A.L. (2000) Topology of evolving networks: local events and universality. Phys. Rev. Lett. 85, 5234–5237.
29. Broder, A. *et al.* (2000) Graph structure in the web. Comput. Netw. 33, 309–320.
30. Wasserman, S. and Faust, K. (1994) Social Network Analysis: Methods and Applications. Cambridge University Press, Cambridge.
31. Abello, J., Buchsbaum, A., and Westbrook, J. (1998) A functional approach to external graph algorithms. Lect. Notes Comput. Sci. 1461, 332–343.
32. Strogatz, S.H. (2001) Exploring complex networks. Nature 410, 268–276.
33. Williams, R.J. and Martinez, N.D. (2000) Simple rules yield complex food webs. Nature 404, 180–183.
34. Bahlla, U.S. and Iyengar, R. (1999) Emergent properties of networks of biological signalling pathways. Science 283, 381–387.
35. Li, R.D. and Erneux, T. (1992) Preferential instability in arrays of coupled lasers. Phys. Rev. A 46, 4252–4260.
36. Fabiny, L., Colet, P., Roy, R., and Lenstra, D. (1993) Coherence and phase dynamics of spatially coupled solid state lasers. Phys. Rev. A 47, 4287–4296.
37. Kourtchatov, S.Y., Kikhanskii, V.V., Naparotovich, A.P., Arechi, F.T., and Lapucci, A. (1995) Theory of phase locking of globally coupled laser arrays. Phys. Rev. A 52, 4089–4094.

38. Kozyreff, G., Vladimorov, A.G., and Mandel, P. (2000) Global coupling with time-delay in an array of semiconductor lasers. Phys. Rev. Lett. 85, 3809–3812.
39. Kuramoto, Y. (1984) Chemical Oscillations, Waves and Turbulence. Springer Verlag, Berlin.
40. Wiesenfeld, K., Colet, P., and Strogatz, S.H. (1998) Frequency locking in Josephson arrays: connections with the Kuramoto model. Phys. Rev. E 57, 1563–1569.
41. Turcote, D.L. (1997) Fractals and Chaos in Geology and Geophysics (second edition). Cambridge University Press, Cambridge.
42. May, R.M. (1973) Stability and Complexity in Model Ecosystems. Princeton University Press.
43. Levin, S.A., Greefell, B.T., Hastings, A., and Perelson, A.S. (1997) Mathematical and computational challenges in population biology and ecosystem science. Science 275, 334–343.
44. Arbib, M. (ed.) (1997) The Handbook of Brain Theory and Neural Networks. MIT Press, Cambridge, MA.
45. Stein, D.L. (ed.) (1989) Lectures in the Sciences of Complexity. Addison-Wesley, Reading, MA.
46. Van Wiggeren, G.D. and Roy, R. (1958) Communication with chaotic lasers. Science 279, 1198–1200.
47. Abbot, L.F. and Van Vreeswijk, C. (1993) Asynchronous states in neural networks of pulse-coupled oscillators. Phys. Rev. E 48, 1483–1490.
48. Winfree, A.T. (1967) Biological rythms and the behaviour of populations of coupled oscillators. J. Theor. Biol. 16, 15–42.
49. Winfree, A.T. (1980) The Geometry of Biological Time. Springer Verlag, New York.
50. Strogatz, S.H. (2000) From Kuramoto to Crawdoed: exploring the onset of synchronization in populations of coupled oscillators. Physica D 143, 1–20.
51. Amaral, L.A.N., Scala, A., Barthelemy, M., and Stanley, H.E. (2000) Classes of behaviour of small-world networks. Proc. Natl. Acad. Sci. USA 97, 11149–11152.

J. Ricard *Emergent Collective Properties, Networks and Information in Biology*

DOI: 10.1016/S0167-7306(05)40004-6

CHAPTER 4

# Information and communication in living systems

J. Ricard

Living systems are able to communicate, i.e., they can exchange information. Communication processes are based upon an organization which is common to both man-made and biological systems. A first purpose of this chapter is to derive a number of concepts, such as that of entropy or self-information, defined as the monovariate moment of a distribution of the log of a probability of occurrence of an event. In this chapter, are also discussed the concepts of joint and conditional entropies, as well as that of mutual information, with the aim of showing that the very concept of information is based on the ability of a system (for instance a communication channel) to associate signs in a specific manner so as to allow communication of a message. The relations between communication and mapping, the subadditivity principle, and the laws of coding are also discussed. The subadditivity principle is of particular interest. It states that the joint entropy of the channel is at most equal (and in general smaller) to the sum of the entropies of the source and the destination. This principle prevents the channel from generating its own information. The last part of this chapter is devoted to the analysis of information transfer between DNA and proteins viewed as a communication channel.

*Keywords:* Additivity and monotonocity of uncertainty functions, central dogma, code word, coding, codons, communication channel between DNA and proteins, communication and mapping, communication system, conditional entropy, entropy, genetic code, joint entropy, mean uncertainty functions and Stieltjes integral, mutual information, nonextensive entropies, prefix condition, subadditivity principle, Szilard–Kraft inequalities, triplet, uncertainty functions.

Living systems are able to communicate, i.e., they can exchange information. This property occurs at all levels of biological systems, cellular, multicellular, organismic, and populational levels. In this chapter, we shall only consider the deoxyribonucleic acid (DNA)–protein communication system. True communication processes are based upon an organization which is common to both man-made and biological systems. The theoretical analysis of these systems allows one to derive a number of concepts, such as those of entropy and information, that are rather general and, as we shall see later, not restricted to communication processes.

## *1. Components of a communication system*

Figure 1 shows a communication system, like those commonly used by engineers. The main feature of a communication device is to transmit messages, emitted by a source through a channel, to a destination. The source generates a sequence of signs, or letters, through a Markov process. The message is therefore encoded in the sequence of symbols that constitute an alphabet. During the communication process, random alterations of the message usually occur. These errors are called the noise of the system. The message altered by the noise is decoded to a new alphabet and transmitted to the destination. Hence, communication systems display three main features:

- in practice they often act in one direction, although they could, in principle, display potential symmetry;
- they possess a code that allows correspondence between the alphabets of the source and of the destination;
- from a mathematical viewpoint, the code is the mapping of the letters of an alphabet of a probability space, $\Omega_X$, onto the letters of a different alphabet of another probability space, $\Omega_Y$.

From a functional viewpoint, the components of the DNA–protein communication system, as well as its design, is basically the same as that of systems used by communication engineers and already referred to. The genetic message is stored in the DNA sequence of four base pairs, namely adenine–thymine, thymine–adenine, cytosine–guanine, and guanine–cytosine. This message is then transferred from the DNA alphabet to the messenger ribonucleic acid (mRNA) alphabet. This is the so-called transcription process in which a message written in the DNA alphabet is expressed in the RNA alphabet. The letters of this new alphabet involve the bases adenine, uracil, cytosine, and guanine. mRNA which, by contrast to double-stranded DNA, is a single-stranded molecule, plays the part of the channel that conveys the genetic message to the ribosomes. These organelles can be considered decoders of the message. Whereas both the DNA and mRNA messages are expressed in a well-known 64-letter alphabet, the decoding leads to proteins, which are the translation of this message expressed in a 20-letter alphabet, namely the 20 aminoacids of natural proteins. The genetic code, which will be discussed later, is therefore the mapping of mRNA code words onto protein code words. Hence, as required for an onto type of mapping, the genetic code is degenerate, i.e., several code words should correspond to the same protein code word. Aminoacylated transfer RNAs (tRNAs) are transported to the ribosomes and recognize the code

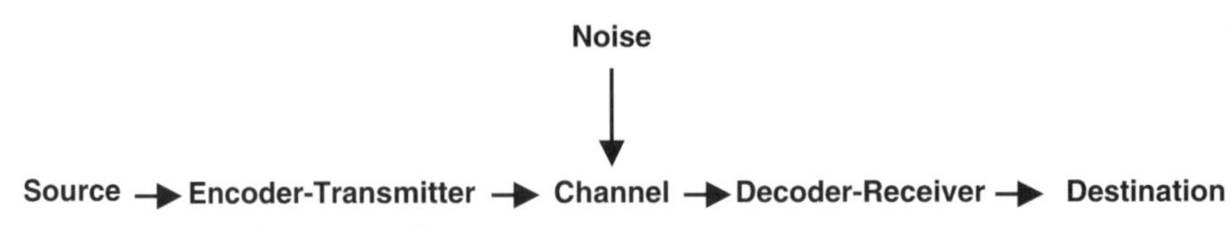


Fig. 1. A communication system. See text.

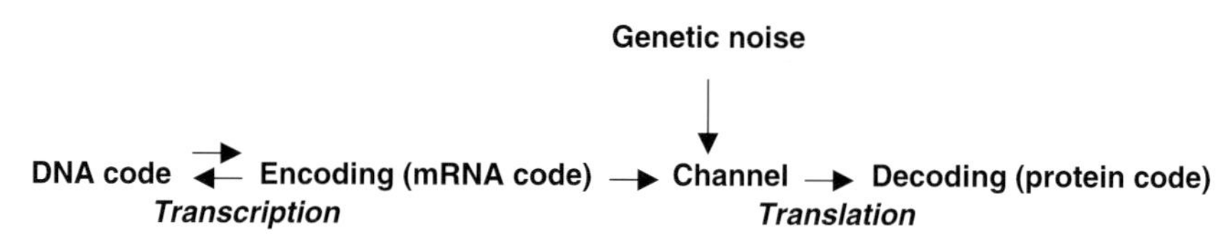


Fig. 2. The DNA–protein communication system. See text.

words of the mRNA message. The aminoacids polymerize in sequence to form a polypeptide chain that dissociates from the transfer RNAs. The genetic message is then expressed in protein code. The genetic noise takes place at all levels of the communication process. Thus, for instance, point mutations alter the sequences of both DNA and mRNA. Hence transfer RNAs may possibly be mischarged with wrong aminoacids. Some of the mistakes can, however, be corrected. The DNA–protein communication system is shown in Fig. 2.

## 2. *Entropy and information*

The concept of information is related to that of uncertainty. The larger the uncertainty of an event, the larger is its potential information. This statement appears, at first sight, somewhat paradoxical. But the paradox is only apparent. One can easily realize that the smaller the probability of occurrence of an event, the larger is its potential information, if the event occurs [1–8]. Alternatively, if the occurrence of an event is certain it brings us no information when it occurs. Let us consider a Markov process that generates a sequence of discrete events $x_1, \ldots, x_i, \ldots, x_N$ that can be considered the letters of an alphabet. Each of these events is granted a certain probability $p(x_1), \ldots, p(x_i), \ldots, p(x_N)$. The corresponding probability space is therefore characterized by an alphabet $X$ and a probability vector $p_X$.

A function aimed at measuring the degree of uncertainty should be a function of the probability of occurrence, $p(x_i)$, of the various events, $x_i$. Moreover, it should meet two axioms: that of monotonicity and that of additivity. By monotonicity it is meant that the uncertainty of the sequence $x_1, \ldots, x_i, \ldots, x_N$ increases as the number of events of this sequence becomes larger. This axiom implies that the uncertainty, $h(x_i)$, of each event should be a function of $1/p(x_i)$, i.e.,

$$h(x_i) = f\{1/p(x_i)\} \tag{1}$$

hence the mean uncertainty of the sequence should be

$$H(X) = \sum_i p(x_i)h(x_i) \tag{2}$$

The axiom of additivity implies that if we have two independent events $x_i$ and $y_j$ pertaining to different alphabets, the joint uncertainty, $h(x_i, y_j)$, of the

where, as usual for continuous functions (Chapter 2), $dF(x)$ and $dF(y)$ are defined as

$$\begin{aligned} dF(x) &= p(x)dx \\ dF(y) &= p(y)dy \end{aligned} \tag{15}$$

Under these forms, $H(X)$ and $H(Y)$ are called Shannon's entropies, or self-information. They are in fact the first monovariate moments of the distributions of $-\log_2 p(x)$ and $-\log_2 p(y)$. One can now raise the question of the maximum possible value of the entropy [10]. If all the probabilities of an event are equal and have the same value, $1/M$, one has

$$p(x_i) = \frac{1}{M} \tag{16}$$

and

$$H(X) = -\sum_{i=1}^{M} p(x_i)\log_2 p(x_i) = -\frac{1}{M}M\log_2\frac{1}{M} = \log_2 M \tag{17}$$

One can now demonstrate that $\log_2 M$ is the maximum value of entropy. In order to do so, one has to demonstrate that

$$H(X) - \log_2 M \leq 0 \tag{18}$$

Hence, one has

$$H(X) - \log_2 M = -\sum_{i=1}^{M} p(x_i)\log_2 p(x_i) - \sum_{i=1}^{M} p(x_i)\log_2 M \tag{19}$$

and

$$H(X) - \log_2 M = \sum_{i=1}^{M} p(x_i)\log_2\frac{1}{p(x_i)M} \tag{20}$$

Moreover, for any positive number $x$, one has $\ln x \leq x - 1$. Hence it follows that

$$\log_2\frac{1}{p(x_i)M} \leq \frac{1}{\ln 2}[1/p(x_i)M - 1] \tag{21}$$

Therefore one has

$$H(X) - \log_2 M \leq \frac{1}{\ln 2}\sum_{i=1}^{M} p(x_i)[1/p(x_i)M - 1] \tag{22}$$

and since

$$\frac{1}{\ln 2}\sum_{i=1}^{M} p(x_i)\left[\frac{1}{p(x_i)M} - 1\right] = \frac{1}{\ln 2}\sum_{i=1}^{M}\left[\frac{p(x_i)}{p(x_i)M} - p(x_i)\right] = 0 \tag{23}$$

it follows that

$$H(X) - \log_2 M \leq 0 \tag{24}$$

and $\log_2 M$ is the maximum entropy value that can be reached when the probabilities of the events are all the same.

One can also define a mean joint uncertainty function, or joint entropy, $H(X, Y)$, as

$$H(X, Y) = -\sum_i \sum_j p(x_i, y_j) \log_2 p(x_i, y_j) \tag{25}$$

or, in a more general form

$$H(X, Y) = -\int_{-\infty}^{+\infty} \int_{-\infty}^{+\infty} \log_2 p(x, y) dF(x, y) \tag{26}$$

where, for continuous functions

$$dF(x, y) = p(x, y) dx dy \tag{27}$$

Under this form, it becomes obvious that Shannon joint entropy is the first bivariate moment of the distribution of $-\log_2 p(x_i, y_j)$. One can easily demonstrate that if the events $x_i$ and $y_j$ are mutually independent, then the joint entropy $H(X, Y)$ is equal to the sum of individual entropies $H(X)$ and $H(Y)$. If the events $x_i$ and $y_j$ are independent, then

$$p(x_i, y_j) = p(x_i) p(y_j) \tag{28}$$

and $H(X, Y)$ assumes the form

$$H(X, Y) = -\sum_i \sum_j p(x_i, y_j)[\log_2 p(x_i) + \log_2 p(y_j)] \tag{29}$$

Moreover, one has

$$-\sum_i \sum_j p(x_i) p(y_j) \log_2 p(x_i) = -\sum_i p(x_i) \log_2 p(x_i) \tag{30}$$

for $\sum p(y_j) = 1$. Similarly, one finds

$$-\sum_i \sum_j p(x_i) p(y_j) \log_2 p(y_j) = -\sum_j p(y_j) \log_2 p(y_j) \tag{31}$$

for $\sum p(x_i) = 1$. Comparing expressions (30) and (31) to Eq. (7) yields

$$H(X, Y) = H(X) + H(Y) \tag{32}$$

If the events $x_i$ and $y_j$ are not mutually independent, then

$$p(x_i, y_j) = p(x_i)p(y_j/x_i) = p(y_j)p(x_i/y_j) \tag{33}$$

which is the well-known Bayes relationship. One can also define conditional entropies, $H(X/Y)$ and $H(Y/X)$, in different but equivalent ways, namely

$$\begin{aligned} H(X/Y) &= -\sum_i \sum_j p(x_i, y_j) \log_2 p(x_i/y_j) \\ H(Y/X) &= -\sum_i \sum_j p(x_i, y_j) \log_2 p(y_j/x_i) \end{aligned} \tag{34}$$

or

$$\begin{aligned} H(X/Y) &= -\sum_i \sum_j p(x_i, y_j) \log_2 \frac{p(x_i, y_j)}{p(y_j)} \\ H(Y/X) &= -\sum_i \sum_j p(x_i, y_j) \log_2 \frac{p(x_i, y_j)}{p(x_i)} \end{aligned} \tag{35}$$

or

$$\begin{aligned} H(X/Y) &= -\sum_i \sum_j p(y_j)p(x_i/y_j) \log_2 p(x_i/y_j) \\ H(Y/X) &= -\sum_i \sum_j p(x_i)p(y_j/x_i) \log_2 p(y_j/x_i) \end{aligned} \tag{36}$$

Making use of the first expression (35) yields

$$H(X/Y) = -\sum_i \sum_j p(x_i, y_j) \log_2 p(x_i, y_j) + \sum_i \sum_j p(x_i, y_j) \log_2 p(y_j) \tag{37}$$

The second term of the right-hand side can be rewritten as

$$\sum_i \sum_j p(x_i, y_j) \log_2 p(y_j) = \sum_i \sum_j p(y_j)p(x_i/y_j) \log_2 p(y_j) \tag{38}$$

Since $\sum_i p(x_i/y_j) = 1$, the above expression reduces to

$$\sum_i \sum_j p(x_i, y_j) \log_2 p(y_j) = \sum_j p(y_j) \log_2 p(y_j) \tag{39}$$

The first term of the right-hand side of Eq. (37) is equal to $H(X, Y)$ and the second (Eq. (39)) to $-H(Y)$. Hence

$$H(X, Y) = H(Y) + H(X/Y) \tag{40}$$

Following the same reasoning, one can demonstrate that

$$H(X,Y) = H(X) + H(Y/X) \tag{41}$$

If the variables $X$ and $Y$ are not independent, then

$$\begin{aligned} H(X/Y) &\neq H(X) \\ H(Y/X) &\neq H(Y) \end{aligned} \tag{42}$$

Under these conditions, $H(X, Y)$ should be different from the sum $H(X) + H(Y)$. The information associated with the communication process, called mutual information, is then defined as

$$I(X:Y) = H(X) + H(Y) - H(X, Y) \tag{43}$$

or

$$H(X, Y) + I(X:Y) = H(X) + H(Y) \tag{44}$$

and is expressed in bits. The expression for $I(X:Y)$ is indeed reminiscent of a covariance. An important question is to know whether information can be positive, negative, or both. This question will be addressed in a forthcoming section of this chapter.

A different, but equivalent, mathematical definition of information can also be given. If, in (43), expression $H(X, Y)$ is replaced by the one obtained from either Eq. (40) or (41), one finds

$$I(X:Y) = H(X) - H(X/Y) \tag{45}$$

or

$$I(X:Y) = H(Y) - H(Y/X) \tag{46}$$

These expressions show that mutual information is a symmetrical function. They can also be rewritten in a different and useful form. Equation (45) above is equivalent to

$$I(X:Y) = -\sum_i p(x_i) \log_2 p(x_i) + \sum_i \sum_j p(x_i, y_j) \log_2 p(x_i/y_j) \tag{47}$$

Moreover one has

$$p(x_i) = \sum_j p(x_i, y_j) \tag{48}$$

Hence, the expression of mutual information assumes the form

$$I(X:Y) = -\sum_i \sum_j p(x_i, y_j) \log_2 p(x_i) + \sum_i \sum_j p(x_i, y_j) \log_2 p(x_i/y_j) \tag{49}$$

and this expression can be rewritten as

$$I(X:Y) = \sum_i \sum_j p(x_i, y_j) \log_2 \frac{p(x_i/y_j)}{p(x_i)} \tag{50}$$

Moreover since

$$\frac{p(x_i/y_j)}{p(x_i)} = \frac{p(y_j/x_i)}{p(y_j)} \tag{51}$$

Eq. (50) is equivalent to

$$I(X:Y) = \sum_i \sum_j p(x_i, y_j) \log_2 \frac{p(y_j/x_i)}{p(y_j)} \tag{52}$$

Expressions (50) and (52) show the potential symmetry of the information transfer between the two alphabets $X$ and $Y$.

### *3. Communication and mapping*

It appears from a previous chapter that communication of a message requires a mapping of the elements of one set onto the elements of another. More precisely, communication between a source and a destination implies that the code words, or the alphabet, $X$, of the probability space $\Omega_X$ associated with the source be mapped onto the alphabet, $Y$, of the probability space $\Omega_Y$ associated with the destination. These two probability spaces will be considered isomorphic if there is a one-to-one mapping, or a bijection, between the letters of the two alphabets $X$ and $Y$ (Fig. 4).

This situation allows, without difficulty, the possibility of communication between two probability spaces. One-to-one mapping, however, is far from universal in both man-made communication devices and living systems. The genetic code, for instance, is the mapping of the mRNA code words, or codons, onto the code words of the protein space. However, as there are 20 different aminoacids in proteins and only four different bases in mRNAs, there cannot be a one-to-one correspondence

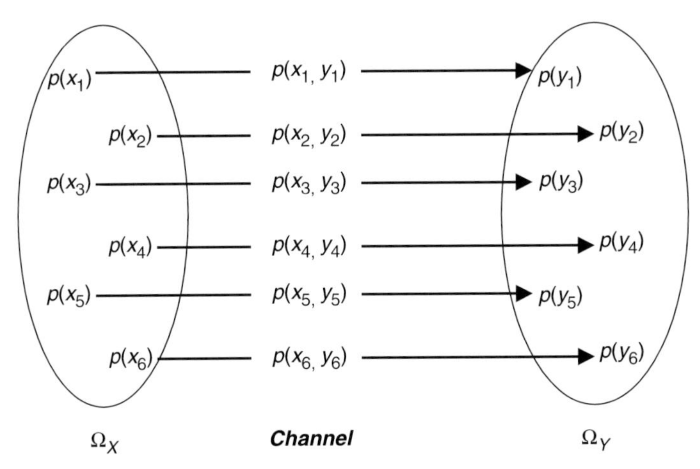


Fig. 4. Communication and mapping. The elements of a probability space $\Omega_X$ are mapped onto the elements of another probability space $\Omega_Y$. From a mathematical viewpoint, communication is a one-to-one mapping of $\Omega_X$ onto $\Omega_Y$.

between the bases of the mRNA space and the aminoacids of the protein space. The code words of the mRNA alphabet contain three bases. But this situation generates in turn 64 different codons, i.e., an excess of messenger code words relative to protein code words. mRNA and protein probability spaces are therefore not isomorphic. The laws that define the correspondence, or mapping, between the letters of the two alphabets are the code.

An important mathematical condition that is required in order to allow the communication of a message between two probability spaces $\Omega_X$ and $\Omega_Y$ is

$$\sum_i p(x_i) = \sum_j p(y_j) = \sum_i \sum_j p(x_i, y_j) = 1 \tag{53}$$

This condition is of necessity fulfilled for isomorphic probability spaces but may, or may not, be satisfied for non-isomorphic spaces. This is illustrated in Figs. 4 and 5. Relationships (53) are fulfilled for probability spaces of Fig. 4 but not for the probability spaces of Fig. 5.

## 4. *The subadditivity principle*

Communication theory is based on a principle called the subadditivity principle the validity of which relies upon expressions (53). We know that for all $x > 0$, then $\ln x \leq x - 1$. Let us consider the function $I^*(X : Y)$ defined as

$$I^*(X : Y) = -3.3219 \log e \sum_i \sum_j p(x_i, y_j) \left[ \frac{p(x_i)}{p(x_i/y_j)} - 1 \right] \tag{54}$$

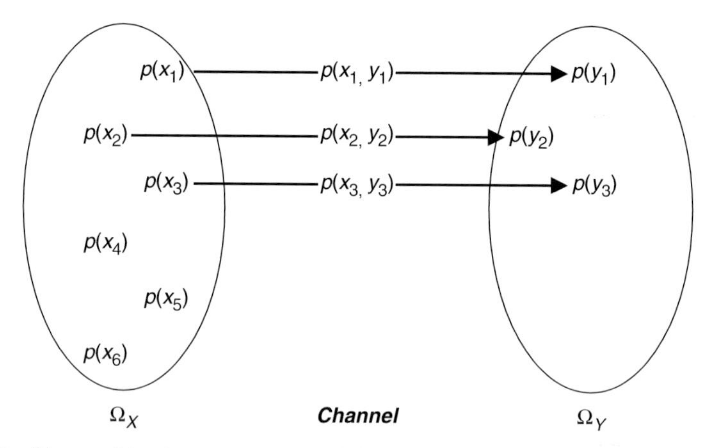


Fig. 5. Lack of isomorphism between two probability spaces. See text.

This expression is to be compared to the expression of $I(X:Y)$ (Eq. (50)) and one has of necessity

$$3.3219 \log e\left[\frac{p(x)_i}{p(x_i/y_j)} - 1\right] \geq \log_2 \frac{p(x_i)}{p(x_i/y_j)} \tag{55}$$

It follows that

$$I(X:Y) \geq I^*(X:Y) \tag{56}$$

Hence, if it is possible to demonstrate that $I^*(X:Y) = 0$, this will imply that $I(X:Y) \geq 0$. Equation (54) can be rewritten as

$$I^*(X:Y) = -3.3219 \log e \sum_i \sum_j \left[\frac{p(x_i, y_j)p(x_i)}{p(x_i/y_j)} - p(x_i, y_j)\right] \tag{57}$$

which reduces to

$$I^*(X:Y) = 3.3219 \log e\left[\sum_i \sum_j p(x_i, y_j) - \sum_i p(x_i) \sum_j p(y_j)\right] \tag{58}$$

If relationships (53) are fulfilled the expression in brackets in Eq. (58) above vanishes and $I^*(X:Y) = 0$. This leads to the conclusion that $I(X:Y) \geq 0$.

The conclusion that information is of necessity positive in a communication process implies that

$$H(X, Y) \leq H(X) + H(Y) \tag{59}$$

which is the mathematical expression of the subadditivity principle. Hence this principle can be formulated in the following way: the sum of individual entropies in a communication process is of necessity larger than, or at most equal to, the joint entropy. This principle implies that

$$\begin{aligned} H(X) &\geq H(X/Y) \\ H(Y) &\geq H(Y/X) \end{aligned} \tag{60}$$

The relationships between the various functions $H(X)$, $H(Y)$, $I(X:Y)$, $H(X/Y)$, and $H(X/Y)$, are illustrated in Fig. 6.

It is now of interest to come back to the significance and the role of these mathematical functions in a communication process, and to understand the intuitive meaning of the subadditivity principle. $H(X)$ and $H(Y)$ represent the uncertainty, hence the self-information, of the source and destination, respectively. $H(X, Y)$ is the uncertainty, or the joint entropy, of the $XY$ pairs during the communication process. $I(X:Y)$ is the information channelled from the source to the destination. $H(X/Y)$ measures the entropy of the source given the entropy of the destination. It is called equivocation. Conversely, $H(Y/X)$ is the entropy of the destination given that of the source, and is termed ambiguity.

One can easily see that if $H(X/Y) = H(Y/X) = 0$, then one has

$$I(X:Y) = H(X) = H(Y) = H(X, Y) \tag{61}$$

In this case, the communication process takes place without any noise. This is precisely why $H(X/Y)$ and $H(Y/X)$ have been called equivocation and ambiguity. Under these conditions, the communication channel is a noiseless channel. Conversely, if $H(X/Y)$ and $H(Y/X)$ are both different from zero, the communication

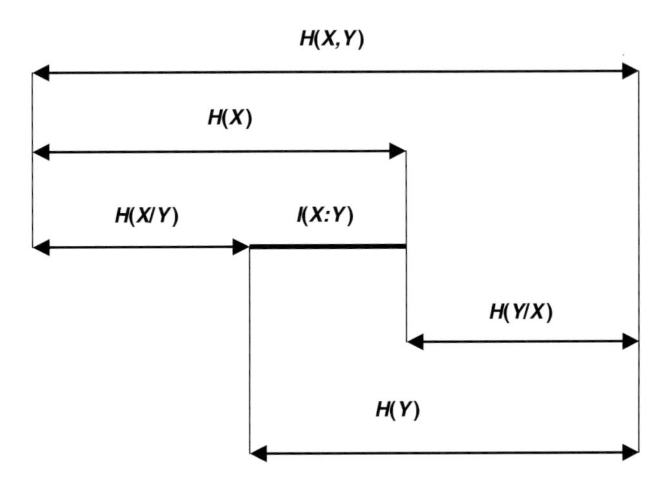


Fig. 6. Relationships between individual, joint, conditional entropies and mutual information. This schematic mode of representation is due to Quastler [32].

channel is a noisy channel. Hence conditional entropies $H(X/Y)$ and $H(Y/X)$ represent a measure of the noise in the channel. Communication through a channel can take place only if the noise is kept at a low level and this is precisely the intuitive meaning of the subadditivity principle. As $H(X)$ depends upon the source only, whereas $H(X/Y)$ relies upon both the source and the noise, the entropy of the source is of necessity larger than its conditional entropy.

## 5. *Nonextensive entropies*

Classical Shannon entropies are indeed extensive functions. It may be of interest to define and use entropies that are both nonextensive and more general than classical entropies discussed so far [1–8]. Tsallis [11–13] has defined such nonextensive entropies as

$$H_q(X) = \frac{1 - \sum_i p(x_i)^q}{q - 1} \tag{62}$$

where $q$ is a real number. $H_q(X)$ is more general than classical entropy $H(X)$ but becomes identical to this function when $q \to 1$. As a matter of fact, one has

$$p(x_i)^q = p(x_i) \exp[(q - 1) \ln p(x_i)] \tag{63}$$

This becomes evident upon rewriting this expression (63) as

$$p(x_i)^{q-1} = \exp[(q - 1) \ln p(x_i)] \tag{64}$$

which in turn becomes

$$(q - 1) \ln p(x_i) = (q - 1) \ln p(x_i) \ln e = (q - 1) \ln p(x_i) \tag{65}$$

Hence relation (62) can be rewritten as

$$H_q(X) = \frac{1 - \sum p(x_i) \exp[(q - 1) \ln p(x_i)]}{q - 1} \tag{66}$$

As $q \to 1$

$$\exp[(q - 1) \ln p(x_i)] = 1 + (q - 1) \ln p(x_i) \tag{67}$$

and one has

$$H_q(X) = \frac{1 - \sum_i p(x_i) - (q - 1) \sum_i p(x_i) \ln p(x_i)}{q - 1} \tag{68}$$

Moreover as

$$\sum_i p(x_i) = 1 \tag{69}$$

expression (68) reduces to

$$H_q(X) = -\sum_i p(x_i) \ln p(x_i) \tag{70}$$

which is the classical expression for an entropy. The same reasoning could indeed hold if ln were to be replaced by $\log_2$. A similar definition of the nonextensive entropy of another alphabet $Y$ would be

$$H_q(Y) = \frac{1 - \sum_j p(y_j)^q}{q - 1} \tag{71}$$

and one can see that [14].

$$H_q(X + Y) = H_q(X) + H_q(Y) + (q - 1)H_q(X)H_q(Y) \tag{72}$$

It thus appears that the entropies are nonextensive unless $q = 1$. This interesting matter will not be discussed any further here.

## 6. *Coding*

Let us consider two probability spaces $\Omega_X$ and $\Omega_Y$. The first is associated with a source and the second with a destination. The alphabet of the source is $X$ and that of the destination is $Y$. A mapping of the letters of the alphabet $X$ onto the letters of the alphabet $Y$ is called a code. The simplest possible situation is obtained when the two alphabets $X$ and $Y$ have the same number of letters. The code then consists in a one-to-one correspondence between these symbols. But this situation is not often encountered in man-made and living systems. A classical type of code consists in the coding of the 26 letters of the latin alphabet by the binary alphabet 0 and 1.

### 6.1. *Code words of identical length*

If the letters are used with the same frequency, it is obvious that one needs several binary symbols to specify each letter of the alphabet. If, for the sake of simplicity, we assume a message to be written in a four-letter alphabet A, B, C, and D, these letters can be specified, in binary symbols as 11, 10, 01, and 00, respectively (Table 1). Similarly, with three binary symbols, one specifies 8 letters A, B, C, D, E, F, G, and H (Table 1).

Table 1
Coding of a four- and eight-letter alphabet with two symbols. See text.

| Four-letter alphabet | | Eight-letter alphabet | |
|---|---|---|---|
| A | 11 | A | 111 |
| B | 10 | B | 110 |
| C | 01 | C | 011 |
| D | 00 | D | 101 |
| | | E | 100 |
| | | F | 010 |
| | | G | 001 |
| | | H | 000 |

One can notice that, in these two cases, the number of letters of the two alphabets are $2^2 = 4$ and $2^3 = 8$, respectively. More generally, coding in binary digits will be both simple and straightforward if the number of letters to be coded is of the form $2^n = N$, where $N$ is the total number of letters of the alphabet (4 and 8) and $n$ the number of binary symbols that have to be associated in order to specify each letter. The number of binary symbols required to form a code word is therefore $n = \log_2 N$. This is precisely why the entropies are usually expressed in logarithms to the base two. One can easily see that, in the two previous examples, we have chosen code words of length $n = 2$ and $n = 3$, i.e., $\log_2 4 = 2$ and $\log_2 8 = 3$.

This situation can easily be extended to the case in which more than two symbols are used to code a message. This is precisely what occurs for the genetic message. This message, written in the mRNA molecule, uses four symbols, namely uracil (U), cytosine (C), adenine (A), and guanine (G). If the code words were made up of two symbols, one could specify $4^2 = 16$ aminoacids only. Hence the code words of the genetic code, or codons, have to be made up of three symbols in order to be able to specify the 20 aminoacids present in protein molecules. But then we have $4^3 = 64$ codons, i.e., an excess of codons relative to the number of aminoacids to be specified. This question will be discussed later.

### *6.2. Code words of variable length*

Many codes use code words of variable length and, in order to decrease the number of symbols used for coding, it is advantageous to assign short code words to the most frequent letters of the alphabet [8,15–21]. Let us consider an alphabet $X$ made up of $N$ independent symbols, the probability of occurrence of the sequence

$$X = [x_1, x_2, \ldots, x_i, \ldots, x_N] \tag{73}$$

is

$$p(X) = \prod_{n=1}^{N} p(x_n) \tag{74}$$

In order to be efficient, coding requires that no code word be the same as the initial part, called prefix, of another code word. Such a code is called a prefix code. Let us call $X_1, X_2, \ldots, X_J$ all the sequences of length $N$ produced by the source. A necessary and sufficient condition for the existence of a prefix code in binary digits is

$$\sum_{j=1}^{N} 2^{-n_j} \leq 1 \tag{75}$$

where the $n_j$ are the lengths of the various code words. If the size of the alphabet is $M$, then expression (75) becomes

$$\sum_{j=1}^{N} M^{-n_j} \leq 1 \tag{76}$$

Expressions (75) and (76), which are known as the Szilard–Kraft inequalities, can be justified easily. Let us consider expression (75) and the binary sequences $X_1, X_2, \ldots, X_j$ that fulfil the prefix condition [8], they can be viewed as binary expansions of real numbers between 0 and 1, namely

$$1011 = 1 \times \left(\frac{1}{2}\right)^1 + 0 \times \left(\frac{1}{2}\right)^2 + 1 \times \left(\frac{1}{2}\right)^3 + 1 \times \left(\frac{1}{2}\right)^4 \tag{77}$$

If the code word $v_j$ has length $n_j$, no other code word can fall in the interval

$$v_j \leq v_j + 2^{-n_j} \tag{78}$$

without violating the prefix condition.

One can also ask whether there is a relationship between the entropy of the source and the mean length of the code words. The mean number of binary digits per source sequence is defined as

$$\langle n \rangle = \sum_{j=1}^{J} p(X_j) n_j \tag{79}$$

and the entropy of the source as

$$H(X) = -\sum_{j=1}^{J} p(X_j) \log_2 p(X_j) \tag{80}$$

Let us consider the difference between these two functions. One has

$$H(X) - \langle n \rangle = -\sum_{j=1}^{J} p(X_j) \log_2 p(X_j) - \sum_{j=1}^{J} p(X_j) n_j \tag{81}$$

As

$$n_j = -\log_2 2^{-n_j} \tag{82}$$

The source of entropy $H(X)$ and the mean number of binary digits $\langle n \rangle$ have the following values

$$H(X) = 0.5728$$
$$\langle n \rangle = 1.1475 \qquad (99)$$

and the condition

$$\langle n \rangle - H(X) = 0.5747 > 0 \qquad (100)$$

is met. Moreover

$$H(X) + 1 = 1.5727 \qquad (101)$$

and, as expected, this value is larger than $\langle n \rangle$.

## *7. The genetic code and the Central Dogma*

The genetic code can be considered a correspondence between the sequence of bases in DNA, or its transcript mRNA, and the sequence of aminoacids in proteins. There are four bases in DNA, namely adenine (A), thymine (T), guanine (G), and cytosine (C). With the exception of thymine, which is replaced by uracil (U), the same bases exist in RNA. As everyone knows today, DNA is double-stranded whereas RNA is not. In the DNA molecule, the bases are associated by pairs, viz. A...T, T...A, C...G, G...C, through hydrogen bonds. Twenty different aminoacids are present in proteins. If each of the four base pairs in DNA (A...T, T...A, C...G, G...C) and each of the four bases in RNA (A, U, G, C) were to code for a single specific aminoacid, the genetic code would then involve only four aminoacids whereas there are up to twenty different aminoacids in proteins. If two base pairs in DNA, or two bases in RNA, code for one aminoacid, only $4^2 = 16$ aminoacids could be specified. Hence we are led to the conclusion that the genetic code requires three base pairs in DNA, and three bases in mRNA, to code for an aminoacid [22,23]. But then the genetic code can specify $4^3 = 64$ aminoacids. This implies either that several triplets of bases code for the same aminoacid or that some triplets do not specify any aminoacid or, last but not least, that both possibilities occur.

The genetic code could *a priori* be overlapping or nonoverlapping. Examples of an overlapping and a nonoverlapping code are shown in Fig. 7. In these schemes A, B,

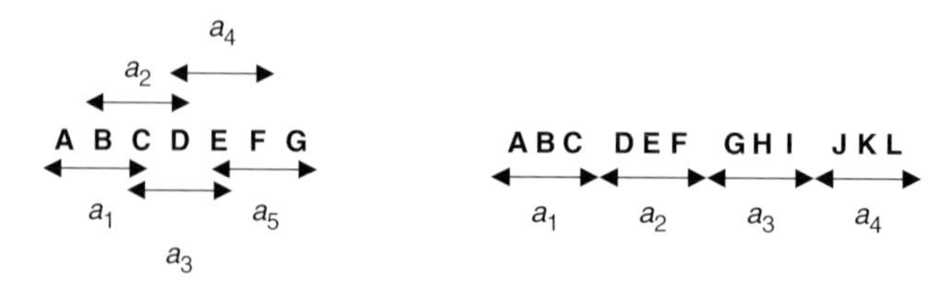


Fig. 7. Overlapping and nonoverlapping codes. See text.

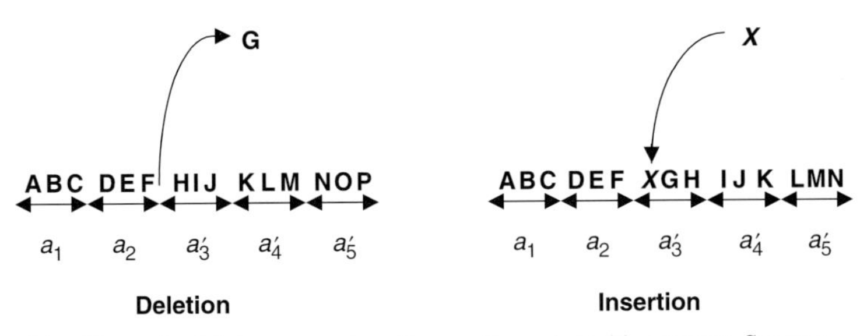


Fig. 8. Possible effects of a deletion or an insertion on the aminoacid sequence. See text.

C, . . . and $a_1, a_2, a_3, \ldots$ represent bases and aminoacids, respectively. Simple genetic observations and experiments can allow one to distinguish between these possibilities. Some mutations correspond to the substitution of one base by another and, with an overlapping code, should result in the substitution of two or three consecutive aminoacid residues in the polypeptide chain. This is usually not the case. Many mutations result in the change of one aminoacid only. This implies that the code is of the nonoverlapping type. Moreover one could expect that adjacent triplets be separated by bases playing the part of a "comma". This, again, is not the case. As a matter of fact, mutations corresponding to base deletion or insertion, as shown in Fig. 8, result in a frame shift that changes the whole aminoacid sequence starting at the point where insertion or deletion has taken place [24]. The existence of frame shifts leads to the possibility that the same DNA sequence be read in different manners leading in turn to the synthesis of different polypeptide chains [25].

The genetic code was broken, at least in part, through the use of synthetic trinucleotides, polyribonucleotides, and copolymers of ribonucleotides of defined sequence [26,27]. These polymers can be obtained by polymerizing ribonucleoside bisphosphates to RNAs thanks to the enzyme polynucleotide phosphorylase. As polymerization occurs randomly, one can obtain copolymers of different composition by simply changing the proportions of ribonucleotides. One can then use these copolymers as models of mRNA that can be translated in a cell-free extract. However, the use of copolymers as templates reveals the composition of codons not their sequence. A major step in breaking the genetic code was achieved in 1964 by Nirenberg who observed that synthetic trinucleotides promote the binding of specific tRNAs to ribosomes even in the absence of protein synthesis. An immediate consequence of this result is that synthetic trinucleotides mimic a triplet of bases of ribosome-bound RNA that is itself associated with a specific tRNA for which it is a code word. Moreover as it is known that aminoacids can be bound to the 3′-end of tRNAs, these RNAs play the part of adaptors that define the exact position of an aminoacid residue in the polypeptide chain. The triplet of bases complementary to a given codon of mRNA is located on a loop of tRNA and is called the anticodon (Fig. 9).

All sixty-four trinucleotides have been synthesized and the binding of a specific tRNA has been assayed for each trinucleotide. As these tRNAs can bind aminoacids at their 3′-end, one can define a correspondence between the sequence of trinucleotides and the tRNA-bound aminoacids. This correspondence defines

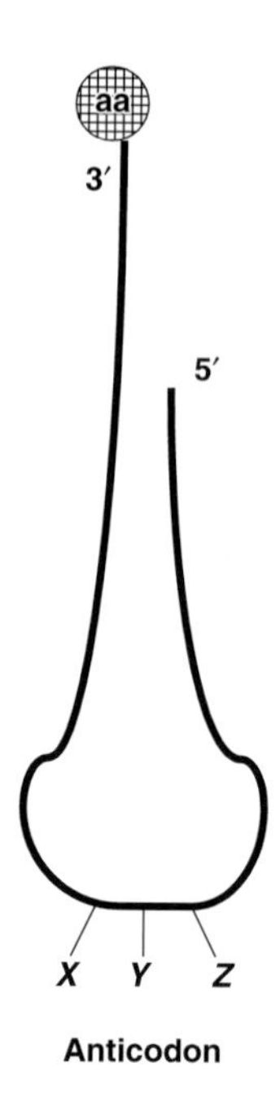


Fig. 9. tRNA with its anticodon. See text.

what is called the genetic code. This technique together with the use of copolymers with a defined sequence allowed biologists to break the genetic code.

The correspondence between the sixty-one codons and the twenty aminoacids was then established. Three triplets (UAA, UAG, and UGA) do not code for any aminoacid but rather for chain termination. The genetic code is then highly degenerate and most of the aminoacids are coded by several codons (Fig. 10). Only two of them, viz. tryptophan and methionine, are each coded for by one triplet. With the exception of mitochondrial and chloroplast DNA, the genetic code is universal.

From a mathematical viewpoint, the genetic code can be considered a mapping of a 64-letter alphabet onto a 20-letter alphabet. Hence the two probability spaces are not isomorphic. This situation is illustrated in Fig. 10 and has two important consequences that have been mentioned earlier. Owing to the lack of isomorphism, the code is degenerate and three codons (UAA, UAG, and UGA) do not code for any aminoacid and allow termination of the genetic message. It is on the third nucleotide of the codons that the variation of the nucleotides is the largest.

A first consequence of the degeneracy of the code is that it limits the impact of mutations. If, for instance, six different codons code for the same aminoacid, different changes of the code word will not affect the aminoacid sequence. The second consequence is that the genetic code is associated with a so-called Central Dogma [28]. The Central Dogma has been formulated by Crick as a biological property defined in the specific context of genetic expression. It is in fact a general property of *any* code in which a source alphabet is larger than a destination alphabet.

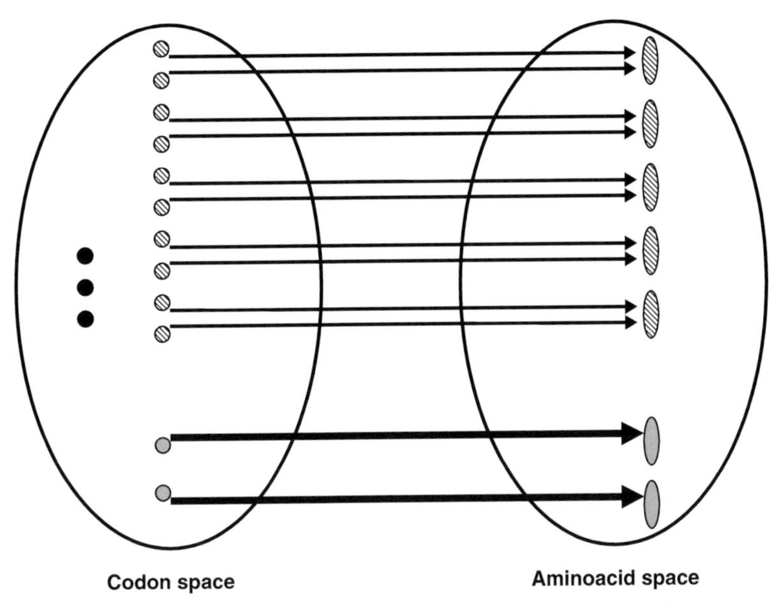


Fig. 10. Most aminoacids can be specified by different triplets. In the codon space three codons (the three black dots) do not code for any aminoacid, two codons (shown at the bottom of the codon space) code for one aminoacid each, and most of the aminoacids are coded by several codons. In order to make the figure easy to read, two codons only pertain to each aminoacid. This is indeed pure convention.

In other words, the Central Dogma reflects the lack of isomorphism between two probability spaces, and this defines a direction of the information flow. If we consider the gene and mRNA spaces, they are isomorphic. Hence, information can flow in either direction, i.e., from DNA to mRNA, or from mRNA to DNA. This is in fact what has been found experimentally. The information flow from mRNAs to proteins is indeed allowed, but it is strictly forbidden in the opposite direction. As outlined by Yokey [7], this is not, as thought initially [28], a specific biological property, but a property of *any* code in which the two probability spaces are not isomorphic. By the same token, this principle does not forbid information transfer from protein to protein and this conclusion is indeed at variance with the classical formulation of the Central Dogma [28]. It seems that prions, i.e., proteinaceous infective agents devoid of nucleic acids, can transfer an information to other prion molecules, therefore allowing propagation of the scrapie disease. Likewise, it will be shown in a forthcoming chapter that induced conformation changes that occur in protein edifices could correspond in fact to real information transfer between polypeptide chains.

## *8. Accuracy of the communication channel between DNA and proteins*

It has already been outlined that noise takes place during emission, conduction, and reception of a message. This is mathematically expressed by the result that the information transferred from a source to a destination is of necessity smaller than the source entropy. The loss of entropy during emission, conduction, and reception of a message is then quantitatively expressed by the corresponding conditional entropy. Since the loss of entropy is of necessity smaller than the source entropy, the subaddivity principle must hold in a communication process. Although this principle is of necessity met in communication processes, it is of interest to discuss briefly what the significance could be of lack of subadditivity. Such a situation would imply that the communication channel generates its own information or, put in other words, there is *emergence* of novel information in the communication channel. This is indeed impossible during a communication process, but we shall see in the next chapter that networks can display such an unexpected property. In this perspective, the lack of subadditivity can be considered a criterion of emergence in a system. Last but not least, communication of a message requires a specific arrangement of signs, or of molecular signals, in the communication channel. The concept of association of molecular signals is therefore common to communication and organization processes.

These general considerations indeed apply to the information flow between DNA and proteins. During this communication process, the amplitude of the noise can, to a certain extent, be limited to a low level thanks to editing mechanisms [29]. Thus an important step in protein synthesis is the aminoacylation of transfer RNA (tRNA), i.e., the binding of an aminoacyl residue to a RNA which is then able to recognize, in a specific manner, a given codon of the mRNA. The overall process must be highly specific since any misacylation of the tRNA would result in a scrambling of the aminoacid sequence, even to the suppression of the corresponding protein. In order to limit noise during the communication process, misacylated tRNA can be hydrolyzed before incorporation of the "wrong" aminoacid in the polypeptide chain.

The problem of the communication between genes and proteins raises another question. Roughly speaking, a gene can be approximated to a linear entity whereas a protein usually has a globular structure. One can therefore wonder whether the four base pairs in DNA, or the codons in mRNAs, are sufficient to define, in terms of information theory the three-dimensional structure of proteins. Since the work of Anfinsen [30] it has often been claimed that the aminoacid sequence of a protein is sufficient to define *one* globular structure only. This is no doubt true in a number of cases. However, proteins in aqueous media display internal movements and often exist under different conformation states. It is therefore difficult to imagine that there is the same entropy in a sequence of code words and in the dynamic three-dimensional structure of a protein. If this remark were sound, this would imply that the relationship between DNA, or mRNA, and proteins would be more complex than a mere communication process defined by the genetic code. This conclusion is made even more obvious if one recalls that nascent protein chains quite often undergo post-translational transformations such as specific protein cleavages, methylation, etc....

A last question, which is perhaps worth mentioning now, is to know whether gene DNA can be assimilated to a computer program. This view has been implicitly, or explicitly, adopted by most biologists. It has, however, been outlined by Atlan [31] that this identification is no more than a metaphor and that there is no more reason to believe that DNA is analogous to a computer program rather than to the data of another program. If this view were correct, it would imply that this program is the enzymatic machinery that allows treatment of the data stored in the DNA. In spite of its philosophical interest, this debate is beyond the scope of the present chapter.

## *References*


1. Shannon, C.E. (1948) A mathematical theory of communication. Bell System Technical Journal 27, 379–423.
2. Shannon, C.E. (1948) A mathematical theory of communication. Bell System Technical Journal 27, 623–656.
3. Shannon, C.E. (1949) The Mathematical Theory of Communication. University of Illinois Press, Urbana.
4. Kullback, S. (1959) Information Theory and Statistics. Wiley and Sons, New York.
5. Fano, R.M. (1961) The Transmission of Information. MIT Press and Wiley and Sons, New York.
6. Cover, T.M. and Thomas, J.A. (1991) Elements of Information Theory. Wiley and Sons, New York.
7. Yokey, H.P. (1992) Information Theory and Molecular Biology. Cambridge University Press.
8. Callager, R.G. (1964) Information theory. In: Margenau, H. and Murphy, H. (eds.) The Mathematics of Physics and Chemistry. Vol. II, 190–248.
9. Kendall, M.G. (1948) The Advanced Theory of Statistics. Vol. 1 Charles Griffin and Company, London.
10. Atlan, H. (1972) L'Organisation Biologique et la Théorie de l'Information. Herman, Paris.
11. Tsallis, C. (1988) Possible generalization of the Boltzmann-Gibbs statistics. J. Stat. Phys. 52, 479–487.
12. Tsallis, C. (1998) The role of constraints within generalized nonextensive statistics. Physica A 261, 534–554.
13. Tsallis, C. (2001) On mixing and metaequilibrium in nonextensive statistics. Physica A 302, 187–192.
14. Yamano, T. (2001) Information theory based on nonadditive information content. Physical Review E 63, 046105-1–046105-7.
15. McMillan, B. (1953) The basic theorems of information theory. Ann. Math. Stat. 24, 196–219.
16. Wolfowitz, J. (1961) Coding Theorems of Information Theory. Springer Verlag, Berlin and Prentice-Hall, Englewood Cliffs, NJ.
17. Feinstein, A. (1958) Foundations of Information Theory. McGraw-Hill Book Company, New York.
18. Peterson, W. (1961) Error Correcting Codes. MIT Press and John Wiley and Sons, New York.
19. Wozencraft, J.M. and Reiffen, B. (1961) Sequential Decoding. MIT Press and John Wiley and Sons, New York.
20. Callager, R.G. (1963) Low Density Parity Check Codes. MIT Press, Cambridge.
21. Massey, J.L. (1963) Threshold Decoding. MIT Press, Cambridge.
22. Crick, F.H.C., Barnett, L., Brenner, S., and Watts-Tobin, R.J. (1961) General nature of the genetic code for proteins. Nature 192, 1227–1232.
23. Crick, F.H.C. (1966) The genetic code III. Sci. Amer. 215(4), 55–62.
24. Garen, A. (1968) Sense and nonsense in the genetic code. Science 160, 149–159.
25. Farabaugh, P.J. (1996) Programmed translational frameshifting. Microbiol. Rev. 60, 103–134.
26. Nirenberg, M.W. (1963) The genetic code II. Sci. Amer. 208(3), 80–94.
27. Nirenberg, M.W. (1968) The genetic code. Nobel Lectures: Physiology or Medicine (1963–1970), 341–369 American Elsevier.

28. Crick, F.H.C. (1970) Central dogma of molecular biology. Nature 227, 561–563.
29. Hopfield, J.J. (1974) Kinetic proofreading: a new mechanism for reducing errors in biosynthetic processes requiring high specificity. Proc. Natl. Acad. Sci. USA 71, 4135–4139.
30. Anfinsen, C.B. (1973) Principles that govern the folding of protein chains. Science 181, 223–230.
31. Atlan, H. (1990) The cellular computer DNA: program or data? Bull. Mathem. Biol. 52, 335–348.
32. Quastler, H. (1958) A primer on information theory. In: Yokey, H.P., Platzman, R.L., and Quastler, H. (eds.) Symposium on Information Theory in Biology, Pergamon Press, New York.

J. Ricard *Emergent Collective Properties, Networks and Information in Biology*

DOI: 10.1016/S0167-7306(05)40005-8

CHAPTER 5

# Statistical mechanics of network information, integration, and emergence

J. Ricard

In the previous chapter, it was shown that the concept of information is based on the ability of a system to associate signs in a specific manner so as to allow a function. This is precisely what many biological networks do. Thus, for instance, enzymes specifically associate chemical substances (the "substrates" of the chemical reaction) as to form reaction products. Hence, from that respect, a chemical reaction possesses information. From a general viewpoint, many networks, whatever their physical nature and function, possess the ability to bring together, to associate, different chemicals and therefore possess information. In this chapter, is discussed how the organization of a protein network can be associated with information. Mathematical functions have been derived that allow one to express the degree of information associated with the organization of the system. Moreover subadditivity does not necessarily apply to protein networks. In fact, a network behaves as a coherent whole, as a true system, if its information is different from the total information of its component sub-systems considered in isolation. Under these conditions, network information can be larger or smaller than the total information of its components. In the first case the system is integrated, in the second case it is called emergent. This means that the global collective properties of a network cannot be reduced to those of its sub-networks. Hence, contrary to communication channels, emergent networks can generate their own information.

*Keywords:* Boltzmann statistics and emergence, competition and mutual information, emergence in protein networks, image, information and organization, mutual information of integration, organization of protein networks, probability sub-spaces, self-information of integration, subadditivity in protein networks.

Most biochemical networks involve the association of different molecular signals. These molecular signals can be, for instance, two different ligands $x$ and $y$ bound to a macromolecule. A protein that binds several molecules, $x$ and $y$, at specific sites tends to form a network whose nodes are the protein states that have bound several molecules $x$ and $y$. Multiple ligand binding equilibria can be viewed as simple models of networks [1–3]. The molecular signals we are referring to could also be protein conformations in a multimolecular complex. Thus, for instance, complex protein edifices that can bind different ligands inducing different protein conformations that spread over the whole complex protein edifice [4,5] represent examples of biochemical networks that we are going to study later

(Chapter 9). Whatever be their physical nature and biological function, these networks all have in common the ability to bring together, to associate, different ligands or protein conformations.

There is no doubt that the concept of information is of major importance in biology and in many other fields of rational knowledge as well. As we have seen in the previous chapter, the very concept of information is rooted in the study of communication processes such as the communication of the genetic message to protein structure and function. From a mathematical viewpoint, the concept of information relies upon the mapping of the elements of the set **X** onto the elements of the set **Y**, i.e., $f : X \rightarrow Y$. Moreover, in this kind of mapping, to each element $x \in X$ is associated a *unique image* $y \in Y$ defined as

$$fX = \{y \in Y; y = fx, x \in X\} \tag{1}$$

This rule implies the existence of the subadditivity principle, i.e.,

$$H(X) + X(Y) \geq H(X, Y) \tag{2}$$

Hence, like any other type of relation $xRy$, a mapping can be defined as a specific subset of the cartesian product of two sets **X** and **Y**. In contrast, a network is a set of connected nodes that include a relation $xRy$ in which the image is not necessarily unique.

In the present chapter, we wish to address the following four questions.

First, is it possible to derive a new vision of information, not from the concept of mapping, but from that, more general, of relation?

Second, if this is possible, could one expect the subadditivity principle to apply?

Third, if this is not the case, what could be the physical significance of a lack of subadditivity?

Fourth, since a network can be considered a subset of the cartesian product **XY** of two sets **X** and **Y**, is it possible to predict the properties of the subset from the independent study of **X** and **Y**?

The answers to these questions can be of great interest. Thus, for instance, if the answer to the first question is positive, it would imply that it is possible to derive the mathematical expression of information involved in an organization process. In other words, this means that cell information is not restricted to the sequence information of the genome.

If, for this putative new type of information, subadditivity is not required and, in particular, if the sign of inequality (2) is reversed, it would imply that the network generates its own information and therefore displays *emergence* of information.

The last point raises an even more general philosophical question. If it is possible to predict the properties of the whole network from those of the two sets **X** and **Y**, it means it is possible to reduce the properties of a system to those of its component sub-systems.

The aim of this chapter is to answer these questions.

## 1. Information and organization of a protein network

### 1.1. Subsets of the protein network

There exists many different types of protein networks. In this chapter, we shall be concerned about two of them. The first is a multi-sited protein that binds two different types of ligands, $x$ and $y$, and therefore exists under different interconnected states. These states correspond to the nodes of the network. Some of the protein states have bound neither $x$ nor $y$, some have bound $x$, others have bound $y$, and some others have bound both $x$ and $y$. Hence the network constitutes a lattice [6]. If $p(N_{\kappa,\lambda})$, with $\kappa, \lambda \in Z^+$, is the probability of occurrence of a node associated with $\kappa$ molecules of $x$ and $\lambda$ molecules of y ($\forall\kappa, \lambda = 0, 1, 2, 3, \ldots$), the axiomatic definition of a probability implies that

$$\sum_{\kappa}\sum_{\lambda} p(N_{\kappa,\lambda}) = 1 \quad (\kappa, \lambda \in Z^+) \tag{3}$$

If $\kappa = \lambda = 0$, the protein state pertaining to the corresponding node has bound neither $x$, nor $y$. If $\kappa \geq 1$ and $\lambda = 0$ the protein states of the corresponding nodes have bound molecules of $x$ but no molecule of $y$. If, conversely $\kappa = 0$ and $\lambda \geq 1$ the protein states of the corresponding nodes have bound no molecule of $x$ but have bound molecules of $y$. Last, if $\kappa \geq 1$ and $\lambda \geq 1$ the protein states have bound both $x$ and $y$. The same reasoning can be applied to a lattice of proteins that change their conformation in succession. This succession is now well documented and is called conformational spread [4]. This matter is discussed at length in Chapter 9. Hence, whatever the physical nature of the network, it can be associated with a probability space $\Omega_N$ that collects the probabilities of occurrence of all the nodes. Moreover the topology of the network is defined by a certain rule, for instance that both $\kappa$ and $\lambda$ have to be smaller than, or equal to, a certain number $n$, or that their sum has to be smaller than, or equal to, $n$. In the first case, it is assumed that the protein bears two classes of $n$ sites, each class being able to accommodate ligand $x$, or ligand $y$. In the second case one postulates there exists a single class of $n$ sites in such a way that the ligands compete for the same sites. One has [7,8], for the network of Fig. 1A,

$$\Omega_N = \{p(N_{\kappa,\lambda}); \kappa, \lambda \in Z^+, \kappa, \lambda \leq n\} \tag{4}$$

and for the network of Fig. 1B

$$\Omega_N = \{p(N_{\kappa,\lambda}); \kappa, \lambda \in Z^+, \kappa + \lambda \leq n\} \tag{5}$$

According to the definition of this type of network, one can distinguish three different types of subsets (Fig. 1).

The first type of subset, $\Omega_0$, usually consists of the probability of occurrence of a node the corresponding protein state of which has bound neither $x$ nor $y$, i.e.,

$$\Omega_0 = \{p(N_{0,0})\} \tag{6}$$

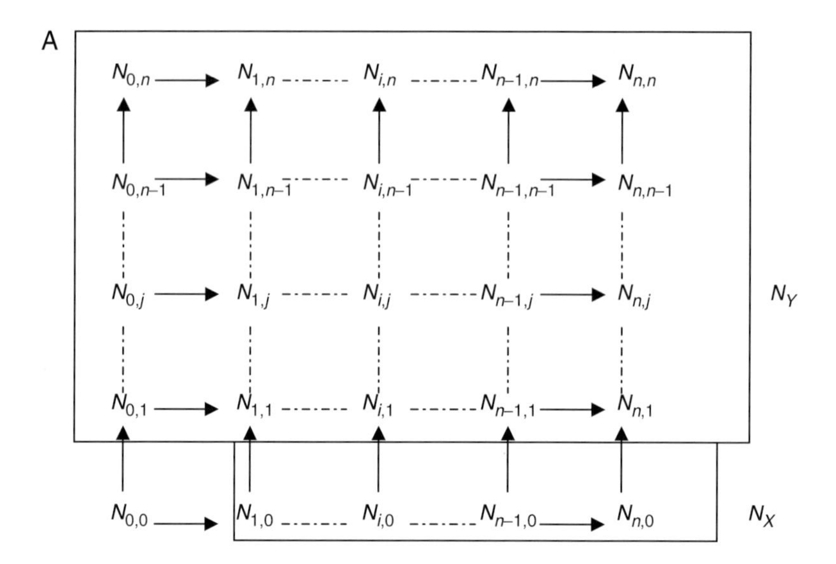


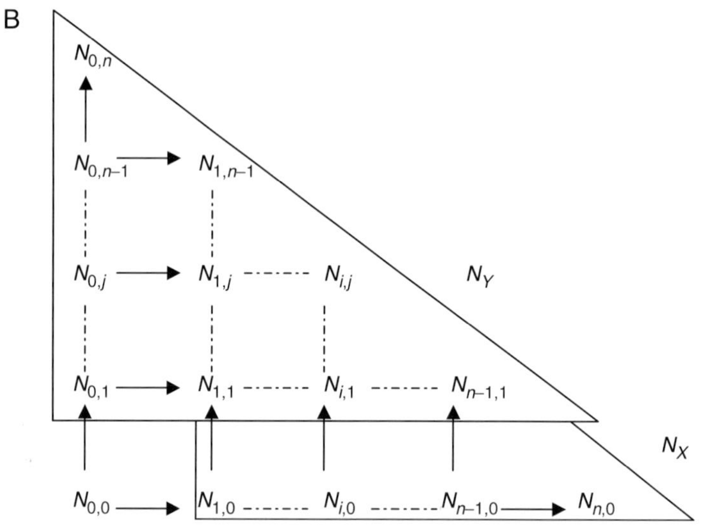


Fig. 1. Two different types of protein networks. (A) A square lattice is made up of a protein that bears two classes of $n$ sites each. The sites belonging to the first class accomodate ligand $x$ whereas the sites belonging to the second class accomodate ligand $y$. The nodes of the lattice represent the states of the protein. There exist three different types of nodes: a node that has bound neither $x$ nor $y$ ($N_{0,0}$), a population of nodes that have bound $x$ and *possibly* $y$ ($N_X$) and a population of nodes that have bound $y$ and *possibly* $x$ ($N_Y$). Each node can be granted a certain probability of occurrence. The rule that defines the topology of the lattice is thus $\kappa, \lambda \leq n$ ($\kappa, \lambda \in Z^+$). (B) In this case, the protein bears $n$ sites only in such a way that ligands $x$ and $y$ compete for the same sites. As for Fig. 1 A, there exist three different types of nodes: a node that has bound neither $x$ nor $y$ ($N_{0,0}$), nodes that have bound $x$ and *possibly* $y$ ($N_X$) and nodes that have bound $y$ and *possibly* $x$ ($N_Y$). The rule that defines the topology of the network is therefore different from the previous one viz. $\kappa + \lambda \leq n$ ($\kappa, \lambda \in Z^+$).

It can happen that this subset collects several nodes but in most cases it has only one. The second subset, $\Omega_{N_X}$, contains probabilities of occurrence of nodes whose protein states are all associated with $x$ (and possibly with $y$) i.e.,

$$\Omega_{N_X} = \{p(N_{i,\lambda});\, i \in N, \lambda \in Z^+, i, \lambda \leq n\} \tag{7}$$

for the network of Fig. 1A and

$$\Omega_{N_X} = \{p(N_{i,\lambda});\, i \in N, \lambda \in Z^+, i + \lambda \leq n\} \tag{8}$$

for the network of Fig. 1B. Similarly, the third subset, $\Omega_{N_Y}$, gets together the probabilities of occurrence of nodes whose protein states are all associated with $y$, regardless of whether they have bound $x$ or not. Hence one has

$$\Omega_{N_Y} = \{p(N_{\kappa,j});\, \kappa \in Z^+, j \in N, \kappa, j \leq n\} \tag{9}$$

for the model of Fig. 1A and

$$\Omega_{N_Y} = \{p(N_{\kappa,j});\, \kappa \in Z^+, j \in N, \kappa + j \leq n\} \tag{10}$$

for the model of Fig. 1B. The intersection, $\Omega_{N_{XY}}$, of the subsets $\Omega_{N_X}$ and $\Omega_N$ can then be expressed as

$$\Omega_{N_{XY}} = \Omega_{N_X} \cap \Omega_{N_Y} = \{p(N_{i,j});\, i, j \in N, i, j \leq n\} \tag{11}$$

for model of Fig. 1A and

$$\Omega_{N_{XY}} = \{p(N_{i,j});\, i, j \in N, i + j \leq n\} \tag{12}$$

for model of Fig. 1B. Expressions (11) and (12) represent the probabilities of occurrence of the nodes of the relation $x_i R y_j$ with $R = i, j \leq n$ for Eq. (11) and $R = i + j \leq n$ for Eq. (12). If there exist in the network several nodes, $N_{i,\lambda}$, associated with the same number, $i$, of molecules of $x$, whether or not these nodes are also associated with molecules of $y$, the probability that the protein has bound $i$ molecules of $x$ is

$$p(x_i) = \sum_{\lambda} p(N_{i,\lambda}) \quad (i \in N, \lambda \in Z^+) \tag{13}$$

Similarly, one can define $p(y_i)$, i.e., the probability that the protein has bound $j$ molecules of $y$. One has

$$p(y_j) = \sum_{\kappa} p(N_{\kappa,j}) \quad (j \in N; \kappa \in Z^+) \tag{14}$$

Hence one can define two sets, $\Omega_X$ and $\Omega_Y$, that collect the values of $p(x_i)$ and $p(y_j)$, respectively. These two sets are different from $\Omega_{N_X}$ and $\Omega_{N_Y}$. They can be defined as

$$\begin{aligned}\Omega_X &= \{p(x_i);\, i \in N\} \\ \Omega_Y &= \{p(y_j);\, j \in N\}\end{aligned} \tag{15}$$

The corresponding states, $x_i$ and $y_j$, define two sets **X** and **Y** whose cartesian product is **XY**. The probability space of this system **XY** is

$$\Omega_{XY} = \Omega_{N_{XY}} = \{p(x_i, y_j);\, i, j \in N\} \tag{16}$$

and pertains to the relation $x_i R y_j$. However, depending on the topology of the network, some of the states of the cartesian product may be forbidden. This is precisely what is occurring in the case of the model of Fig. 1B. The probability space of the network can therefore be expressed under two equivalent forms

$$\begin{aligned}\Omega_N &= \Omega_0 + \Omega_{N_X} + \Omega_{N_Y} - \Omega_{N_{XY}} \\ \Omega_N &= \Omega_0 + (\Omega_{N_X} - \Omega_{N_Y}) \cup (\Omega_{N_Y} - \Omega_{N_X}) + \Omega_{N_{XY}}\end{aligned} \tag{17}$$

In the latter form, it appears that a network can be defined as a relation between $x_i$ and $y_j$ included in a set of connected nodes that bear $x$ or $y$ but never both of them. This situation is clearly at variance with a communication process that can be expressed by an onto type of mapping.

### *1.2. Definition of self- and mutual information of integration*

Exactly as there exist mathematical functions that express the information content of a message, one can easily imagine that there exist mathematical functions that can express the information content of an organized system. One can define the mutual information involved in an organized system as the capacity of the system to specifically associate variables $x$ and $y$, for instance molecular signals or elements, according to certain rules in order to generate a structure and a function. The respective sizes of the subsets $\Omega_0$, $\Omega_X$, $\Omega_Y$, and $\Omega_{XY}$ that define the global organization of the network depend on this ability. If there exist several nodes, $N_{i,\lambda}$, that are associated with the same number, $i$, of molecules of $x$, the probability of occurrence, $p(x_i)$, of these nodes in the network is given by the relationship (13) above. Exactly as in communication theory, the smaller the value of $p(x_i)$, the larger is its uncertainty of occurrence, which can be expressed as

$$h(x_i) = -\log_2 p(x_i) \tag{18}$$

A similar definition should hold for the uncertainty of occurrence of the nodes bearing $j$ molecules of $y$ viz.

$$h(y_j) = -\log_2 p(y_j) \tag{19}$$

It is clear that neither $h(x_i)$ nor $h(y_j)$ takes account of a possible organization of the network for the corresponding probabilities do not specify whether the nodes pertaining to protein states that have bound $i$ molecules of $x$ have also bound molecules of $y$, and conversely. Last, one can also define an uncertainty of occurrence of the pair $p(x_i, y_j)$, namely

$$h(x_i, y_j) = -\log_2 p(x_i, y_j) \tag{20}$$

Since these definitions of entropies refer to protein states that have bound $x$ or $y$, or both $x$ and $y$, the protein state which is unliganded is of necessity excluded from the above definition of uncertainties. It is also of interest to stress that whereas $h(x_i)$ and $h(y_j)$ do not take account of any kind of organization in the binding of $x$ and $y$ to the protein, $h(x_i, y_j)$ does. This important matter will be discussed later. One can also define global uncertainties for the whole network viz.

$$\begin{aligned} H(X) &= \sum_i h(x_i) = -\sum_i \log_2 p(x_i) \\ H(Y) &= \sum_j h(y_j) = -\sum_j \log_2 p(y_j) \\ H(X, Y) &= \sum_i \sum_j h(x_i, y_j) = -\sum_i \sum_j \log_2 p(x_i, y_j) \end{aligned} \tag{21}$$

Whereas $H(X)$ and $H(Y)$ define the total uncertainty of linear sequences of $x_i$ and $y_j$ values, respectively, $H(X, Y)$ defines how the organization of $x$ relative to $y$ affects the uncertainty of the overall system. The numerical values of $H(X), H(Y)$, and $H(X, Y)$ depend on the number of nodes of the network. If we wish to have functions that are independent of the number of nodes and that can be compared, one has to define uncertainties per node of the global network, viz.

$$\begin{aligned} H(X)_N &= -\sum_i p(x_i) \log_2 p(x_i) \\ H(Y)_N &= -\sum_j p(y_j) \log_2 p(y_j) \\ H(X, Y)_N &= -\sum_i \sum_j p(x_i, y_j) \log_2 p(x_i, y_j) \end{aligned} \tag{22}$$

Although these expressions seem formally identical to Shannon entropies, i.e., moments of monovariate and bivariate distributions, in fact they are not, because

$$\begin{aligned} &\sum_i p(x_i) < 1 \\ &\sum_j p(y_j) < 1 \\ &\sum_i \sum_j p(x_i, y_j) < 1 \end{aligned} \tag{23}$$

As we shall see in the next section, these inequalities are the consequence of the first relation (17). In fact the probabilities $p(x_i)$, $p(y_j)$, and $p(x_i, y_j)$ refer to subsets of the probability space $\Omega_N$. In expressions (22) and (23) the running indices $i$ and $j$ cannot take zero value, i.e., $i,j \in N$. In the following, the expressions $H(X)_N$, $H(Y)_N$, and $H(X, Y)_N$ are called self- and joint information per node. These expressions, as well as $H(X), H(Y)$, and $H(X, Y)$, take account of the relative sizes of the subsets $\Omega_0$, $\Omega_X$, $\Omega_Y$, and $\Omega_{XY}$. Although the node involved in subset $\Omega_0$ does not appear explicitly in these functions, it nevertheless plays a role in the values of the probabilities $p(x_i), p(y_j)$, and $p(x_i, y_j)$. The larger the size of $\Omega_0$, the smaller these probabilities.

In order to express this idea of organization quantitatively, one can compare probabilities $p(x_i)$ with corresponding conditional probabilities $p(x_i/y_j)$. As already outlined, $p(x_i)$ is the probability of occurrence of nodes bearing $i$ molecules of $x$ regardless of whether they also bear $y$. Hence $p(x_i)$ is totally independent of the local organization of the system and can even assume the same value whether or not the protein binds $y$. On the other hand, conditional probabilities $p(x_i/y_j)$ express a certain type of local organization of a protein state, which has already bound a fixed number, $j$, of $y$ molecules and binds $i$ molecules of another ligand $x$. The local degree of organization of the system is therefore expressed by the local mutual information of integration i.e.,

$$I(x_i : y_j) = \log_2 \frac{p(x_i/y_j)}{p(x_i)} = \log_2 \frac{p(y_j/x_i)}{p(y_j)} = \log_2 \frac{p(x_i, y_j)}{p(x_i)p(y_j)} \tag{24}$$

One can also define the total mutual information of integration of the network as

$$I(X : Y) = \sum_i \sum_j \log_2 \frac{p(x_i, y_j)}{p(x_i)p(y_j)} = \sum_i \sum_j \log_2 \frac{p(x_i/y_j)}{p(x_i)} \quad (i,j \in N) \tag{25}$$

As previously, the summation is effected over the nodes belonging specifically to $\Omega_X$ and $\Omega_Y$. Indeed the resulting expression depends on the number of nodes of the network. In order to avoid this difficulty, one has to define a mutual information of integration per node of the network, i.e.,

$$I(X : Y)_N = \sum_i \sum_j p(x_i, y_j) \log_2 \frac{p(x_i, y_j)}{p(x_i)p(y_j)} = \sum_i \sum_j p(x_i, y_j) \log_2 \frac{p(x_i/y_j)}{p(x_i)} \tag{26}$$

This relationship expresses the information per node of the network regardless of whether the nodes belong to subsets $\Omega_0, \Omega_X$, or $\Omega_Y$. Hence, it is evident that this expression takes account of the sizes of subsets $\Omega_X$ and $\Omega_Y$. But it also takes account of the size of subset $\Omega_0$, for $p(x_i, y_j)$ and $p(x_i)$ are computed from all the nodes of the network. Moreover, an important feature of expression (26) is the number of the elements of $\Omega_X$ (or $\Omega_Y$) relative to that of $\Omega_{XY}$, which is a distinctive feature of mean mutual organizing information from classical Shannon mutual entropy. Although they may have different numerical values, expressions (25) and (26) usually display the same sign, positive or negative, depending on the respective values of $p(x_i/y_j)$ and $p(x_i)$, but this is not compulsory. It may occur for instance that $I(X:Y)$ be negative whereas $I(X:Y)_N$ is positive. Hence, just as for classical information in communication theory, mutual information of integration per node of the network can be defined as the difference between either $H(X)_N$ and $H(X/Y)_N$ or $H(Y)_N$ and $H(Y/X)_N$ i.e.,

$$I(X:Y)_N = H(X)_N - H(X/Y)_N = H(Y)_N - H(Y/X)_N \tag{27}$$

where $H(X/Y)_N$ and $H(Y/X)_N$ are conditional uncertainties of $X$ and $Y$ expressed per node of the network. Whereas $H(X)_N$ is the uncertainty of occurrence of nodes bearing $x$ without reference to any type of organization, $H(X/Y)_N$ is the uncertainty of occurrence of nodes bearing $x$ in a strictly organized system, i.e., in a network bearing both ligands $x$ and $y$ according to a certain type of organization. The same reasoning can be followed for $H(Y)_N$ and $H(Y/X)_N$.

## 2. *Subadditivity and lack of subadditivity in protein networks*

We have previously seen that the subaddivity principle requires that

$$\sum_i \sum_j p(x_i, y_j) - \sum_i p(x_i) \sum_j p(y_j) = 0 \tag{28}$$

When the sums of the probabilities of occurrence $p(x_i)$, $p(y_j)$, and $p(x_i, y_j)$ are independently equal to one, this relationship is of necessity fulfilled. As already pointed out, these conditions cannot apply to a protein network. Hence one can expect that the subadditivity principle may not apply to protein networks. The validity of this conclusion can easily be demonstrated. From the first relation (17) one has

$$p(N_{0,0}) + \sum_i \sum_\lambda p(N_{i,\lambda}) + \sum_\kappa \sum_j p(N_{\kappa,j}) - \sum_i \sum_j p(N_{i,j}) = 1 \quad (\kappa, \lambda \in Z^+; i, j \in N) \tag{29}$$

which can be written as

$$p(N_{0,0}) + \sum_i p(N_{i,0}) + \sum_j p(N_{0,j}) + \sum_i \sum_j p(N_{i,j}) = 1 \tag{30}$$

This expression is equivalent to the second Eq. (17). Setting for simplicity

$$\begin{aligned} p(N_{0,\lambda})_T &= p(N_{0,0}) + \sum_j p(N_{0,j}) \\ p(N_{\kappa,0})_T &= p(N_{0,0}) + \sum_i p(N_{i,0}) \end{aligned} \tag{31}$$

it follows that

$$\sum_i \sum_j p(N_{i,j}) = \sum_i \sum_j p(x_i, y_j) = 1 - \left\{p(N_{0,\lambda})_T + p(N_{\kappa,0})_T - p(N_{0,0})\right\} \tag{32}$$

In these equations, $p(N_{0,\lambda})_T$ is the probability that protein states have bound no $x$ molecule. Similarly, $p(N_{\kappa,0})_T$ is the probability that protein states have bound no $y$ molecule. Moreover one has

$$\begin{aligned} \sum_i p(x_i) &= \sum_i p(N_{i,0}) + \sum_i \sum_j p(x_i, y_j) \\ \sum_j p(y_j) &= \sum_j p(N_{0,j}) + \sum_i \sum_j p(x_i, y_j) \end{aligned} \tag{33}$$

which can be rewritten as

$$\begin{aligned} \sum_i p(x_i) &= p(N_{\kappa,0})_T - p(N_{0,0}) + \sum_i \sum_j p(x_i, y_j) \\ \sum_j p(y_j) &= p(N_{0,\lambda})_T - p(N_{0,0}) + \sum_i \sum_j p(x_i, y_j) \end{aligned} \tag{34}$$

From the definition of $p(N_{0,\lambda})_T$ and $p(N_{\kappa,0})_T$ one has

$$\begin{aligned} \sum_i p(x_i) &= 1 - p(N_{0,\lambda})_T \\ \sum_j p(y_j) &= 1 - p(N_{\kappa,0})_T \end{aligned} \tag{35}$$

which leads to

$$\sum_i \sum_j p(x_i, y_j) - \sum_i p(x_i) \sum_j p(y_j) = p(N_{0,0}) - p(N_{0,\lambda})_T p(N_{\kappa,0})_T \tag{36}$$

Now the probability of occurrence of the node $N_{0,0}$ is equal to the product of $p(N_{\kappa,0})_T$ by the conditional probability $p\left\{(N_{0,\lambda})_T/(N_{\kappa,0})_T\right\}$ or, alternatively, equal to the product of $p(N_{0,\lambda})_T$ by $p\left\{(N_{\kappa,0})_T/(N_{0,\lambda})_T\right\}$. Equation (36) can then be written as

$$\sum_i \sum_j p(x_i, y_j) - \sum_i p(x_i) \sum_j p(y_j) = p(N_{\kappa,0})_T \left[ p\left\{(N_{0,\lambda})_T/(N_{\kappa,0})_T\right\} - p(N_{0,\lambda})_T \right] \tag{37}$$

or as

$$\sum_i \sum_j p(x_i, y_j) - \sum_i p(x_i) \sum_j p(y_j)$$
$$= p(N_{0,\lambda})_T \left[ p\{(N_{\kappa,0})_T / (N_{0,\lambda})_T\} - p(N_{\kappa,0})_T \right] \quad (38)$$

Equations (37) and (38) can adopt positive or negative values depending on the sign of their right-hand side members. Hence it appears that subadditivity is not a necessary prerequisite for protein networks. One can therefore expect mutual information of integration to take positive or negative values.

## 3. *Emergence and information of integration of a network*

If we compare the elements of the subsets $\Omega_{N_X}$, $\Omega_{N_Y}$, and $\Omega_{N_{XY}}$ it appears that all the probabilities of occurrence, $p(N_{i,j})(\forall i, j \in N)$, appear in the expression of $p(x_i)$, $p(y_j)$, and $p(x_i, y_j)$. However, the probabilities $p(N_{i,0})(\forall i \in N)$ and $p(N_{0,j})(\forall j \in N)$ specifically appear in the expressions of $p(x_i)$ and $p(y_j)$, respectively. Hence it follows that

$$p(x_i) > p(x_i, y_j) \quad \text{and} \quad p(y_j) > p(x_i, y_j) \quad (39)$$

The differences $p(x_i) - p(x_i, y_j)$ and $p(y_j) - p(x_i, y_j)$ can be such that

$$p(x_i) p(y_j) > p(x_i, y_j) \quad (40)$$

If this inequality holds for all the $x_i$ and $y_j$ values, one has

$$I(X:Y) = \sum_i \sum_j \log_2 \frac{p(x_i, y_j)}{p(x_i) p(y_j)} < 0 \quad (41)$$

Since this situation is possible, mutual information of integration per node can be negative. This is not compulsory however. If the differences $p(x_i) - p(x_i, y_j)$ and $p(y_j) - p(x_i, y_j)$ are not large, the organized region of the network that associates ligand $x$ and $y$, i.e., the intersection $\Omega_{XY}$, has less information than $\Omega_X$ and $\Omega_Y$.

It is of interest to understand on an intuitive basis the reason for this situation. If

$$\Omega_0 + (\Omega_{N_X} - \Omega_{N_Y}) \cup (\Omega_{N_Y} - \Omega_{N_X}) = \emptyset \quad (42)$$

the network is equivalent to a relation $x_i R y_j$, subadditivity should of necessity be satisfied, and the network reduces to the cartesian product of systems **X** and **Y**, or to a subset of this cartesian product. This matter will be discussed later. In the case of models such as those of Fig. 1A and B the topology of the network is set by a rule such as $i, \lambda \leq n$ or $i + \lambda \leq n$ (Eqs. 7 and 8). Hence it appears that emergence of self-information can possibly arise in the network when some nodes associate $x$ and $y$

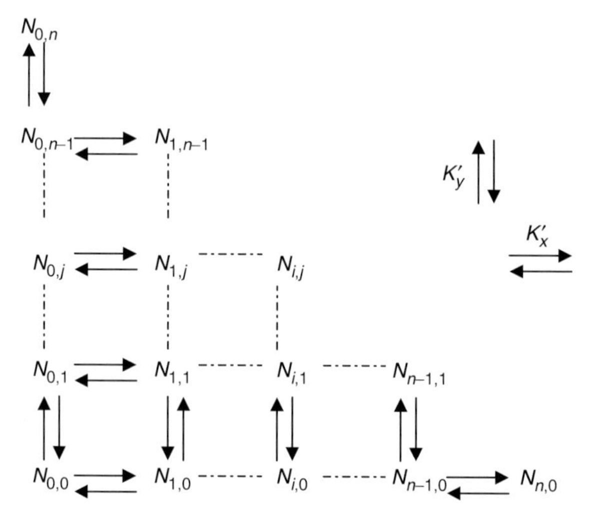


Fig. 2. A network made up of a protein bearing $n$ sites able to bind ligand $x$ or ligand $y$. The present situation is equivalent to that described in Fig. 1B, but the system is assumed to be in thermodynamic equilibrium. Under these conditions, it is possible to calculate the probabilities of occurrence of the various nodes.

according to a strictly defined relation whereas others are associated with $x$ or $y$, but not with both of them, as shown in the second Eq. (17). If this is the case, the system can be defined as complex.

This situation can be illustrated with the simple model networks in which the rule $i+\lambda \leq n$ or $\kappa + j \leq n$ apply. As shown for the model of Fig. 1B, we assume that a protein has $n$ sites that each of which binds either ligand $x$ or ligand $y$ in such a way there is competition between the two. Moreover, we assume there is equilibrium between the various protein states (Fig. 2). Owing to the competition between $x$ and $y$ there is a negative correlation in the two binding processes. This means that if a sub-population of protein molecules binds a large number of $x$ molecules, it will, of necessity, bind a small number of $y$ molecules. Moreover it is assumed that the binding of $x$ or $y$ to the protein does not alter the other binding processes.

The total "concentration", or "density", of protein is

$$N_T = N_{0,0}\left(1 + K'_x x + K'_y y\right)^n \tag{43}$$

where $N_{0,0}$ is the "density" of the node that has bound neither x nor y, $K'_x$ and $K'_y$ the microscopic binding constants of $x$ and $y$ to the protein sites. In the probability space $\Omega_N$ the probability of occurrence of the node $N_{\kappa,\lambda}$ is

$$p(N_{\kappa,\lambda}) = \frac{\binom{n}{\lambda}\binom{n-\lambda}{\kappa} K'^{\kappa}_x x^{\kappa} K'^{\lambda}_y y^{\lambda}}{\left(1 + K'_x x + K'_y y\right)^n} \quad (\forall \kappa, \lambda \in Z^+) \tag{44}$$

and one can check that

$$\sum_{\kappa}\sum_{\lambda}\frac{\binom{n}{\lambda}\binom{n-\lambda}{\kappa}K_x'^{\kappa}x^{\kappa}K_y'^{\lambda}y^{\lambda}}{\left(1+K_x'x+K_y'y\right)^n}=1 \quad (\forall\kappa,\lambda\in Z^+) \tag{45}$$

The elements, $p(N_{i,\lambda})$, of subset $\Omega_{N_X}$ are

$$p(N_{i,\lambda})=\frac{\binom{n}{\lambda}\binom{n-\lambda}{i}K_x'^{i}x^{i}K_y'^{\lambda}y^{\lambda}}{\left(1+K_x'x+K_y'y\right)^n} \quad (\lambda\in Z^+, i\in N) \tag{46}$$

and one can derive from these expressions the value of $p(x_i)$, i.e., the probability that the protein has bound $i$ molecules of $x$ regardless the number of molecules of $y$ it has also bound. One finds

$$p(x_i)=\sum_{\lambda}\frac{\binom{n}{\lambda}\binom{n-\lambda}{i}K_x'^{i}x^{i}K_y'^{\lambda}y^{\lambda}}{\left(1+K_x'x+K_y'y\right)^n}=\frac{\binom{n}{i}K_x'^{i}x^{i}(1+K_y'y)^{n-i}}{\left(1+K_x'x+K_y'y\right)^n} \tag{47}$$

Similarly, the elements $p(N_{\kappa,j})$ of the subset $\Omega_{N_Y}$ are expressed as

$$p(N_{\kappa,j})=\frac{\binom{n}{j}\binom{n-j}{\kappa}K_x'^{\kappa}x^{\kappa}K_y'^{j}y^{j}}{\left(1+K_x'x+K_y'y\right)^n} \quad (\kappa\in Z^+, j\in N) \tag{48}$$

and from this expression one can derive the value of $p(y_j)$, i.e., the probability that the protein has bound $j$ molecules of $y$ regardless of the number of molecules $x$ it has bound. One has

$$p(y_j)=\sum_{\kappa}\frac{\binom{n}{j}\binom{n-j}{\kappa}K_x'^{\kappa}x^{\kappa}K_y'^{j}y^{j}}{\left(1+K_x'x+K_y'y\right)^n}=\frac{\binom{n}{j}K_y'^{j}y^{j}(1+K_x'x)^{n-j}}{\left(1+K_x'x+K_y'y\right)^n} \tag{49}$$

Last, the elements, $p(N_{i,j})$ or $p(x_i, y_j)$, of the subset $\Omega_{XY}$ are

$$p(N_{i,j})=p(x_{i,j})=\frac{\binom{n}{j}\binom{n-j}{i}K_x'^{i}x^{i}K_y'^{j}y^{j}}{\left(1+K_x'x+K_y'y\right)^n} \tag{50}$$

One can also define the probability, $p(N_{0,\lambda})_T$, that the protein has bound no $x$ and the probability, $p(N_{\kappa,0})_T$, that it has bound no $y$. One finds

$$p(N_{0,\lambda})_T = \frac{(1 + K'_y y)^n}{\left(1 + K'_x x + K'_y y\right)^n} \tag{51}$$

and

$$p(N_{\kappa,0})_T = \frac{(1 + K'_x x)^n}{\left(1 + K'_x x + K'_y y\right)^n} \tag{52}$$

One can also derive the probability that the protein has bound no $x$ and no $y$, i.e.,

$$p(N_{0,0}) = \frac{1}{\left(1 + K'_x x + K'_y y\right)^n} \tag{53}$$

Hence, one can check that

$$\begin{aligned} p(N_{0,\lambda})_T + \sum_i p(x_i) = 1 \\ p(N_{\kappa,0})_T + \sum_j p(y_j) = 1 \end{aligned} \tag{54}$$

One can also calculate conditional probabilities involved in the system, for instance the probability that the protein binds $i$ molecules of $x$ given it has already bound $j$ molecules of $y$. One has

$$p(x_i/y_j) = \frac{\binom{n-j}{i} K'^i_x x^i}{(1 + K'_x x)^{n-j}} \quad (\forall i, j \in N) \tag{55}$$

As this expression is different from Eq. (47), one can conclude that the network contains mutual information of integration.

It is now possible to derive the expression of $p(N_{0,0}) - p(N_{0,\lambda})_T p(N_{\kappa,0})_T$ and one finds

$$p(N_{0,0}) - p(N_{0,\lambda})_T p(N_{\kappa,0})_T = \frac{\left(1 + K'_x x + K'_y y\right)^n - (1 + K'_x x)^n \left(1 + K'_y y\right)^n}{\left(1 + K'_x x + K'_y y\right)^n} \tag{56}$$

One can see that this expression is of necessity negative if $K'_x x$ and $K'_y y$ are different from zero. Hence the subadditivity principle does not appear to be a necessary prerequisite of the system. The mutual information of integration of a node, $I(x_i : y_j)$, total information, $I(X : Y)$, and mutual information of integration per node, $I(X : Y)_N$ can then be derived. One finds

$$I(x_i : y_j) = \log_2 \frac{\binom{n-j}{i}\left(1 + K'_x x + K'_y y\right)^n}{\binom{n}{i}(1 + K'_x x)^{n-j}(1 + K'_y y)^{n-i}}$$

$$I(X : Y) = \sum_i \sum_j \log_2 \frac{\binom{n-j}{i}\left(1 + K'_x x + K'_y y\right)^n}{\binom{n}{i}(1 + K'_x x)^{n-j}(1 + K'_y y)^{n-i}} \quad (\forall i, j \in N) \tag{57}$$

$$I(X : Y)_N = \sum_i \sum_j \frac{\binom{n}{j}\binom{n-j}{i} K'^i_x x^i K'^j_y y^j}{\left(1 + K'_x x + K'_y y\right)^n} \log_2 \frac{\binom{n-j}{i}\left(1 + K'_x x + K'_y y\right)^n}{\binom{n}{i}(1 + K'_x x)^{n-j}\left(1 + K'_y y\right)^{n-i}}$$

One can easily see that there are conditions in which $I(X : Y)$ and $I(X : Y)_N$ have negative values. These conditions are obvious. If $K'x$ and $K'y$ are small enough, the mutual information of the $N_{i,j}$ node is

$$I(x_i : y_j) \approx \log_2 \frac{\binom{n-j}{i}\left(1 + nK'_x x + nK'_y y\right)}{\binom{n}{i}\{1 + (n-j)K'_x x\}\left\{1 + (n-i)K'_y y\right\}} \tag{58}$$

Hence, whatever be the values of $i$ and $j$ $(\forall i, j \in N)$ one has

$$\begin{aligned} &\binom{n}{i} > \binom{n-j}{i} \\ &(n-j)\binom{n}{i} \geq n\binom{n-j}{i} \\ &(n-i)\binom{n}{i} \geq n\binom{n-j}{i} \end{aligned} \tag{59}$$

It thus appears that, whatever be the values of $i$ and $j$, the denominator of expression (58) is larger than the numerator. It follows that under these conditions, the mutual information of integration per node of the network is negative.

Table 1

The network is the one shown in Fig. 1B. The protein bears three binding sites. Any of these can accommodate either ligand $x$ or ligand $y$ which are in mutual competition. The system is assumed to be in thermodynamic equilibrium. The results illustrate the view that, depending the concentration of $x$ and $y$, mutual information per node can take negative values

| | | |
|---|---|---|
| $K'_x x = K'_y y = 0.1$ | $n = 3$ | |
| $p(x_1) = 0.210$ | $p(x_1/y_1) = 0.165$ | $p(x_1, y_1) = 0.034$ |
| $p(x_2) = 0.019$ | $p(x_1/y_2) = 0.090$ | $p(x_1, y_2) = 0.001$ |
| | $p(x_2/y_1) = 0.008$ | $p(x_2, y_1) = 0.001$ |
| $I(X : Y)_N = -0.016$ bit/node | | |
| $K'_x x = K'_y y = 0.25$ | $n = 3$ | |
| $p(x_1) = 0.347$ | $p(x_1/y_1) = 0.320$ | $p(x_1, y_1) = 0.111$ |
| $p(x_2) = 0.069$ | $p(x_1/y_2) = 0.200$ | $p(x_1, y_2) = 0.013$ |
| | $p(x_2/y_1) = 0.040$ | $p(x_2, y_1) = 0.013$ |
| $I(X : Y)_N = -0.035$ bit/node | | |
| $K'_x x = K'_y y = 0.5$ | $n = 3$ | |
| $p(x_1) = 0.281$ | $p(x_1/y_1) = 0.444$ | $p(x_1, y_1) = 0.187$ |
| $p(x_2) = 0.140$ | $p(x_1/y_2) = 0.333$ | $p(x_1, y_2) = 0.046$ |
| | $p(x_2/y_1) = 0.111$ | $p(x_2, y_1) = 0.046$ |
| $I(X : Y) = +0.123$ bit/node | | |

This situation is indicative of a phenomenon of emergence which is generated by the competition of binding of the two ligands $x$ and $y$ to the same sites of the protein. This situation is illustrated by simulation in Table 1.

## 4. *The physical nature of emergence in protein networks*

If one assumes that the probabilities of occurrence $p(x_i)$, $p(y_j)$ and $p(x_i, y_j)$ are distributed according to Boltzmann statistics [6], one has

$$\begin{aligned} p(x_i) &= \exp\{-(E_i - E_0)/k_B T\} \\ p(y_j) &= \exp\{-(E_j - E_0)/k_B T\} \\ p(x_i, y_j) &= \exp\{-(E_{i,j} - E_0)/k_B T\} \end{aligned} \tag{60}$$

where $E_i$, $E_j$, $E_{i,j}$, and $E_0$ are the free energy levels of the states $x_i$, $y_j$, $x_i y_j$, and $N_{0,0}$, respectively. As usual, $k_B$ and $T$ are the Boltzmann constant and the absolute temperature, respectively. Setting

$$\begin{aligned} \varepsilon(x_i) &= E_i - E_0 \\ \varepsilon(y_j) &= E_j - E_0 \\ \varepsilon(x_i, y_j) &= E_{i,j} - E_0 \end{aligned} \tag{61}$$

it follows that

$$H(X)_N = -\{\log_2 p(x)\}_N = \frac{\varepsilon(x)_N}{1.4426k_B T}$$

$$H(Y)_N = -\{\log_2 p(y)\}_N = \frac{\varepsilon(y)_N}{1.4426k_B T}$$

$$H(X,Y)_N = -\{\log_2 p(x,y)\}_N = \frac{\varepsilon(x,y)_N}{1.4426k_B T} \tag{62}$$

In these expressions, $\{\log_2 p(x)\}_N$, $\{\log_2 p(y)\}_N$, and $\{\log_2 p(x,y)\}_N$ represent the log of probabilities per node of the network. Similarly, $\varepsilon(x)_N$, $\varepsilon(y)_N$, and $\varepsilon(x,y)_N$ are the corresponding energy levels expressed per node of the network.

These expressions show that the self-information of each system **X**, **Y**, and **XY** are proportional to the corresponding energy level per node. Moreover one can derive from these equations the mutual information per node of the system. One finds

$$I(X:Y)_N = \frac{\varepsilon(x)_N + \varepsilon(y)_N - \varepsilon(x,y)_N}{1.4426k_B T} \tag{63}$$

and the system **XY** displays emergence of self-information relative to **X** and **Y** if the mean energy level of the states of the **XY** system is larger than the mean energy levels of the states of systems **X** and **Y**, i.e.,

$$\varepsilon(x,y)_N > \varepsilon(x)_N + \varepsilon(y)_N \tag{64}$$

Under these conditions, the increase of energy level of the nodes associated with both $x$ and $y$ originate from the kinetic energy spent in the collisions between the protein, and the two ligands. The situation expressed by Eq. (64) is depicted in Fig. (3). Under the conditions of Eq. (64), a system **XY** displays emergence both of energy and of self-information relative to **X** and **Y**, considered in isolation. Such a system can be considered complex. As we shall see later, emergence of self-information in a system is probably the best definition of complexity. Indeed inequality (64) can be reversed and then the system is *integrated* and behaves as a coherent whole. As we shall see later, it is tempting to assume that the increase of self-information and energy can be used by the system to perform a certain function.

## 5. *Reduction and lack of reduction of biological systems*

The problems we are now dealing with belong to the famous philosophical problem of reduction, i.e., whether it is legitimate or not to *reduce* the properties of a system to the properties of component sub-systems. This problem of the feasibility of reduction which we have mentioned in the first chapter of this book will now be more specifically considered in the case of a network. If we consider the organized

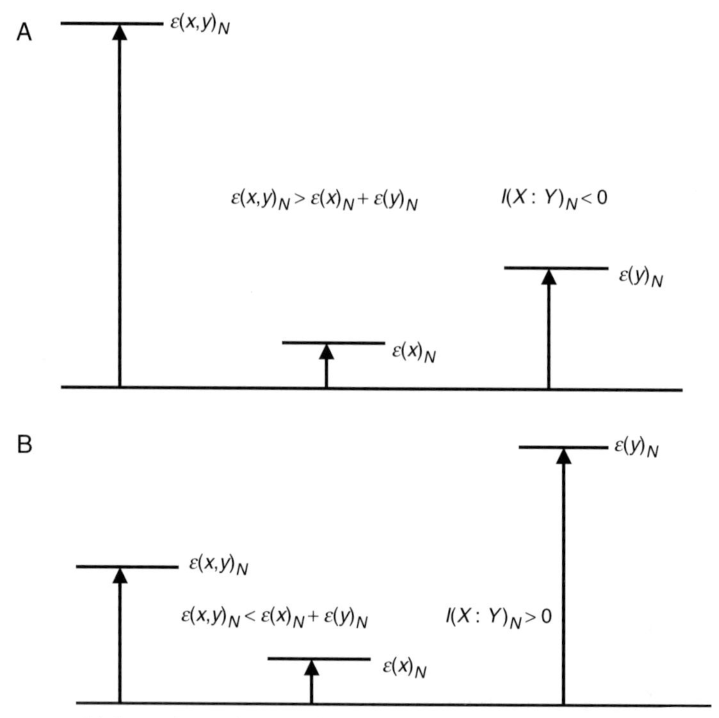


Fig. 3. Emergence of information and energy levels. (A) In this figure $\varepsilon(x, y)_N > \varepsilon(x)_N + \varepsilon(y)_N$ and this relation implies that $I(X : Y)_N < 0$. Hence the system displays emergence of information. **(B)** Contrary to the previous situation $\varepsilon(x, y)_N < \varepsilon(x)_N + \varepsilon(y)_N$ and this implies that $I(X : Y)_N > 0$. The system is integrated and does not display emergence of information.

region of a protein network, i.e., the set of nodes that associate a definite number of ligand $x$ with a definite number of ligand $y$, this region has already been called the system **XY** which defines a relation between $x_i$ and $y_j$. The problem is to know whether it is possible to reduce the properties of this system to the properties of the sub-systems **X** and **Y**. Or, put in other words, whether it is possible to predict the probabilities of occurrence of states $x_i y_j$ once we know the probabilities of occurrence of states $x_i$ (of sequential sub-system **X**) and $y_j$ (of sequential sub-system **Y**). It is obvious that such a situation will occur if **XY** is the cartesian product of **X** and **Y**. This problem can be given a more general solution by using self-information functions already defined. Let us consider the two protein networks already referred to. The protein involved in the first network bears $2n$ independent sites. A first class of $n$ sites bind specifically ligand $x$ and a second class of $n$ sites bind ligand $y$ (Fig. 4A). The second network is that of Fig. 4B and the corresponding protein bears $n$ sites that can bind, with different affinity constants, the two ligands $x$ and $y$ which are then in mutual competition. For the network of Fig. 4A, the total "concentration", or "density" of protein, $N_T$, is

$$N_T = N_{0,0}(1 + K'_x x)^n (1 + K'_y y)^n \tag{65}$$

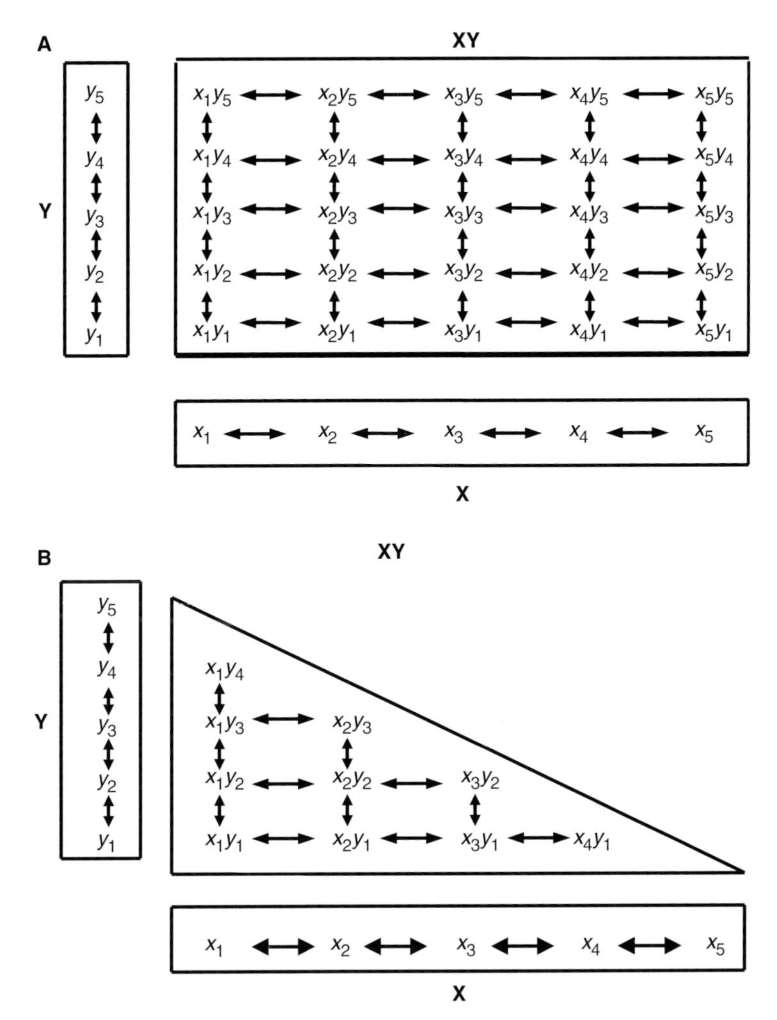


Fig. 4. Reduction of a network to its component sub-systems. (A) The lattice is the one shown in Fig. 1A under equilibrium conditions. System **XY** is then the cartesian product of **X** by **Y**. Under these conditions, properties of system **XY** can then be deduced from the individual properties of **X** and **Y**. Reduction of the system to its component sub-systems is then possible. (B) The network is the one shown in Fig. 1B under equilibrium conditions. System **XY** is not the cartesian product of **X** by **Y**. Reduction of the system to its component sub-systems is impossible.

where $K'_x$ and $K'_y$ are the intrinsic binding constants. One can calculate the probability that nodes be associated with $i$ molecules of $x$ or $j$ molecules of $y$. One finds

$$p(x_i) = \frac{\binom{n}{i} K'^{i}_{x} x^i}{(1 + K'_x x)^n} \tag{66}$$

and

$$p(y_j) = \frac{\binom{n}{j} K_y'^j y^j}{(1 + K_y' y)^n} \tag{67}$$

Similarly, one can calculate the probability that a node be associated with $i$ molecules of $x$ and $j$ molecules of $y$. One then has

$$p(x_i, y_j) = \frac{\binom{n}{i}\binom{n}{j} K_x'^i x^i K_y'^j y^j}{(1 + K_x' x)^n (1 + K_y' y)^n} \tag{68}$$

Comparing Eqs. (66), (67), and (68) leads to

$$p(x_i, y_j) = p(x_i)\, p(y_j) \tag{69}$$

As the expression of $p(x_i)$ is identical to the probability that a protein state has bound $i$ molecules $x$ in the sequential system **X** (Fig. 4), and $p(y_j)$ is the probability that a protein state of the sequential system **Y** (Fig. 4) has bound $j$ molecules of $y$, it follows that the organized system **XY** of the network can be *reduced* to the individual properties of sequential systems **X** and **Y** of Fig. 4. Comparison of expressions (66–68) leads to the conclusion that

$$H(X, Y)_N = H(X)_N + H(Y)_N \tag{70}$$

This equation formalizes the process of reduction [7,9–23]. Moreover, Eq. (69) shows that the system **XY** of Fig. 4 is the cartesian product of **X** and **Y**.

However, in the case of the network of Fig. 5, a comparison of Eqs. (47), (49), and (50) shows that

$$p(x_i, y_j) \neq p(x_i)\, p(y_j) \tag{71}$$

Hence, system **XY** cannot be the cartesian product of **X** and **Y** and one has either

$$H(X, Y)_N < H(X)_N + H(Y)_N \tag{72}$$

or

$$H(X, Y)_N > H(X)_N + H(Y)_N \tag{73}$$

In the case of expression (72), the system **XY** is *integrated* i.e., its degree of freedom is smaller than that of its component sub-systems **X** and **Y**, whereas in the case of expression (73), the system is *emergent* or *complex*. As we shall see in the next chapter, it is striking that emergence, or complexity, can be generated by so simple a process as competition of two ligands for the same site.

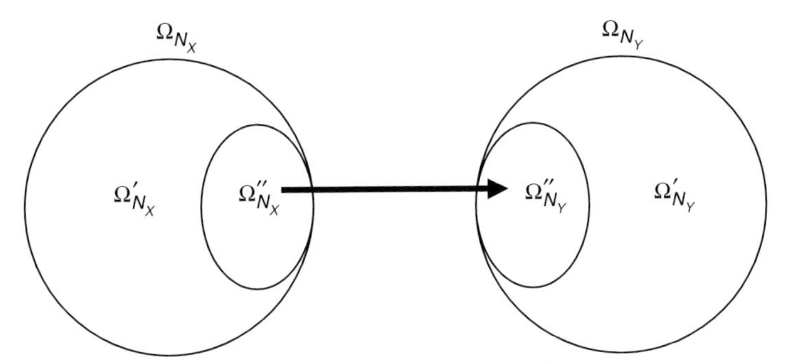


Fig. 5. The organization of the probability space that generates integration or emergence. The condition that generates emergence of mutual information is $|\Omega_{N_{XY}}| < |\Omega'_{N_X} \cup \Omega'_{N_Y}|$. Conversely, integration of the system is obtained if $|\Omega_{N_{XY}}| > |\Omega'_{N_X} \cup \Omega'_{N_Y}|$. See text.

## 6. *Information, communication, and organization*

It has become evident in recent years that no obvious correlation exists between the number of genes of living organisms and their apparent complexity. Moreover it appears now that a part, perhaps a large part, of cell information is epigenetic in nature. We have seen, in the present chapter, that a very simple protein network can store information. Although synthesis of the protein itself requires a certain amount of genetic information, the total information stored in the network is larger than the corresponding genetic information. The topology of the lattice, as expressed by the probabilities of occurrence of the various protein states, possesses information that has nothing to do with the classical genetic message. Moreover the kind of epigenetic information we are dealing with is adaptive, i.e., it can be adjusted in response to external signals, for instance the concentrations of ligands $x$ and $y$.

In classical Shannon communication theory, the very idea of information stems from the purely mathematical concept of mapping of the elements of a probability space $\Omega_X$ onto the elements of another probability space $\Omega_Y$, i.e.,

$$f : \Omega_X \to \Omega_Y \tag{74}$$

Information transferred in the communication channel is quantitatively expressed by the mutual information $I(X : Y)$, which is itself defined from Shannon entropies $H(X)$, $H(Y)$, and $H(X, Y)$. As seen in the previous chapter, these entropies are in fact monovariate and bivariate moments of the frequency distributions of the uncertainties $h(x_i)$, $h(y_j)$ and $h(x_i, y_j)$. It follows from these definitions that $I(X : Y)$ is the covariance of the bivariate distribution. Owing to the mathematical expression of a communication channel, the joint entropy $H(X, Y)$ should be less than, or at most equal to, the sum of $H(X)$ and $H(Y)$. This result supports the so-called subadditivity principle which serves as a basis for any communication process.

Common sense, however, tells us that the organization of any system requires some sort of information. As we have seen in this chapter, it is possible to develop a variant of Shannon theory that allows one to grasp and to express quantitatively the main features of system organization [24]. The ideal protein networks studied in this chapter can be represented by a relation between *some* elements of two probability spaces $\Omega_{N_X}$ and $\Omega_{N_Y}$(which are in fact subspaces of $\Omega_N$). One can split $\Omega_{N_X}$ into two subsets $\Omega'_{N_X}$ and $\Omega''_{N_X}$. Hence one has

$$\begin{aligned}\Omega'_{N_X} &= \{p(N_{x0});\ x \in N\} \\ \Omega''_{N_X} &= \{p(N_{xy});\ x, y \in N\}\end{aligned} \tag{75}$$

Similarly, $\Omega_{N_Y}$ can be split into two subsets $\Omega'_{N_Y}$ and $\Omega''_{N_Y}$

$$\begin{aligned}\Omega'_{N_Y} &= \{p(N_{0y});\ y \in N\} \\ \Omega''_{N_Y} &= \{p(N_{xy});\ x, y \in N\}\end{aligned} \tag{76}$$

It then follows that

$$\Omega''_{N_X} = \Omega''_{N_Y} = \Omega_{N_{XY}} \tag{77}$$

and the elements of this set fulfil the relation

$$p(x_i)Rp(y_j) = p(x_i y_j) \tag{78}$$

Moreover one has

$$\begin{aligned}\Omega'_{N_X} &= \Omega_{N_X} - \Omega_{N_Y} \\ \Omega'_{N_Y} &= \Omega_{N_Y} - \Omega_{N_X}\end{aligned} \tag{79}$$

Hence one can see that the conditions for global, collective, behavior rely upon the comparison of expressions (77) and (79). If

$$|\Omega_{N_{XY}}| > |(\Omega_{N_X} - \Omega_{N_Y}) \cup (\Omega_{N_Y} - \Omega_{N_X})| \tag{80}$$

the system behaves as an integrated global entity. Alternatively if

$$|\Omega_{N_{XY}}| < |(\Omega_{N_X} - \Omega_{N_Y}) \cup (\Omega_{N_Y} - \Omega_{N_X})| \tag{81}$$

the network can be a global emergent system. Expressions (80) and (81) can also be written as

$$\begin{aligned}|\Omega_{N_{XY}}| &> |\Omega'_{N_X} \cup \Omega'_{N_Y}| \\ |\Omega_{N_{XY}}| &< |\Omega'_{N_X} \cup \Omega'_{N}|\end{aligned} \tag{82}$$

The first expression (82) is the condition for integration, the second the condition for emergence. It is therefore evident that mutual information in the Shannon sense and mutual information of organization are based on two different mathematical concepts. Information transfer in a communication channel is based upon a mapping of the elements of the probability space $\Omega_X$ onto the elements of another probability space $\Omega_Y$. Mutual information of organization of a network relies upon a relation $p(x_i)Rp(y_j)$ between *some* elements only of set $\Omega_{N_X}$ and *some* elements only of set $\Omega_{N_Y}$. This situation is illustrated in Fig. 5.

Moreover, whereas Shannon entropies are monovariate and bivariate moments of uncertainty distributions, the expressions for self- and joint information per node of the global network, $H(X)_N$, $H(Y)_N$, and $H(X;Y)_N$ are not. As a matter of fact, the probabilities $p(x_i)$, $p(y_j)$, and $p(x_iy_j)$ are not the probabilities of occurrence of states $x_i, y_j$, and $x_iy_j$ in subsets $\Omega_X$, $\Omega_Y$, and $\Omega_{XY}$, respectively, but the probabilities of occurrence of these states in the "global" set $\Omega_N$. As the entropies used to express quantitatively the information transferred between a source and a destination are different from the entropies involved in the expression of the degree of integration or emergence of a system, one can wonder whether these two types of functions have something in common. In fact, both of them express the ability of a system to associate in a specific manner signs, or molecular signals, $x_i$ and $y_j$, so as to generate a function whatever, its nature. This is, perhaps, the most general definition of biological information.

The study of networks, or lattices, has also allowed us to raise an even more general question. This is the philosophical problem of reduction of the properties of a system to the properties of the component sub-systems. If we consider a system **XY** made up of two component sub-systems **X** and **Y**, it will be possible to reduce the properties of the system **XY** to the individual properties of its components only if **XY** is the cartesian product of **X** by **Y**. If this is not the case, the system possesses global, collective properties that are present neither in **X** nor in **Y**.

The last point of interest is to know the origin of integration and emergence in protein networks and lattices. In fact in the simple case of a protein that binds two ligands $x$ and $y$, integration or emergence appear if the binding of a ligand modifies the binding of the other one. The simplest situation one can think of is obtained when the two ligands compete for the same sites. We shall see in the next chapter that different molecular events can lead to the same result. Whatever that may be, it is interesting to point out that simple and trivial events in a population of protein states can generate global, collective, properties.

## *References*


1. Wyman, J. (1964) Linked functions and reciprocal effects in haemoglobin: a second look. Adv. Prot. Chem. 19, 223–286.
2. Wyman, J. (1967) Allosteric linkage. J. Am. Chem. Soc. 89, 2202–2218.
3. Wyman, J. (1968) Regulation in macromolecules as illustrated by haemoglobin. Q. Rev. Biophys. 1, 35–80.

4. Duke, T.A.J., Le Novère, N., and Bray, D. (2001) Conformational spread in a ring of proteins: a stochastic approach to allostery. J. Mol. Biol. 308, 541–553.
5. Bray, D. and Duke, T. (2004) Conformational spread: the propagation of allosteric states in large multiprotein complexes. Annu. Rev. Biophys. Biomol. Struct. 33, 53–73.
6. Hill, T.L. (1987) Statistical Mechanics. Principles and Selected Applications. Dover Publications Inc., New York.
7. Ricard, J. (2004) Reduction, integration and emergence in biochemical networks. Biology of the Cell 96, 719–725.
8. Ricard, J. (2005) Statistical mechanics of organization, information and emergence in protein networks. J. Nonequilibrium Thermo. in press.
9. Descartes, R. (1959, third edition) Règles pour la Direction de l'Esprit. Traduction et notes Sirven J. Vrin, Paris.
10. Descartes, R. (re-issue) Discours de la Méthode. Flammarion, Paris.
11. Medawar, P. (1974) A geometric model of reduction and emergence. In: Ayala, F.J. and Dobzhansky, T. (eds.) Studies in the Philosophy of Biology, Macmillan, London, pp. 57–63.
12. Goodfield, J. (1974) Changing strategies: a comparison of reductionist attitudes in biological and medical research in the nineteenth and twentieth centuries. In: Ayala, F.J. and Dobzhansky, T. (eds.) Studies in the Philosophy of Biology, Macmillan, London, pp. 65–86.
13. Popper, K.R. (1974) Scientific reduction and the essential incompleteness of all science. In: Ayala, F.J. and Dobzhansky, T. (eds.) Studies in the Philosophy of Biology, Macmillan, London, pp. 259–284.
14. Nagel, E. (1961) The Structure of Science. Harcourt, Brace and World, New York.
15. Nagel, E. (1998) Reductionism and anitireductionism. Novartis Foundation Symposium, John Wiley and Sons, Chichester and New York, pp. 3–14.
16. Nurse, P. (1998) Reductionism and explanation in cell biology, Novartis Foundation Symposium, John Wiley and Sons, Chichester and New York, pp. 93–105.
17. May, R. (1998) Levels of organization in ecological systems, Novartis Foundation Symposium, John Wiley and Sons, Chichester and New York, pp. 193–202.
18. Bock, G.R. and Goode, J.A. (eds.) (1998) The Limits of Reductionism in Biology. Novartis Foundation Symposium, John Wiley and Sons, Chichester and New York.
19. Robinson, J.D. (1986) Reduction, explanation and the quests of biological research. Philosophy of Science 53, 333–353.
20. Thorpe, W.H. (1974) Reductionism in biology. In: Ayala, F.J. and Dobzhansky, T. (eds.) Studies in the Philosophy of Biology, Macmillan, London, pp. 109–138.
21. Beckner, M. (1974) Reduction, hierarchies and organization. In: Ayala, F.J. and Dobzhansky, T. (eds.) Studies in the Philosophy of Biology, Macmillan, London, pp. 163–177.
22. Ricard, J. (2001) Complexity, reductionism and the unity of science. 97–105. In: Agazzi, E. and Faye, J. (eds.) The Problem of the Unity of Science, World Scientific, New Jersey, London, pp. 95–105.
23. Ricard, J. (2001) Reduction, integration emergence and complexity in biological networks. In: Agazzi, E. and Montecocco, L. (eds.) Complexity and Emergence, World Scientific, New Jersey, London, pp. 101–112.
24. Ricard, J. (2003) What do we mean by biological complexity? C. R. Biologies 326, 133–140.

J. Ricard *Emergent Collective Properties, Networks, and Information in Biology*

DOI: 10.1016/S0167-7306(05)40006-X

CHAPTER 6

# On the mechanistic causes of network information, integration, and emergence

J. Ricard

As seen in the previous chapter, the concept of mutual information applies equally well to the communication of a message and to the organization of a system. One can wonder, however, about the mechanistic causes of network information, integration and emergence. The aim of the present chapter is to describe the mechanistic causes of these events. In fact, they can have different origins. The first, and the simplest one, is a negative correlation between two binding processes. Thus, for instance, if a protein bears identical sites each able to bind either ligand $x$ or ligand $y$, the competition between these ligands is sufficient to generate integration or emergence of the network. The second cause of these events is a physical interaction that may possibly exist between two binding processes. If a protein, for instance, has two sites, one specific for the binding of $x$, the other one specific for the binding of $y$, interaction between these events can generate integration or emergence. Last, a certain type of global network topology can also generate the same events. If a graph is made up of several modules that do not, *per se*, possess any information aggregation of these modules can generate a system that possesses information.

*Keywords:* information and statistical correlation between events, information and physical interaction between events, topological information.

In the previous examples of equilibrium networks, we have shown that a negative correlation between two binding processes in competition can generate emergence and complexity. Common sense, however, tells us that mutual information of integration may not be restricted to a statistical correlation between discrete variables, but can be extended to physical interaction between events.

## 1. *Information and physical interaction between two events*

Let us consider, for instance, a protein that, under equilibrium conditions, binds two ligands at two specific sites in interaction (Fig. 1). This situation may imply that the binding of one ligand affects the association of the other ligand with the protein, and conversely. Hence the variables $X$ and $Y$ (Chapters 4 and 5) assume one value each, $x_a$ and $y_b$, and there cannot exist any correlation between them. If, however, the

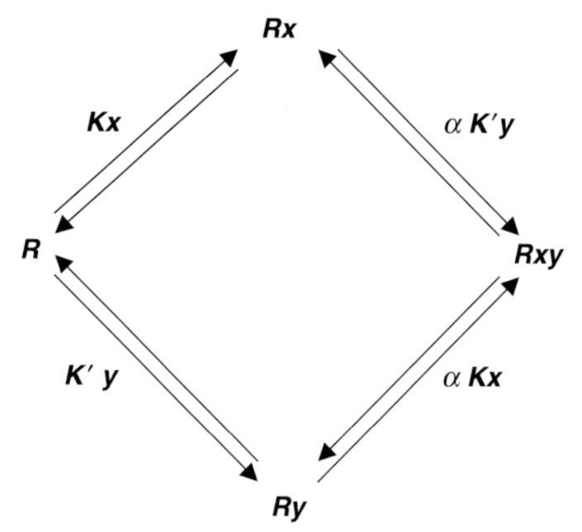


Fig. 1. Mutual information of integration between two binding sites of a protein. A protein $R$ bears two sites. One of these sites binds ligand $x$ whereas the other binds ligand $y$. The corresponding affinity constants are $K$ and $K'$, respectively. Depending on the value of the coefficient $\alpha$ the binding of $x$ facilitates or hinders that of $y$ and conversely. See text.

corresponding probabilities $p(x_a)$ and $p(x_b)$ express some kind of mutual physical dependence between two events, one can define the corresponding uncertainty functions as [1]

$$\begin{aligned} H(X) &= -\log_2 p(x_a) \\ H(Y) &= -\log_2 p(y_b) \\ H(X, Y) &= -\log_2 p(x_a, y_b) \\ H(X/Y) &= -\log_2 p(x_a/y_b) \\ H(Y/X) &= -\log_2 p(y_b/x_a) \end{aligned} \tag{1}$$

From these expressions, one can calculate a positive or a negative mutual information if

$$p(x_a/y_b) \neq p(x_a) \quad \text{and} \quad p(y_b/x_a) \neq p(y_b) \tag{2}$$

In fact, the expression for the corresponding mutual information of integration is

$$I(X:Y) = \log_2 \frac{p(x_a/y_b)}{p(x_a)} = \log_2 \frac{p(y_b/x_a)}{p(y_b)} \tag{3}$$

Let us consider a protein, a receptor for instance, that binds two ligands $x$ and $y$ at two specific sites (Fig. 1). It is quite possible that the binding of $x$ facilitates, or hinders, that of $y$, and conversely. In this case there is no correlation between the two binding processes, but a physical interaction between them. One can express the situation by stating that information "propagates" from one binding site to another, or, in other words, there exists a positive or a negative cooperativity between the sites [2–5]. One can express this effect by a constant factor, $\alpha$, that can assume different, but fixed values, depending on the interaction occurring between the physically interacting events [6]. If $\alpha = 1$ the two processes do not interfere. If $\alpha > 1$ the binding of $x$ facilitates that of $y$, and conversely. Last, if $0 < \alpha < 1$ the binding of $x$ hinders that of $y$, and conversely.

One can calculate the probabilities that the protein has bound either $x$ or $y$ or both $x$ and $y$. One finds

$$
\begin{aligned}
p(x) &= \frac{Kx(1+\alpha K'y)}{1+Kx+K'y+\alpha KK'xy} \\
p(y) &= \frac{K'y(1+\alpha Kx)}{1+Kx+K'y+\alpha KK'xy} \\
p(x,y) &= \frac{\alpha KK'xy}{1+Kx+K'y+\alpha KK'xy} \\
p(x/y) &= \frac{\alpha Kx}{1+\alpha Kx} \\
p(y/x) &= \frac{\alpha K'y}{1+\alpha K'y}
\end{aligned}
\tag{4}
$$

In these expressions $K$ and $K'$ are the equilibrium binding constants of $x$ and $y$ to the protein, respectively. One can note that if $\alpha = 1$

$$
\begin{aligned}
p(x/y) &= p(x) = \frac{Kx}{1+Kx} \\
p(y/x) &= p(y) = \frac{K'y}{1+K'y}
\end{aligned}
\tag{5}
$$

Hence mutual information in this simple system can be expressed as

$$
I(X:Y) = \log_2 \frac{\alpha(1+Kx+K'y+\alpha KK'xy)}{(1+\alpha Kx)(1+\alpha K'y)} \tag{6}
$$

From this expression, it appears that if $\alpha = 1$ the network does not contain any information, if $\alpha > 1$ mutual information is of necessity positive whereas if $0 < \alpha < 1$ this information is negative. The latter situation implies that the system displays emergence of information and energy. This means that the ternary complex $Rxy$ contains more energy than $Rx$ or $Ry$ and can potentially have novel properties, such as catalysis of a chemical reaction, if the protein is an enzyme. This matter will be discussed in the following chapter.

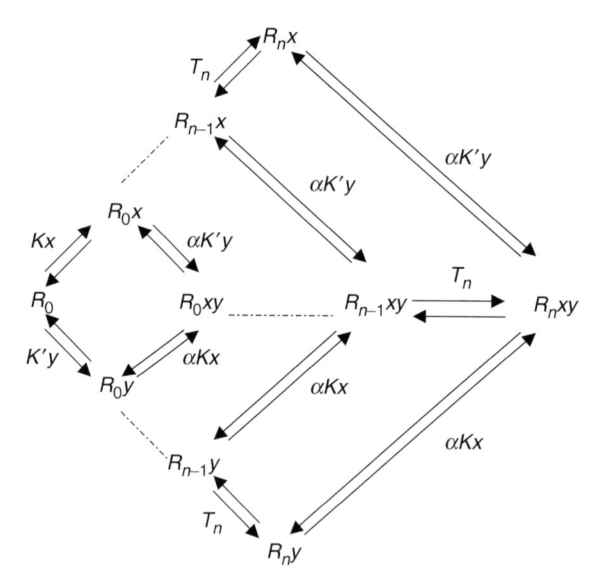


Fig. 2. Mutual information of integration between two binding sites of a protein that undergoes multiple conformation changes. As for Fig. 1. the two binding constants for ligands $x$ and $y$ are $K$ and $K'$, respectively. See text.

## 2. *Emergence and topological information*

The network considered in the previous section was extremely simple and more complicated networks involving the existence of physical interactions between two binding processes can also display emergent properties. Let us consider, as an example of this situation, the ideal network shown in Fig. 2. Here we assume, as previously, that a protein, a hormone receptor for instance, binds either one or two ligands, $x$ and $y$, at specific sites. As in the previous case, we postulate that the network occurs under thermodynamic equilibrium conditions. The two binding processes may, or may not, physically interact, i.e., the binding of $x$ may, or may not, facilitate or hinder that of $y$, and conversely. The additional idea, which is introduced in this model, is that the binding of $x$, as well as that of $y$, induces a sequence of conformational changes of the protein. For simplicity, we assume that the equilibrium constants of these isomerizations, $T_1, T_2, T_3 \ldots$, are the same whether the protein has bound $x$, $y$ or both $x$ and $y$. All these states originate from the initial protein–ligand(s) complexes that undergo subsequent conformation changes. Although the states do not play any role in the binding process *per se* they nevertheless contribute to the topology of the network.

The total protein concentration, or the total "node density", $R_T$, is

$$R_T = R_0\{1 + (Kx + K'y + \alpha KK'xy)(1 + T_1 + T_1T_2 + \cdots + T_1T_2 \ldots T_n)\} \qquad (7)$$

where $R_0$ is the free concentration of the protein. $\alpha$ measures the strength of the interaction that may possibly exist between the binding of $x$ and that of $y$ to their specific sites. $\alpha$ will therefore be equal to one if the occupancy of one site does not interfere with that of the other. It will be larger than one if the occupancy of a site increases the apparent affinity of the other for the corresponding ligand, and smaller than one in the opposite case [6,7]. One can define the following partition function [9]

$$\Pi(n) = 1 + T_1 + T_1 T_2 + \cdots + T_1 T_2 \ldots T_n \tag{8}$$

with

$$\Pi(0) = T_0 = 1 \tag{9}$$

and Eq. (7) assumes the form

$$R_T = R_0 \{1 + (Kx + K'y)\Pi(n) + \alpha KK'xy\Pi(n)\} \tag{10}$$

One can define different types of probabilities that can be used to express different types of information of the network. The probabilities of ligand binding to the protein are defined by the ratio of the number of protein molecules that have bound $x$ or $y$ over the total number of protein molecules. One has thus

$$\begin{aligned} p(x) &= \frac{Kx\Pi(n)(1 + \alpha K'y)}{1 + (Kx + K'y)\Pi(n) + \alpha KK'xy\Pi(n)} \\ p(y) &= \frac{K'y\Pi(n)(1 + \alpha Kx)}{1 + (Kx + K'y)\Pi(n) + \alpha KK'xy\Pi(n)} \end{aligned} \tag{11}$$

One can, in the same way, define the probability that the protein be in state $R_k$ and bear ligand $x$, or ligand $y$. One has then

$$\begin{aligned} p(R_k x) &= \frac{KxT_0 T_1 \ldots T_k(1 + \alpha K'y)}{1 + (Kx + K'y)\Pi(n) + \alpha KK'xy\Pi(n)} \\ p(R_k y) &= \frac{K'yT_0 T_1 \ldots T_k(1 + \alpha Kx)}{1 + (Kx + K'y)\Pi(n) + \alpha KK'xy\Pi(n)} \end{aligned} \tag{12}$$

It is also possible to define the conditional probability that the protein has bound $x$ given it has already bound $y$, and one finds

$$p(x/y) = \frac{\alpha Kx\Pi(n)}{\Pi(n) + \alpha Kx\Pi(n)} = \frac{\alpha Kx}{1 + \alpha Kx} \tag{13}$$

This expression is equal to the conditional probability that the protein be in state $R_k x$ given the value $R_k y$, i.e.,

$$p(R_k x/R_k y) = \frac{\alpha Kx}{1 + \alpha Kx} \tag{14}$$

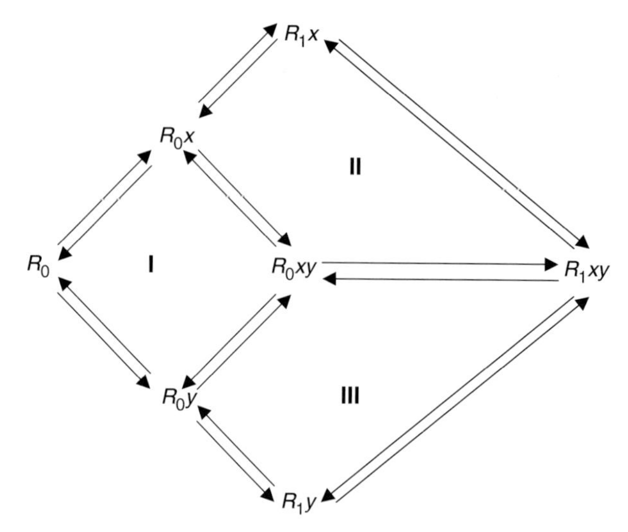


Fig. 3. Topological information of a network. The network is made up of three aggregated sub-networks I, II, and III. These sub-networks considered in isolation do not contain information. When they are aggregated, however, the overall system has negative information.

One can then derive the overall information of the network from the first expression (11) and from Eq. (13) above. One finds

$$I(X:Y) = \log_2 \frac{p(x/y)}{p(x)} = \log_2 \frac{\alpha\{1 + (Kx + K'y)\Pi(n) + \alpha KK'xy\Pi(n)\}}{(1 + \alpha Kx)(1 + \alpha K'y)\Pi(n)} \tag{15}$$

It is obvious from this equation that the overall mutual information thus defined can be positive, nil, or negative. Of particular interest is the situation where $I(X:Y) < 0$. This situation appears as a consequence of an interplay between two factors: the value of the $\alpha$ factor, and the topology of the network, as expressed by the partition function $\Pi(n)$. The interesting result that comes out from this equation is that even if $\alpha = 1$ the mutual information $I(X:Y)$ assumes negative values. This conclusion is illustrated in Fig. 3. The overall network shown in this figure possesses three cycles I, II, and III. When $\alpha = 1$, none of these cycles considered in isolation has any mutual information. However, when aggregated as a network, the whole system contains information which, at least in the present case, is negative. This information, which is associated with the overall network topology, can be called topological information (Table 1). Indeed this overall negative information becomes even more negative if $0 < \alpha < 1$.

$I(X:Y)$ is the overall information of the network. One can also derive the expression of the information per protein conformer. This expression can be derived from

$$\frac{p(R_kx/R_ky)}{p(R_kx)} = \frac{\alpha\{1 + (Kx + K'y)\Pi(n) + \alpha KK'xy\Pi(n)\}}{T_0T_1 \ldots T_k(1 + \alpha Kx)(1 + \alpha K'y)} \tag{16}$$

Table 1

Negative topological information of a cycle made up of three sub-cycles. The sub-cycles I, II, and III do not contain information when considered in isolation. The overall cycle, however, does. The resulting topological information is negative

| $\alpha = 1$ | $Kx = 1$ | $K'y = 2$ | $T_1 = 3$ |
|---|---|---|---|

Cycle I

$$I_1(X:Y) = \log_2 \frac{1+1+2+1x2}{(1+1)(1+2)} = 0\,\text{bit}$$

Cycle II

$$I_2(X:Y) = \log_2 \frac{1+3+2+3x2}{(1+3)(1+2)} = 0\,\text{bit}$$

Cycle III

$$I_3(X:Y) = \log_2 \frac{1+1+3+1x3}{(1+1)(1+3)} = 0\,\text{bit}$$

Overall cycle

$$I(X:Y) = \log_2 \frac{1+(1+2)(1+3)+1x2(1+3)}{(1+3)(1+1)(1+2)} = -0.192\,\text{bit}$$

and one has

$$\sum_k \log_2 \frac{p(R_k x/R_k y)}{p(R_k x)} = \sum_k \log_2 \frac{\alpha\{1+(Kx+K'y)\Pi(n)+\alpha KK'xy\Pi(n)\}}{T_0 T_1 \ldots T_k(1+\alpha Kx)(1+\alpha K'y)} \tag{17}$$

with $k \in Z^+$. As expected, this expression is equivalent to Eq. (15). Moreover the joint probability that a conformer $R_k$ has bound both $x$ and $y$ is

$$p(R_k x, y) = \frac{\alpha KK'xyT_0 T_1 \ldots T_k}{1+(Kx+K'y)\Pi(n)+\alpha KK'xy\Pi(n)} \tag{18}$$

Hence the information per protein conformer can be derived from the general equation

$$I(X:Y)_N = \sum_k p(R_k x, y) \log_2 \frac{p(R_k x/R_k y)}{p(R_k x)} \tag{19}$$

Table 2

Overall information and information per node of a composite network. The overall information of the network is negative whereas the information per node can be positive or negative depending on the $\alpha$ value. The reason for this difference relies upon the partition function $\Pi(n)$ present in both the numerator and denominator of the overall information (15) and absent in the denominator of the logarithmic factor of expressions (17) and (20). In fact, information per node does not take into account the global organization of the network

| $K'y = 2$ | $K'y = 2$ | $T_1 = 3$ | $T_0T_1 = 3$ | $\Pi(n) = 1 + 3 = 4$ |
|---|---|---|---|---|
| $\alpha = 1$ | | | | |
| $I(X:Y) = +0.235$ bit/node | | | | |
| $I(X:Y)_N = -0.192$ bit | | | | |
| $\alpha = 0.5$ | | | | |
| $I(X:Y) = +0.073$ bit/node | | | | |
| $I(X:Y)_N = -0.497$ bit | | | | |
| $\alpha = 0.05$ | | | | |
| $I(X:Y) = -0.058$ bit/node | | | | |
| $I(X:Y)_N = -2.785$ bit | | | | |

which can be rewritten in explicit form as

$$I(X:Y)_N = \sum_k \frac{\alpha KK'xyT_0T_1 \ldots T_k}{1 + (Kx + K'y)\Pi(n) + \alpha KK'xy\Pi(n)} \log_2 \frac{\alpha\{\Pi(n)(Kx + K'y) + \alpha KK'xy\Pi(n) + 1\}}{T_0T_1 \ldots T_k(1 + \alpha Kx)(1 + \alpha K'y)} \tag{20}$$

Some values of the mean mutual information per conformer are given in Table 2.

One can notice from Eqs. (15) and (20) that the system is adaptive, i.e., it can change its mutual information content depending on the values of "external" signals $x$ and $y$. Last but not least, it is worth stressing that the sign of expression (15) is not necessarily the same as that of Eq. (20). Thus one can see from the results of Table 2 that the overall information of the network can be negative whereas the mutual information per node is positive.

## 3. *Organization and the different types of mutual information*

As we have seen in the previous sections above, the concept of mutual information applies equally well to the communication of a message and to the organization of a system. In the latter case, information can be considered a certain improbable configuration of signs or of molecular elements that is associated with the organization of a biological structure endowed with a certain function. We have called this type of information mutual information of integration. There are in fact several types of mutual information of integration: the mutual information associated with a negative correlation between two variables; the mutual information generated by the physical interaction between two events; the mutual information associated with network topology.

### *3.1. Mutual information and negative correlation*

We have considered previously a protein bearing two classes of binding centers. The sites of the first class bind, with the same microscopic binding constant, a ligand $x$ whereas the sites of the second class bind another ligand $y$. If there is no physical interaction between the two binding processes the network cannot have any mutual information because there is neither any correlation between the binding of $x$ and that of $y$, nor any physical interaction between the two binding processes. But if we have only one class of sites on the protein, instead of two, and if two different ligands $x$ and $y$ bind to these identical sites, there exists a correlation between the binding of $x$ and that of $y$. As the number of binding sites on the protein is constant, the larger the number of molecules $x$ that bind to the protein, the smaller is the number of binding sites available for the binding of $y$. Hence there exists a negative statistical correlation between the two binding processes. It is striking to note that such a simple event is sufficient to generate a lack of subadditivity and emergence of self-information in the system.

### *3.2. Mutual information and physical interaction between two binding processes*

Let us consider a protein that bears two binding sites. One site specifically binds $x$ whereas the other specifically binds $y$. If the two processes do not physically interact, i.e., if the binding of $x$ does not modify that of $y$, and conversely, there cannot exist any mutual information in this kind of network. But if the two processes physically interact i.e., if the binding of $x$ facilitates, or hinders, that of $y$, the system possesses positive or negative information. The physical interaction between the two binding processes, which can be considered a sort of cooperativity, is expressed by the parameter $\alpha$ that can take values larger, or smaller, than one depending on whether the binding of $y$ facilitates, or hinders, that of $x$, and conversely [7]. In this case, mutual information is a measure of the positive or negative influence that can be transported from one site to the other one.

### *3.3. Mutual information and network topology*

If a graph is made up of several sub-graphs, the overall topology of the network can generate information. Thus, for instance, in the case of the graph of Fig. 3 describing the binding of two ligands, $x$ and $y$, to a protein that undergoes sequential conformation changes; the whole system can possess mutual information even if component sub-graphs have an $\alpha$ value equal to one, i.e., do not possess information. Each sub-graph can be considered a module [10] devoid of any information. The aggregation of these modules generates information which, in the present case, is negative. It is the topology of the graph that generates information through the partition function $\Pi(n)$.

## 4. *Organization and negative mutual information*

It is striking to note that, with systems as different as those previously studied in Chapters 5 and 6, there can exist emergence of self-information in the network. As a matter of fact, in these graphs one always has three subsets of nodes: nodes that are associated with $x$ and $y$, nodes associated with $x$ but not with $y$, and nodes associated with $y$ but not with $x$. These subsets have been designated $\Omega_{XY}$, $\Omega_{N_X} - \Omega_{N_Y}, \Omega_{N_Y} - \Omega_{N_X}$ and $\Omega_{NX} - \Omega_{NX}$, respectively. Indeed the cardinals of these subsets assume mathematical expressions that depend on the topology of the network considered. Thus, in the case of the network displaying a negative statistical correlation between two ligands that bind to common sites, one has (Chapter 5)

$$
\begin{aligned}
|\Omega_{XY}| &= \sum_i \sum_j p(N_{i,j}) = \sum_i \sum_j \frac{\binom{n}{j}\binom{n-j}{i} K_x'^i K_y'^j x^i y^j}{\left\{1 + K_x' x + K_y' y\right\}^n} = |\Omega_{N_{XY}}| \\
|\Omega_{N_X} - \Omega_{N_Y}| &= \sum_i p(N_{i,0}) = \frac{(1 + K_x' x)^n}{\left\{1 + K_x' x + K_y' y\right\}^n} \\
|\Omega_{N_Y} - \Omega_{N_X}| &= \sum_j p(N_{0,j}) = \frac{(1 + K_y' y)^n}{\left\{1 + K_x' x + K_y' y\right\}^n}
\end{aligned}
\tag{21}
$$

Similarly, for a protein that binds two ligands $x$ and $y$ at two different interacting sites, one has

$$
\begin{aligned}
|\Omega_{XY}| &= p(Rx, y) = \frac{\alpha K K' xy}{1 + Kx + K'y + \alpha K K' xy} = |\Omega_{N_{XY}}| \\
|\Omega_{N_X} - \Omega_{N_Y}| &= p(Rx, 0) = \frac{Kx}{1 + Kx + K'y + \alpha K K' xy} \\
|\Omega_{N_Y} - \Omega_{N_X}| &= p(R0, y) = \frac{K'y}{1 + Kx + K'y + \alpha K K' xy}
\end{aligned}
\tag{22}
$$

and, last but not least, for a protein that binds two ligands at two specific interacting sites and that undergoes conformational changes, one finds

$$
\begin{aligned}
|\Omega_{XY}| &= p(Rx, y) = \frac{\alpha K K' x y \Pi(n)}{1 + (Kx + K'y)\Pi(n) + \alpha K K' x y \Pi(n)} = \left|\Omega_{N_{XY}}\right| \\
\left|\Omega_{N_X} - \Omega_{N_Y}\right| &= p(Rx, 0) = \frac{Kx\Pi(n)}{1 + (Kx + K'y)\Pi(n) + \alpha K K' x y \Pi(n)} \\
\left|\Omega_{N_Y} - \Omega_{N_X}\right| &= p(R0, y) = \frac{K'y\Pi(n)}{1 + (Kx + K'y)\Pi(n) + K K' x y \Pi(n)}
\end{aligned}
\tag{23}
$$

Whatever the physical nature of the system, emergence of self-information implies that the network possesses both organized and non-organized regions. In organized regions protein specifically associate both $x$ and $y$, whereas in non-organized regions it associates either $x$, or $y$ but not both of them. The cardinals $|\Omega_{XY}|$, $|\Omega_{N_X} - \Omega_{N_Y}|$, and $|\Omega_{N_Y} - \Omega_{N_X}|$ express quantitatively the sizes of these regions. As we have seen in the previous chapter, a necessary condition for emergence of self-information is

$$
|\Omega_{XY}| < \left|\Omega_{N_X} - \Omega_{N_Y}\right| \cup \left|\Omega_{N_Y} - \Omega_{N_X}\right| \tag{24}
$$

Although this condition is somewhat less restrictive than the reversal of sub-additivity, it is, perhaps, more intuitive. Classical Quastler representation [10] allows to visualize the entropies $H(X)$, $H(Y)$, $H(X,Y)$, $H(X/Y)$, $H(Y/X)$ as well as $I(X:Y)$ when mutual information is negative (Fig. 4).

As anticipated previously, network information is adaptive, i.e., it varies as a function of external signals. When the concentration of a ligand, or a substrate, changes, the probability of occurrence of certain nodes of the network varies, leading

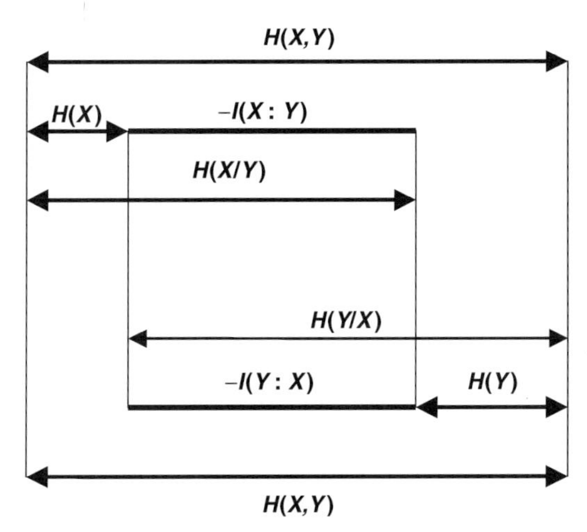


Fig. 4. Quastler diagram of an emergent system. This diagram illustrates the conditions required for the obtainment of negative mutual information viz. $H(X/Y) > H(X)$ and $H(Y/X) > H(Y)$. See text.

in turn to a change of its information content. Hence, contrary to genetic information which is constant, the information content of metabolic networks is not. As a matter of fact, the information content of a biological network can be split into two components: a genetic component, i.e., the genetic information required for the synthesis of the enzymes of the network; a topological component that reflects the probabilities of occurrence of the various nodes and the gross organization of the epigenetic networks.

## *References*


1. Ricard, J. (2003) What do we mean by complexity? C. R. Biologies 326, 133–140.
2. Adair, G.S. (1925) The hemoglobin system: VI. The oxygen dissociation curve of hemoglobin. J. Biol. Chem. 63, 529–545.
3. Levitzki, A. (1978) Quantitative Aspects of Allosteric Mechanisms. Springer-Verlag, Berlin.
4. Tanford, C. (1961) Physical Chemistry of Macromolecules. Wiley, New York.
5. Wyman, J. (1967) Linked functions and reciprocal effects in haemoglobin: a second look. Adv. Prot. Chem. 19, 223–286.
6. Wyman, J. (1967) Allosteric linkage. J. Amer. Chem. Soc. 89, 2202–2213.
7. Laidler, K.J. and Bunting, P.S. (1973) The Chemical kinetics of Enzyme action. Second Edition. Clarendon Press, Oxford.
8. Neet, (1995) Cooperativity in enzyme function: equilibrium and kinetic aspects. In: Purich, D.L. (ed.) Methods in Enzymology, 249(part D), Academic Press, New York, pp. 177–240.
9. Hill, T.L. (1987) Statistical Mechanics. Principles and Selected Applications. Dover Publications, Inc., New York.
10. Hatwell, L.H., Hopfield, J.J., Leibler, S., and Murray, A.W. (1999) From molecular to modular cell biology. Nature 402 Suppl., C47–C52.
11. Quastler, H. (1958) A primer on information theory. In: Yokey, H.P., Platzman, R.L., and Quastler H. (eds.) Symposium on Information Theory in Biology, Pergamon Press, New York.

J. Ricard *Emergent Collective Properties, Networks and Information in Biology*

DOI: 10.1016/S0167-7306(05)40007-1

CHAPTER 7

# Information and organization of metabolic networks

J. Ricard

As enzyme reactions are already networks, any metabolic network, which is made up of many enzyme reactions, can be viewed as a supernetwork, or a network of networks. The aim of this chapter is to study both isolated enzyme reactions, viewed as networks, and metabolic networks. An enzyme that binds two substrates randomly usually possesses mutual information which is a quantitative expression of the corresponding network organization and catalytic efficiency. The more this system is emergent, the more it is catalytically efficient. Moreover departure from pseudo-equilibrium conditions enhances the emergent character of the system. The overall information content of a metabolic network results from the information of individual enzyme reactions and from the topology of the metabolic network. A metabolic network whose individual enzyme reactions are *emergent* tends to be robust to external perturbations. Moreover, by-passing "slow" steps of a metabolic network results in an increase of its information. One may therefore speculate, in a neo-Darwinian perspective, that it might be advantageous for metabolic networks to be of the fuzzy-organized type.

*Keywords:* fuzzy-organized networks, information and departure from equilibrium, metabolic networks as networks of networks, information and nonequilibrium, information of an enzyme reaction, fractionation factors, organization and catalytic efficiency, regular networks, robustness of multienzyme networks.

Metabolic networks are represented by a set of nodes connected according to a certain topology. In this classical perspective, the nodes are chemicals, or metabolites, that undergo enzyme catalyzed transformations whereas the edges between them are the corresponding chemical reactions [1–13]. Although sensible at first sight, this mode of description suffers from several major pitfalls: first, it overlooks the fact that an enzyme reaction is not a link between two nodes, but a true network whose nodes are the various interconvertible states of the same enzyme; second, it does not take into account the fact that most enzyme reactions involve in fact two, or three, substrates, which implies in turn that the enzyme is able to associate these substrates in a specific manner so as to form the products of the reaction. Hence, in the spirit of the previous chapters and in agreement with a previous definition of the concept of information, one should expect that an enzyme reaction can itself possess mutual information of integration. It is clear that the classical representation of networks does not allow us to take into account the potential information stored in a metabolic network. If we want to derive the expression of this information, a metabolic network should be viewed as a kind of supernetwork, or network of networks. As an isolated enzyme reaction can itself be considered a network, it is probably useful to begin this chapter on metabolic networks by a study of simple enzyme reactions.

## *1. Mutual information of individual enzyme reactions*

Let us consider, as an example, a group-transfer enzyme-catalyzed reaction

$$AX + B \xrightarrow{E} A + XB$$

where $E$ is the enzyme. Under "initial" conditions, i.e., before the final equilibrium state is reached, this process can follow either of three possible routes: a sequential and ordered binding of substrates to the enzyme; a sequential and random binding of the two substrates; a substitution mechanism [14–16]. In the first case, the two substrates bind to the enzyme in a strictly ordered fashion before the products are released. One has for instance

$$E \underset{k_{-1}}{\overset{k_1[AX]}{\rightleftarrows}} EAX \underset{k_{-2}}{\overset{k_2[B]}{\rightleftarrows}} EAXB \xrightarrow{k} E + A + XB$$

where the $k$ values are rate constants. But the enzyme can also bind the substrates randomly before releasing the products. One then has

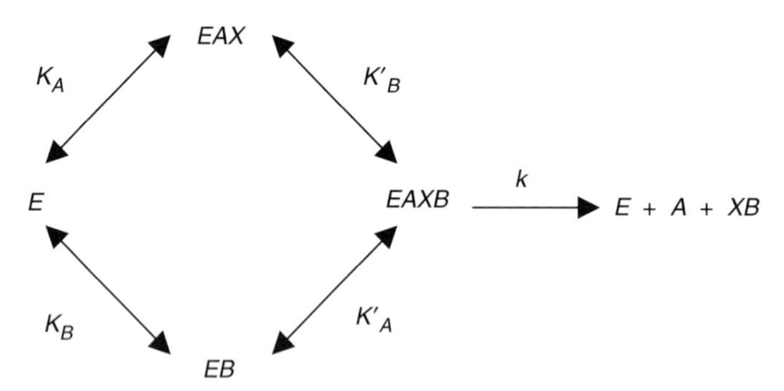


where the $K$'s are the substrate binding constants, i.e., the ratio of the forward and backward rate constants. Last, the enzyme may bind a substrate, for instance $AX$, release the corresponding product $A$, and bind the second substrate $B$, before releasing the last product $XB$. Hence one has

$$E \underset{k_{-1}}{\overset{k_1[AX]}{\rightleftarrows}} EAX \xrightarrow{k_2} EX + A \underset{k_{-3}}{\overset{k_3[B]}{\rightleftarrows}} EXB \xrightarrow{k_4} E + XB$$

The first and the third of these models are assumed to be in steady state, while the second is assumed to be close to thermodynamic equilibrium. This is why equilibrium, or binding, constants have been introduced in this model. Under these conditions the mutual information of integration of the network can be defined and mathematically expressed.

If the substrates bind to the enzyme and if the various enzyme–substrate(s) complexes are close to thermodynamic equilibrium, one can derive the expression for the probabilities that the enzyme has bound either $AX$ or $B$, or both $AX$ and $B$. One finds

$$
\begin{aligned}
p(AX) &= \frac{K_A[AX](1 + K'_B[B])}{1 + K_A[AX] + K_B[B] + K_A K'_B[AX][B]} \\
p(B) &= \frac{K_B[B](1 + K'_A[AX])}{1 + K_A[AX] + K_B[B] + K_A K'_B[AX][B]} \\
p(AX, B) &= \frac{K_A K'_B[AX][B]}{1 + K_A[AX] + K_B[B] + K_A K'_B[AX][B]}
\end{aligned} \qquad (1)
$$

Moreover the substrate binding constants are not independent because thermodynamics demands that

$$K_A K'_B = K'_A K_B \qquad (2)$$

On can also define conditional probabilities $p(AX/B)$ and $p(B/AX)$. One finds

$$
\begin{aligned}
p(AX/B) &= \frac{K'_A[AX]}{1 + K'_A[AX]} \\
p(B/AX) &= \frac{K'_B[B]}{1 + K'_B[B]}
\end{aligned} \qquad (3)
$$

Combining expressions (1) and (3), one can check that

$$p(AX, B) = p(AX/B)p(B) = p(AX)p(B/AX) \qquad (4)$$

thus showing that, as expected, Bayes' theorem applies to the enzyme system. The mutual information of integration of this simple enzyme system can be calculated through the relationship

$$I(AX : B) = I(B : AX) = \log_2 \frac{p(AX/B)}{p(AX)} = \log_2 \frac{p(B/AX)}{p(B)} \qquad (5)$$

and one finds

$$I(AX:B) = I(B:AX) = \log_2 \frac{1 + K_A[AX] + K_B[B] + K_A K'_B[AX][B]}{1 + K'_A[AX] + K'_B[B] + K'_A K'_B[AX][B]} \frac{K'_B}{K_B} \tag{6}$$

This expression is indeed similar to Eq. (6) of Chapter 6. It follows that if $K'_B = K_B$ and $K'_A = K_A$ then $I(AX{:}B) = 0$, and the network does not contain any mutual information of integration;

$K'_B > K_B$ and $K'_A > K_A$ implies that $I(AX:B) > 0$;
$K'_B < K_B$ and $K'_A < K_A$ means that $I(AX:B) < 0$.

In the second and third cases considered above, the network behaves as a real system, as a coherent whole, and the mutual information of integration, whether positive or negative, is a measure of this integration. Whereas in the second case, the system can be considered "integrated", it can be considered "emergent" in the last one. Moreover, Eq. (5) above shows that the sign of the mutual information of integration relies on the respective values of conditional and individual probabilities. As expected, if the corresponding conditional and individual probabilities are equal, i.e., $p(AX) = p(AX/B)$ and $p(B) = p(B/AX)$, the network has no mutual information of integration. If $p(AX) < p(AX/B)$ and $p(B) < p(B/AX)$ the mutual information is positive and if $p(AX) > p(AX/B)$ and $p(B) > p(B/AX)$ the corresponding information is negative.

One must stress again that here information has nothing to do with a correlation between discrete variables, it originates from physical interactions between the subsites of the enzyme that accommodates the two substrates. Hence, if the subadditivity condition does not apply, this means that the binding of $B$ to its corresponding subsite decreases the affinity of $AX$ for the other subsite. Similarly, the binding of $AX$ to the enzyme hinders that of $B$. This situation is reversed if subadditivity applies. The existence of positive or negative information of integration is therefore equivalent to cooperativity between the two subsites of the enzyme's active site.

## 2. *Relationships between enzyme network organization and catalytic efficiency*

When the enzyme reaction follows a compulsory order in the binding of substrates, no organization can be defined through the theoretical concepts already discussed. When the addition of substrates to the protein, however, is random, one can associate the mutual degree of organization of the network to the catalytic power of the enzyme.

The theory of absolute reaction rates associates an energy level with every intermediate of the chemical process [17–26]. The same situation applies to an enzyme catalyzed process. Every binding constant is associated with a free energy increase, or decrease, i.e.,

$$K = e^{-\Delta G/RT} \tag{7}$$

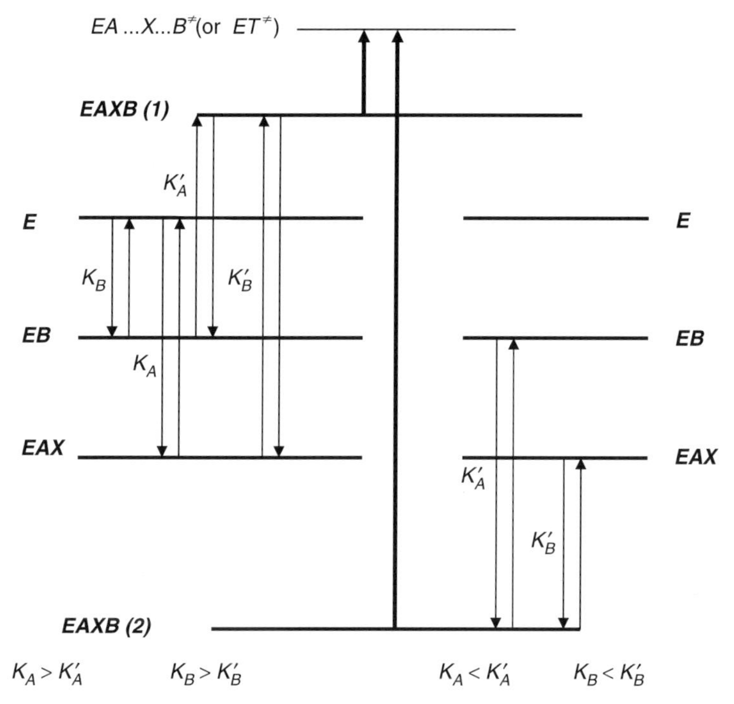


Fig. 1. Catalytic efficiency of an enzyme and mutual information. Left: If $K_A > K'_A$ and $K_B > K'_B$ then $I(AX : B) < 0$ and the energy level of the ternary complex $EAX.B$ is located above the energy levels of $E$, $EAX$, and $EB$. Hence the height of the energy barrier to be overcome in going from $EAX.B$to the transition state $EA \ldots X \ldots B^{\neq}$ is not large and this implies that the catalytic constant is large. Moreover the energy required to go from the states $EAX$ and $EB$ to $EAX.B$ probably originates from the kinetic energy spent during the formation of the ternary complex. Right: If $K_A < K'_A$ and $K_B < K'_B$ then $I(AX : B) < 0$ and the energy level of the ternary complex $EAX.B$ is located below the energy levels of $EAX$ and $EB$. Hence the height of the free energy barrier to be overcome in going from the ternary $EAX.B$ state to the transition state $EA \ldots X \ldots B^{\neq}$ is high and the corresponding catalytic constant is small.

where $\Delta G$ is the free energy increase or decrease, $R$ and $T$ the gas constant and the absolute temperature, respectively. This relationship implies that $0 < K < 1$ if $\Delta G > 0$, $K = 1$ if $\Delta G = 0$, $K > 1$ if $\Delta G < 0$. We have seen previously that $I(AX : B) < 0$ if

$$K_A > K'_A \quad \text{and} \quad K_B > K'_B \tag{8}$$

The corresponding thermodynamic situation is depicted in Fig. 1.

This scheme shows that the free energy decreases when $AX$ or $B$ binds to the enzyme and that the free energy increases when $AX$ or $B$ binds to $E.B$ or to $E.AX$, respectively. This situation is consistent with relationships (8). It also implies that the energy level of the ternary enzyme–substrate complex $E.AX.B$ can possibly be located not too far from the energy level of the enzyme-transition state complex

Table 1
Influence of binding constants $K'_A$ and $K'_B$ on the catalytic constant $k$

| $K'_A$ | $K'_B$ | $\Delta G_{(EAXB-EAX)}$(J) | $k(s^{-1})$ |
|---|---|---|---|
| 0.0002 | 0.0001 | 22,425 | 537 |
| 0.002 | 0.001 | 16,819 | 53.77 |
| 0.02 | 0.01 | 11,212 | 5.37 |
| 0.2 | 0.1 | 5,606 | 0.53 |
| $k_BT/h=10^{13}$; | $T=293°K$; | $R=8.31$ J deg$^{-1}$ mole$^{-1}$; | |
| | $K_A=2$; | $K_B=1$ | |

$ET^{\neq}$, which in turn implies that the reaction can easily overcome the free energy barrier between $E.AX.B$ and $ET^{\neq}$. Let us assume, for instance, it is possible to alter at will, through site-directed mutagenesis $K'_A$ and $K'_B$ without altering $K_A$ and $K_B$ as well as the level of the transition state $ET^{\neq}$. This alteration of binding constants will indeed keep the relationship $K_AK'_B = K'_AK_B$ unchanged. The corresponding free energy changes associated with different values of the binding constant $K'_B$ (or $K'_A$) can be calculated from the classical relationship (Table 1)

$$\Delta G_{(E.AX.B-E.AX)} = -RT\ln K'_B \tag{9}$$

where $R$ and $T$ have their usual significance. The height of the free energy barrier that the reaction should overcome is $\Delta G^{\neq}_{(ET^{\neq}-E.AX.B)}$. Moreover, according to classical Eyring's theory, the corresponding value of the catalytic constant $k$ is expressed as [17–20]

$$k = \frac{k_BT}{h}\exp\left\{-\left[\Delta G^{\neq}_{(ET^{\neq}-E.AX.B)}/RT\right]\right\} \tag{10}$$

where $k_B$ and $h$ are the Boltzmann and Planck constants, respectively.

As shown in Table 1 and Fig. 1, when the value of $K'_B$ decreases $\left(K'_B < K_B\right)$ the catalytic rate constant increases. This is associated with a destabilization of the ground state of the ternary complex $E.AX.B$ relative to what one would have expected if $K_B = K'_B$. Conversely, as the value of $K'_B$ increases $\left(K'_B > K_B\right)$ the catalytic rate constant $k$ decreases and the ground state of the ternary complex $E.AX.B$ becomes stabilized (Fig. 1). If one considers the thought experiment where $K_A$ and $K_B$ have fixed values while $K'_A$ and $K'_B$ are varied, one can calculate the corresponding mutual information $I(AX:B)$ from Eq. (6). It is obvious that changing the values of $K'_A$ and $K'_B$ results in a change of the probabilities of occurrence of $E.AX$, $E.B$, and $E.AX.B$. Therefore the topology of the enzyme network is changed as well as its mutual information (Fig. 2). As a matter of fact, decreasing the mutual information of integration of the network, results in both the emergence of information and of catalytic function. Hence when mutual information of integration is negative, the protein is not only able to bind ligands $AX$ and $B$, but is also able to catalyze their mutual chemical transformation.

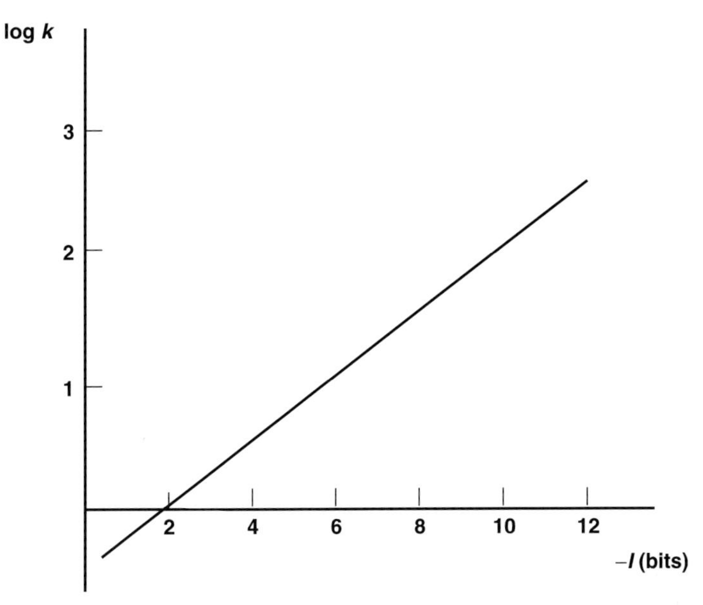


Fig. 2. Variation of the catalytic rate constant $k$ as a function of the mutual information of the enzyme network under pseudo-equilibrium conditions. As emergence of the system increases ($I$ assumes negative values) under pseudo-equilibrium conditions its catalytic constant $k$ becomes larger. Information is computed from the $K_A, K'_B, K_B, K'_B$ values of Table 1.

## *3. Mutual information of integration and departure from pseudo-equilibrium*

In the case of the random binding of the two substrate molecules to the enzyme, it was assumed that the overall system was close to thermodynamic equilibrium. As the system departs from pseudo-equilibrium, however, i.e., when the apparent catalytic constant $k$ increases, the above equations are no longer valid and have to be altered (Fig. 3). One can calculate the probabilities that the enzyme has bound either $AX$ or $B$, or both. One finds

$$\begin{aligned} p(AX) &= \frac{K_1[AX](1 + K_2[B]) + k\delta_{NA}}{1 + K_1[AX] + K_3[B] + K_1K_2[AX][B] + k\delta_D} \\ p(B) &= \frac{K_3[B](1 + K_4[AX]) + k\delta_{NB}}{1 + K_1[AX] + K_3[B] + K_1K_2[AX][B] + k\delta_D} \\ p(AX, B) &= \frac{K_1K_2[AX][B]}{1 + K_1[AX] + K_3[B] + K_1K_2[AX][B] + k\delta_D} \end{aligned} \tag{11}$$

where $\delta_{NA}$, $\delta_{NB}$, and $\delta_D$ are nonlinear functions of the concentration of $AX$ and $B$. They express the departure of the system from pseudo-equilibrium conditions.

$K_1$, $K_2$, $K_3$, and $K_4$ are ratios of rate constants, namely $K_1 = k_1/k_{-1}, \ldots, K_4 = k_4/k_{-4}$. The expressions of these nonlinear functions generated by the departure from equilibrium are

$$
\begin{aligned}
\delta_{NA} &= k_1[AX](k_{-3} + k_4[AX]) \\
\delta_{NB} &= k_3[B](k_{-1} + k_2[B]) \\
\delta_D &= (k_{-1} + k_1[AX] + k_2[B])(k_{-3} + k_4[AX]) + k_3[B](k_{-1} + k_2[B])
\end{aligned}
\tag{12}
$$

Likewise, one can calculate the conditional probabilities $p(AX/B)$ and $p(B/AX)$. One finds

$$
\begin{aligned}
p(AX/B) &= \frac{K_4[AX]}{1 + K_4[AX] + kk_{-3}(k_{-1} + k_2[B])} \\
p(B/AX) &= \frac{K_2[B]}{1 + K_2[B] + kk_{-1}(k_{-3} + k_4[AX])}
\end{aligned}
\tag{13}
$$

It is probably worth stressing again that if the catalytic rate constant $k$ is small, expressions (11) and (13) reduce to Eqs. (1) and (3) above. From Eqs. (11) and (13) one can calculate the ratio $p(AX)/p(AX/B)$, or its reciprocal, as well as the mutual information of the system. One thus finds

$$
I(AX : B) = -\log_2 \frac{\Omega_1 + k\Omega_2 + k^2\Omega_3}{\Omega_4 + k\Omega_5} \frac{K_1}{K_4}
\tag{14}
$$

where the $\Omega$'s are nonlinear functions of the concentrations of $AX$ and $B$, namely

$$
\begin{aligned}
\Omega_1 &= (1 + K_2[B])(1 + K_4[AX]) \\
\Omega_2 &= k_{-3}(1 + K_2[B])(k_{-1} + k_2[B]) + k_{-1}(1 + K_4[AX])(k_{-3} + k_4[AX]) \\
\Omega_3 &= k_1 k_3(k_{-1} + k_2[B])(k_{-3} + k_4[AX]) \\
\Omega_4 &= 1 + K_1[AX] + K_3[B] + K_1 K_2[AX][B] \\
\Omega_5 &= \delta_D = (k_{-1} + k_1[AX] + k_2[B])(k_{-3} + k_4[AX]) + k_3[B](k_{-1} + k_2[B])
\end{aligned}
\tag{15}
$$

Simple inspection of Eq. (14) shows that if the catalytic constant $k$ is very small, $k\Omega_2$, $k\Omega_3$, and $k\Omega_5$ can be neglected and the corresponding mutual information becomes identical to Eq. (6). However, if the value of the catalytic constant $k$ becomes larger, the system drifts from a pseudo-equilibrium to a steady state and its mutual information becomes negative. Departure from thermodynamic equilibrium generates emergence of mutual information. This situation can be observed even if $K_1 = K_4$ and $K_2 = K_3$. Moreover the greater the shift of the system from its initial pseudo-equilibrium, the larger is the emergence of joint entropy (Fig. 4).

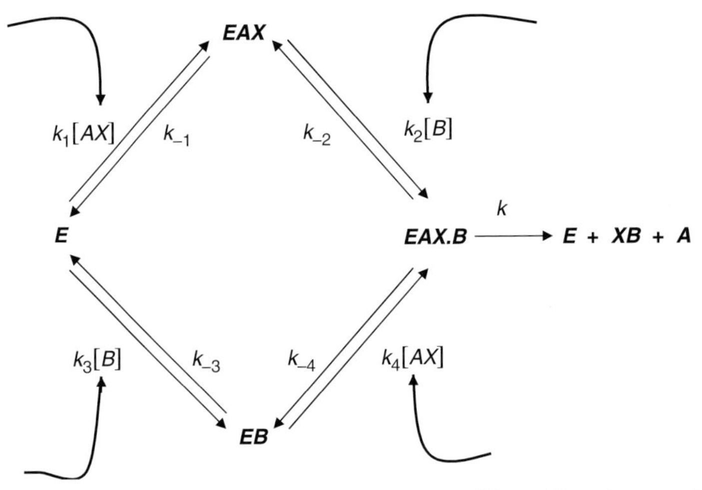


Fig. 3. Two substrate random enzyme network under open conditions. The substrates $AX$ and $B$ are continuously introduced into the medium as to maintain the system under steady state.

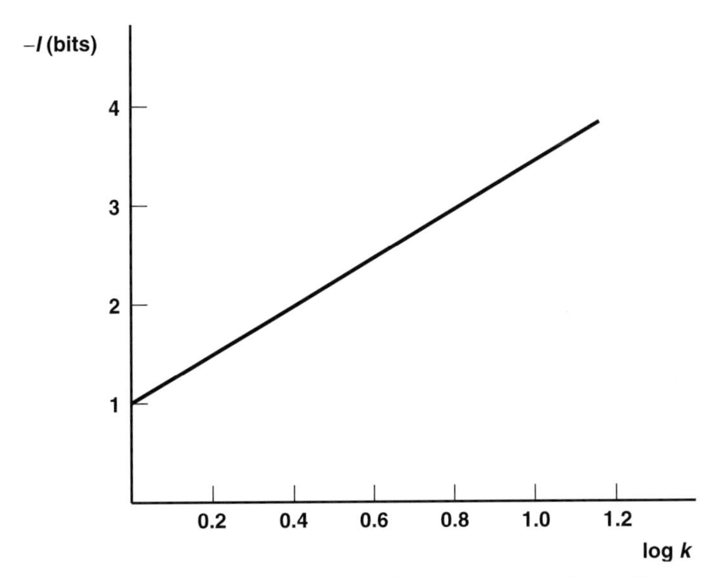


Fig. 4. Mutual information of an enzyme network and departure from pseudo-equilibrium conditions. As the catalytic rate constant $k$ is increased, the system departs from its initial pseudo-equilibrium conditions and stores an increasing amount of negative information. The numerical values of the parameters used to generate the curve are (arbitrary units) $\Omega_1 = 1, \Omega_2 = 2, \Omega_3 = 3, \Omega_4 = 2, \Omega_5 = 4, K_1/K_4 = 2$.

The interesting conclusion of these results is therefore that mutual information of a simple enzyme reaction can have two origins: the physical interaction between two subsites of the enzymes that bind the substrates and the departure of an enzyme reaction from pseudo-equilibrium, thus leading to a system that becomes open and exchanges matter with the external world.

## 4. *Mutual information of integration of multienzyme networks*

### 4.1. *Metabolic networks as networks of networks*

Metabolic processes are in fact multienzyme networks. The graphical representation we are going to adopt consists in defining as nodes individual enzyme reactions. Hence metabolic networks can be considered networks of networks. For reasons that have already been mentioned, this mode of representation is more logical than the classical one which consists in defining as nodes the products of reactions that are taken up as the substrates of the next enzyme process. We shall consider below two classes of multienzyme networks, namely regular and fuzzy-organized networks (Fig. 5). In fuzzy-organized graphs some original edges can be replaced locally by random ones, thus allowing one to bypass certain nodes. Small-world networks, which we have already referred to, belong to this category. Moreover metabolic networks are thermodynamically open. They exchange matter and (or) energy with the external world.

Let us consider an ideal network made up of four nodes. Each node, $Y_1$, $Y_2, \ldots,$ $Y_4$, represents for instance a two-substrate enzyme reaction. This means that each $Y_i$ value is in fact the sum of the concentrations of the various enzyme states of the $i$th chemical process. If this enzyme is able to bind substrates, or reagents, $A_i$ and $B_i$, one has [14–16]

$$Y_i = [E_i] + [E_iA_i] + [E_iB_i] + [E_iA_iB_i] \tag{16}$$

Indeed either $A_i$ or $B_i$, or both of them is (or are) the product(s) of the previous enzyme reaction. However that may be, the four processes are assumed to be in fast equilibrium whereas the overall system may be in steady state. In other words,

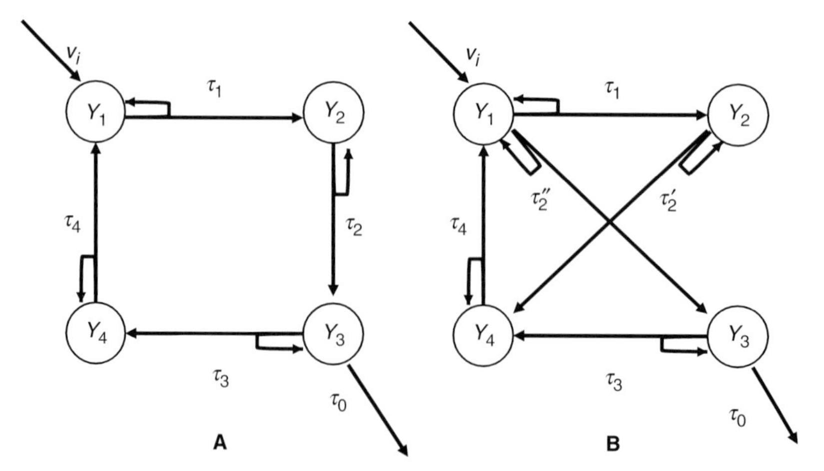


Fig. 5. Two simple ideal open metabolic networks. (A) Regular metabolic network. (B) Fuzzy-organized metabolic network. The nodes $Y_1, \ldots, Y_4$ are enzyme-catalyzed reactions, $\tau_1, \ldots, \tau_4$ are the transition rate constants from one node to another one. $v_1$ is the rate of entry, $\tau_0$ the rate constant of exit.

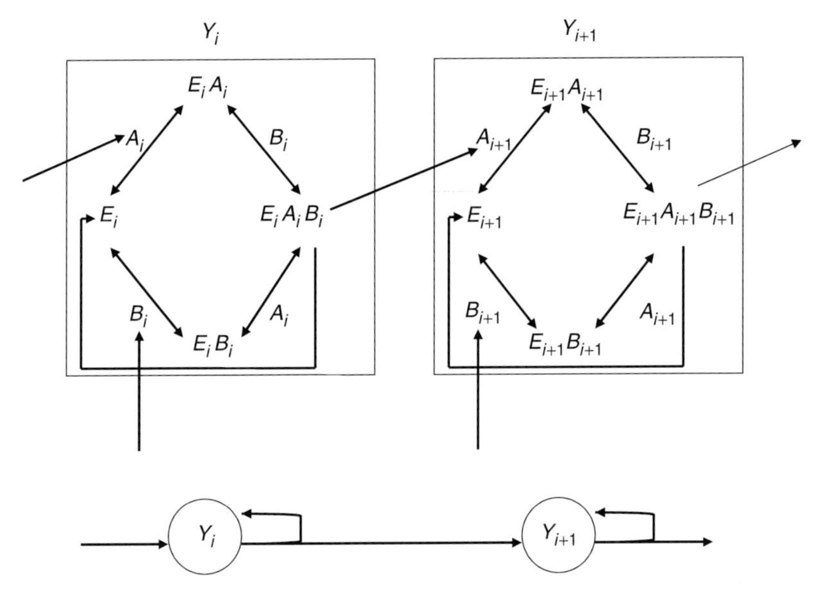


Fig. 6. Two enzyme reactions in a metabolic sequence. The enzymes are $E_i$ and $E_{i+1}$. It is assumed that the processes of substrate binding and release are fast processes relative to the rate of transport of product $A_{i+1}$ from enzyme $E_i$ to enzyme $E_{i+1}$. Chemicals $B_i$ and $B_{i+1}$ are assumed to be taken up in the medium. Each of the enzyme reactions can be considered a node of the network (lower part of the figure).

substrate binding and release are fast processes relative to catalysis and transport of reaction intermediates from one enzyme to another. Owing to this time hierarchy, one can greatly simplify the graph, as shown in Fig. 6. Moreover rate constants such as $k_i^\bullet$ and $k_i^D$ can have different values. In fact, the former is the apparent rate constant of catalysis and free enzyme release whereas the latter is the rate constant of product transport to another enzyme.

### 4.2. *Robustness of multienzyme networks*

An important point in the study of networks is their robustness i.e., their ability to maintain their topology when the external conditions are altered [28]. Networks are robust if the links that associate nodes do not dismantle spontaneously. The strength of these links relies upon the value of the apparent catalytic constant which is in fact the product of the true catalytic constant by the so-called *fractionation factor* [27] viz.

$$k_i^\bullet = k_i f_i = k_i \frac{K_{Ai}K'_{Bi}[A_i][B_i]}{1 + K_{Ai}[A_i] + K_{Bi}[B_i] + K_{Ai}K'_{Bi}[A_i][B_i]} \tag{17}$$

The fractionation factor $f_i$ is in fact the joint probability that enzyme $E_i$ has bound both substrates $A_i$ and $B_i$, i.e., $p(A_i, B_i)$. If, in a thought experiment, we assume that the $E_iA_i$ state can be stabilized, or destabilized, relative to the ternary $E_iA_iB_i$ state

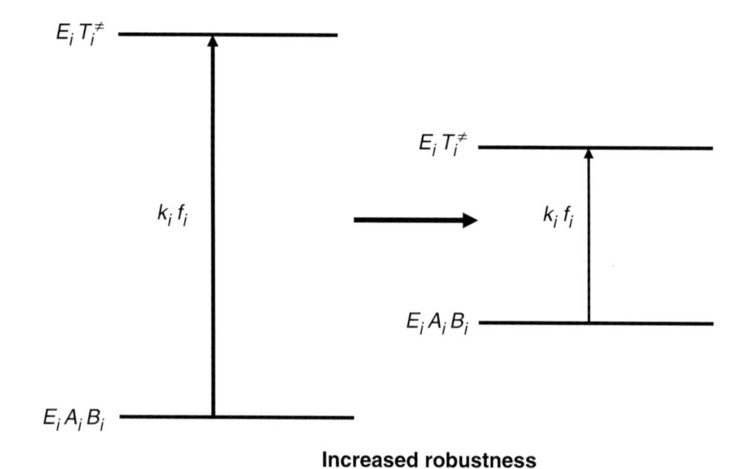


Fig. 7. Destabilization of the ternary states $E_iA_iB_i$ and stabilization of the transition states $E_iT_i^{\neq}$ increase robustness of multienzyme networks. See text.

whereas the energy level of the transition state $E_iT_i^{\neq}$ does not vary, then the product of the catalytic constant $k_i$ by the equilibrium constant $K'_{Bi}$ is constant, i.e.,

$$k_i K'_{Bi} = \lambda \tag{18}$$

where $\lambda$ is a constant (Table 1). It follows that

$$k_i f_i = \frac{\lambda K_{Ai}[A_i][B_i]}{1 + K_{Ai}[A_i] + K_{Bi}[B_i] + K_{Ai}K'_{Bi}[A_i][B_i]} \tag{19}$$

If, in a thought experiment, $E_iA_iB_i$ is subjected to progressive stabilization relative to $E_iA_i$, $K'_{Bi}$ increases and, according to Eq. (19), $k_i f_i$ decreases. Under these conditions, the connection between the nodes $Y_i$ and $Y_{i+1}$ tends to break down. If, conversely, the $E_iA_iB_i$ state undergoes progressive destabilization relative to $E_iA_i$, $K'_{Bi}$ decreases and $k_i f_i$ increases thus leading to a strengthening of the link between the nodes $Y_i$ and $Y_{i+1}$. It follows that if

$$K'_{Ai} \gg K_{Ai} \quad \text{and} \quad K'_{Bi} \gg K_{Bi} \tag{20}$$

then the mini-network that make up the node $Y_i$ is integrated but, as a consequence, the overall network tends to dismantle. Alternatively, if

$$K'_{Ai} \ll K_{Ai} \quad \text{and} \quad K'_{Bi} \ll K_{Bi} \tag{21}$$

the mini-network corresponding to the node $Y_i$ is emergent thus strengthening its link with the node $Y_{i+1}$. More generally speaking, an enzyme network whose nodes are emergent tends to be robust to external perturbations. If there is a functional advantage for a network to be robust, one can speculate there exists a selective pressure that tends to stabilize the transition states $E_iT_i^{\neq}$ ($\forall i = 1, 2, \ldots, n$) of the enzyme reactions and to destabilize the ternary states $E_iA_iB_i$ of the same enzyme processes (Fig. 7). The thermodynamic conditions that generate emergence of nodes tend to increase the robustness of the overall network.

### *4.3. Regular multienzyme networks*

Let us consider the ideally simple network of Fig. 5A. The sum of enzyme concentrations, $Y_T$, involved in this network is indeed

$$Y_T = Y_1 + Y_2 + Y_3 + Y_4 \tag{22}$$

The probabilities of occurrence of the nodes of the graph are

$$\begin{aligned} p(Y_1) &= \frac{Y_1}{Y_T} \\ p(Y_2) &= \frac{Y_2}{Y_T} \\ p(Y_3) &= \frac{Y_3}{Y_T} \\ p(Y_4) &= \frac{Y_4}{Y_T} \end{aligned} \tag{23}$$

The differential equations that describe the dynamics of the regular network are

$$\begin{aligned} \frac{dY_1}{dt} &= v_i + \tau_4 Y_4 - \tau_1 Y_1 \\ \frac{dY_2}{dt} &= \tau_1 Y_1 - \tau_2 Y_2 \\ \frac{dY_3}{dt} &= \tau_2 Y_2 - (\tau_3 + \tau_0) Y_3 \\ \frac{dY_4}{dt} &= \tau_3 Y_3 - \tau_4 Y_4 \end{aligned} \tag{24}$$

In these expressions $v_i$ is the rate of input of matter within the system and the $\tau$'s are the transition rates that connect the nodes. One can also write a conservation Eq. (22). If the system is in steady state, the time derivatives vanish. Moreover the $Y_1$, $Y_2, \ldots$ states can be replaced by their probability of occurrence if each term in Eqs. (24) above is divided by $Y_T$. These probabilities can be obtained by solving the system made up of three steady-state equations plus the conservation equation. One has thus

$$\begin{bmatrix} \tau_1 & -\tau_2 & 0 & 0 \\ 0 & \tau_2 & -(\tau_3+\tau_0) & 0 \\ 0 & 0 & \tau_3 & -\tau_4 \\ 1 & 1 & 1 & 1 \end{bmatrix} \begin{bmatrix} p(Y_1) \\ p(Y_2) \\ p(Y_3) \\ p(Y_4) \end{bmatrix} = \begin{bmatrix} 0 \\ 0 \\ 0 \\ 1 \end{bmatrix} \tag{25}$$

Moreover the steady state requires that

$$\frac{v_i}{Y_T} + \tau_4 p(Y_4) - \tau_1 p(Y_1) = 0 \tag{26}$$

Solving system (25) yields

$$\begin{aligned} p(Y_1) &= \frac{\tau_2\tau_4(\tau_3+\tau_0)}{\tau_1\tau_2(\tau_3+\tau_4)+\tau_4(\tau_1+\tau_2)(\tau_3+\tau_0)} \\ p(Y_2) &= \frac{\tau_1\tau_4(\tau_3+\tau_0)}{\tau_1\tau_2(\tau_3+\tau_4)+\tau_4(\tau_1+\tau_2)(\tau_3+\tau_0)} \\ p(Y_3) &= \frac{\tau_1\tau_2\tau_4}{\tau_1\tau_2(\tau_3+\tau_4)+\tau_4(\tau_1+\tau_2)(\tau_3+\tau_0)} \\ p(Y_4) &= \frac{\tau_1\tau_2\tau_3}{\tau_1\tau_2(\tau_3+\tau_4)+\tau_4(\tau_1+\tau_2)(\tau_3+\tau_0)} \end{aligned} \tag{27}$$

The $\tau$ constants that describe the transitions from node to node are equivalent to apparent rate constants. Thus the $\tau_i$ constant, which is associated with the transition from node $Y_i$ to node $Y_{i+1}$, is equal to the product of the catalytic rate constant, $k_i$, by the rate constant of product transport, $k_i^D$, and by the fractionation factor, $f_i$. Hence one has

$$\tau_i = k_i k_i^D \frac{K_{Ai}K'_{Bi}[A_i][B_i]}{1+K_{Ai}[A_i]+K_{Bi}[B_i]+K_{Ai}K'_{Bi}[A_i][B_i]} \tag{28}$$

Moreover Eqs. (26) and (27) imply that under global steady state

$$\frac{v_i}{Y_T} = p(Y_3)\tau_0 \tag{29}$$

This means that under steady state the input rate should indeed be equal to the output rate. Moreover Eqs. (27) show that the topology of the network and its open character are quantitatively reflected in the values of the probabilities $p(Y_1), p(Y_2), \ldots$. Thus if the open character of the network is increased (by increasing the value of the output constant $\tau_0$), $p(Y_1)$ and $p(Y_2)$ will increase whereas $p(Y_3)$ and $p(Y_4)$ will decrease (Table 2).

Table 2

Influence of the rate constant of exit, $\tau_0$, on the probabilities of occurrence of the nodes of a regular network. One can observe that the probabilities of occurrence of the nodes $Y_3$ and $Y_4$ approach zero for high values of $\tau_0$

| | | | | |
|---|---|---|---|---|
| $\tau_0 = 0.001$ | $p(Y_1)$<br>0.251 | $p(Y_2)$<br>0.251 | $p(Y_3)$<br>0.248 | $p(Y_4)$<br>0.248 |
| $\tau_0 = 0.1$ | $p(Y_1)$<br>0.262 | $p(Y_2)$<br>0.262 | $p(Y_3)$<br>0.238 | $p(Y_4)$<br>0.238 |
| $\tau_0 = 1$ | $p(Y_1)$<br>0.333 | $p(Y_2)$<br>0.333 | $p(Y_3)$<br>0.166 | $p(Y_4)$<br>0.166 |
| $\tau_0 = 10$ | $p(Y_1)$<br>0.458 | $p(Y_2)$<br>0.458 | $p(Y_3)$<br>0.042 | $p(Y_4)$<br>0.042 |
| $\tau_0 = 100$ | $p(Y_1)$<br>0.495 | $p(Y_2)$<br>0.495 | $p(Y_3)$<br>0.000 | $p(Y_4)$<br>0.000 |
| $\tau_1 = \tau_2 = \tau_3 = \tau_4 = 1$ | | | | |

As previously, the probabilities that the enzyme $E_i$ binds either $A_i$ or $B_i$, or both $A_i$ and $B_i$ are

$$
\begin{aligned}
p(A_i)_{Ei} &= \frac{[E_iA_i] + [E_iA_iB_i]}{Y_i} \\
p(B_i)_{Ei} &= \frac{[E_iB_i] + [E_iA_iB_i]}{Y_i} \qquad (30) \\
p(A_i, B_i)_{Ei} &= \frac{[E_iA_iB_i]}{Y_i}
\end{aligned}
$$

But the probability that the network (or the set of the enzymes of the network) binds either $A_i$ or $B_i$, or both $A_i$ and $B_i$ are (if $A_i$ and $B_i$ are specifically bound to one enzyme only of the network)

$$
\begin{aligned}
p(A_i)_N &= \frac{[E_iA_i] + [E_iA_iB_i]}{Y_T} \\
p(B_i)_N &= \frac{[E_iB_i] + [E_iA_iB_i]}{Y_T} \qquad (31) \\
p(A_i, B_i)_N &= \frac{[E_iA_iB_i]}{Y_T}
\end{aligned}
$$

In these expressions and in the following, the index $N$ refers to the whole network, whereas in expressions (30) the index $E_i$ refers to the corresponding specific enzyme. The present notation should not be confused with that used in Chapter 5.

Comparing expressions (30) and (31) leads to

$$
\begin{aligned}
p(A_i)_N &= p(Y_i)p(A_i)_{Ei} \\
p(B_i)_N &= p(Y_i)p(B_i)_{Ei} \\
p(A_i, B_i)_N &= p(Y_i)p(A_i, B_i)_{Ei}
\end{aligned}
\tag{32}
$$

Moreover if, as assumed above, substrate $B_i$ binds only to enzyme $E_i$, the conditional probabilities of $A_i$ binding either to enzyme $E_i$ in isolation, or to the network, given that $B_i$ is already bound, are the same, namely

$$
p(A_i/B_i)_{Ei} = p(A_i/B_i)_N \tag{33}
$$

The corresponding uncertainty functions $H(A_i)_N$ and $H(A_i/B_i)_N$ are thus

$$
\begin{aligned}
H(A_i)_N &= -\log_2\{p(Y_i)p(A_i)_{Ei}\} \\
H(A_i/B_i)_N &= -\log_2 p(A_i/B_i)_{Ei}
\end{aligned}
\tag{34}
$$

and the mutual information of the $i$th step *within the network* is

$$
I(A_i : B_i)_N = -\log_2 \frac{p(A_i)_{Ei}}{p(A_i/B_i)_{Ei}} - \log_2 p(Y_i) \tag{35}
$$

One can notice that the first term of the right-hand side member of the above expression is in fact the mutual information of integration, $I(A_i : B_i)_{Ei}$, of the reaction catalyzed by enzyme $E_i$ and occurring *in isolation*. Hence Eq. (35) reads

$$
I(A_i : B_i)_N = I(A_i : B_i)_{Ei} - \log_2 p(Y_i) \tag{36}
$$

On can also define a mean mutual information per node of the network which can be expressed as

$$
\langle I(A : B)\rangle = -\sum_i p(Y_i)\log_2 \frac{p(A_i)_{Ei}}{p(A_i/B_i)_{Ei}} - \sum_i p(Y_i)\log_2 p(Y_i) \tag{37}
$$

Expressions (35) and (37) can therefore be positive, negative or nil. Relationships (35) and (37) are interesting for several reasons. First, they show that the subadditivity condition is different depending on whether the enzyme is part of a network or in isolation. Second, if $p(A_i)_{Ei} = p(A_i/B_i)_{Ei}$ ($\forall i = 1, 2, \ldots, n$) the corresponding information is not equal to zero. The fact that $E_i$ is not free, but is part of a network, gives this enzyme additional mutual information, which is a form of topological information. Last, it appears from Eq. (37) that if the number of nodes of the network is very large, then the probability of occurrence of $Y_i$, p($Y_i$), is very small but $-\log p(Y_i)$ is large and positive. The second term of the right-hand side of Eq. (37), which we have termed mean topological information, is of necessity positive. From these simple remarks one can conclude there exists an optimal size of the network that leads to the largest possible value of the mean topological information.

*4.4. Fuzzy-organized multienzyme networks*

An ideal four node network displaying fuzzy organization is shown in Fig. 5B. The equations that describe the dynamics of the system are

$$\begin{aligned}\frac{dY_1}{dt} &= v_i + \tau_4 Y_4 - (\tau_1 + \tau_2'')Y_1 \\ \frac{dY_2}{dt} &= \tau_1 Y_1 - \tau_2' Y_2 \\ \frac{dY_3}{dt} &= \tau_2'' Y_1 - (\tau_3 + \tau_0)Y_3 \\ \frac{dY_4}{dt} &= \tau_2' Y_2 + \tau_3 Y_3 - \tau_4 Y_4\end{aligned} \tag{38}$$

Moreover one can write a conservation equation identical to expression (22) and under steady-state conditions the differential equations reduce to

$$\begin{bmatrix} \tau_1 & -\tau_2' & 0 & 0 \\ \tau_2'' & 0 & -(\tau_3 + \tau_0) & 0 \\ 0 & \tau_2' & \tau_3 & -\tau_4 \\ 1 & 1 & 1 & 1 \end{bmatrix} \begin{bmatrix} p(Y_1) \\ p(Y_2) \\ p(Y_3) \\ p(Y_4 \end{bmatrix} = \begin{bmatrix} 0 \\ 0 \\ 0 \\ 1 \end{bmatrix} \tag{39}$$

Solving this system yields

$$\begin{aligned}p(Y_1) &= \frac{\tau_2'\tau_4(\tau_0 + \tau_3)}{(\tau_2'\tau_4 + \tau_1\tau_4 + \tau_2'\tau_1)(\tau_0 + \tau_3) + \tau_2'\tau_2''(\tau_3 + \tau_4)} \\ p(Y_2) &= \frac{\tau_1\tau_4(\tau_0 + \tau_3)}{(\tau_2'\tau_4 + \tau_1\tau_4 + \tau_2'\tau_1)(\tau_0 + \tau_3) + \tau_2'\tau_2''(\tau_3 + \tau_4)} \\ p(Y_3) &= \frac{\tau_2'\tau_2''\tau_4}{(\tau_2'\tau_4 + \tau_1\tau_4 + \tau_2'\tau_1)(\tau_0 + \tau_3) + \tau_2'\tau_2''(\tau_3 + \tau_4)} \\ p(Y_4) &= \frac{\tau_2'\tau_1(\tau_0 + \tau_3) + \tau_2'\tau_2''\tau_3}{(\tau_2'\tau_4 + \tau_1\tau_4 + \tau_2'\tau_1)(\tau_0 + \tau_3) + \tau_2'\tau_2''(\tau_3 + \tau_4)}\end{aligned} \tag{40}$$

As previously, the first Eq. (38) implies, if a steady state is reached, that

$$\frac{v_i}{Y_T} = p(Y_3)\tau_0 \tag{41}$$

Moreover expressions (40) are indeed different from Eqs. (27) for they stem from a different network topology. Of particular interest in the two examples is the result that $p(Y_4)$ decreases for model of Fig. 5A and remains nearly constant

Table 3

Influence of the value of the rate constant of exit, $\tau_0$, on the probabilities of occurrence of the nodes of a fuzzy-organized network. Increasing the value of $\tau_0$ does not result in major changes of the probabilities of occurrence of the nodes $Y_1$, $Y_2$, $Y_4$

| | | | | |
|---|---|---|---|---|
| $\tau_0=0.01$ | $p(Y_1)$ 0.201 | $p(Y_2)$ 0.201 | $p(Y_3)$ 0.198 | $p(Y_4)$ 0.399 |
| $\tau_0=0.1$ | $p(Y_1)$ 0.207 | $p(Y_2)$ 0.207 | $p(Y_3)$ 0.188 | $p(Y_4)$ 0.396 |
| $\tau_0=1$ | $p(Y_1)$ 0.250 | $p(Y_2)$ 0.250 | $p(Y_3)$ 0.125 | $p(Y_4)$ 0.375 |
| $\tau_0=10$ | $p(Y_1)$ 0.314 | $p(Y_2)$ 0.314 | $p(Y_3)$ 0.028 | $p(Y_4)$ 0.343 |
| $\tau_0=100$ | $p(Y_1)$ 0.332 | $p(Y_2)$ 0.332 | $p(Y_3)$ 0.003 | $p(Y_4)$ 0.335 |
| $\tau_1=\tau_2'=\tau_2''=\tau_3=\tau_4=1$ | | | | |

for model of Fig. 5B as the system becomes more open (Table 3). As a matter of fact $p(Y_1)$, $p(Y_2)$, and $p(Y_4)$ has similar or identical values whereas the values of $p(Y_3)$ decrease.

### *4.5. Topological information of regular and fuzzy-organized networks*

Even if the individual enzyme reactions involved in the overall network do not carry any mutual information, which implies that $p(A_i)_{Ei}=p(A_i/B_i)_{Ei}$, the graph still possesses information that reflects its topology and whose mean value per node is

$$\langle I_T \rangle = -\sum_i p(Y_i)\log_2 p(Y_i) \tag{42}$$

This expression is formally equivalent to a self-information. It is therefore of interest to evaluate this topological information content per node for both ideal regular and fuzzy-organized networks. It is well known that the total information of a system is maximum when the probability of occurrence of its elements is about the same. This conclusion should indeed be valid for the nodes of a graph. If a regular network is almost closed, i.e., if the output rate is very small relative to the other transition rates and if these transition rates have identical or similar values, then the probabilities of occurrence of all the nodes are the same, or nearly the same. Hence the mean information per node is maximum (Table 4). In the case of fuzzy-organized networks, however, the degree of certain nodes is increased with respect to what occurs in regular networks. As a consequence, the probability of occurrence of certain nodes can possibly rise, thus maintaining the mean topological information per node at a relatively high value.

In regular networks successions of "slow" and "fast" transition steps results in high and low values of the probabilities of occurrence of the corresponding nodes.

Table 4

Influence of the rate constant of exit $\tau_0$ on the mean information per node in a regular and in fuzzy-organized network. Increasing the value of $\tau_0$ of a regular network results in a marked decrease of information per node. The decrease of information per node is much less apparent for a fuzzy-organized network

| | | | |
|---|---|---|---|
| Regular network | | | |
| $\tau_1 = \tau_2 = \tau_3 = \tau_4 = 1$ | $\tau_0 = 0.01$ | | |
| $p(Y_1)$ | $p(Y_2)$ | $p(Y_3)$ | $p(Y_4)$ |
| 0.251 | 0.251 | 0.248 | 0.248 |
| $\langle I_T \rangle = 1.999$ bits/node | | | |
| $\tau_1 = \tau_2 = \tau_3 = \tau_4 = 1$ | $\tau_0 = 10$ | | |
| $p(Y_1)$ | $p(Y_2)$ | $p(Y_3)$ | $p(Y_4)$ |
| 0.458 | 0.458 | 0.042 | 0.042 |
| $\langle I_T \rangle = 1.413$ bits/node | | | |
| Fuzzy-organized network | | | |
| $\tau_1 = \tau_2' = \tau_2'' = \tau_3 = \tau_4 = 1$ | $\tau_0 = 10$ | | |
| $p(Y_1)$ | $p(Y_2)$ | $p(Y_3)$ | $p(Y_4)$ |
| 0.314 | 0.314 | 0.028 | 0.343 |
| $\langle I_T \rangle = 1.725$ bits/node | | | |

This situation generates a drop of the mean information per node (Table 4). Bypassing a "slow" step allows the graph to hold a probability of occurrence of the nodes at a nearly constant value, thus possibly maintaining at a nearly constant level the mean information per node. Hence it seems there is a functional advantage, for open metabolic networks, to possess a fuzzy-organized topology. One can therefore speculate that a selective pressure has been exerted upon the network in order to favor fuzzy-organized graphs over all others.

## 5. *Enzyme networks and Shannon communication–information theory*

Classical Shannon communication–information theory associates the concept of mutual information with that of a correlation between discrete variables. Moreover, the theory is based on the subadditivity principle, that is, on the idea that the mean joint uncertainty, or entropy, of a pair of discontinuous variables is less than, or at most equal to, the sum of the mean uncertainties of the two variables. In the case of networks, this is in fact equivalent to the measure of the integration of the two separate variables in a coherent entity. The extent of the mutual information of the system is then a measure of its integration.

The view that mutual information, or integration, of a network can only be positive or nil, can be challenged even with simple enzyme reactions. As a matter of fact, most simple enzyme-catalyzed processes can be considered elementary networks that possess both organization and mutual information, for they can associate different chemical reagents in the cleft of the enzyme active site. Here mutual information is an extension of the original Shannon definition. Such apparently simple systems can have no mutual information, or either positive or negative mutual information.

The possible existence of negative mutual information, which implies that the subadditivity principle is violated, is of particular interest for it can be considered a criterion of emergence and complexity [29]. Under these conditions the system has more self-information than the set of individual component sub-systems. This is precisely the situation that occurs with the random binding of two substrates $AX$ and $B$ to an enzyme, when the joint probability of binding both $AX$ and $B$ to the enzyme, is smaller than the product of individual probabilities of binding $AX$ and $B$ alone. From a thermodynamic viewpoint, this means that complexity of a multimolecular system occurs if its energy is high enough and such that the subadditivity relationship is reversed.

In apparently simple systems occurring under pseudo-equilibrium conditions, mutual information results from a physical interaction between the two subsites that bind the substrates. However, if the system departs from pseudo-equilibrium, it gains additional self-information in such a way that its mutual information becomes negative. Hence emergence of information and energy stems from the lack of pseudo equilibrium of the system.

Multienzyme networks result from the association of many enzyme reactions coupled as an open dynamic system. We have studied from a theoretical viewpoint two types of simple graphs, namely regular and fuzzy-organized. In order to be able to study these models, one has to postulate that there exists a time hierarchy within these graphs. The steps of substrate binding involved in every enzyme process is assumed to be fast relative to catalysis and transfer of metabolites to another enzyme. With this assumption in mind, metabolic networks can be modeled by different types of graphs the nodes of which are not metabolites, but individual enzyme reactions. A metabolic network can then be considered a network of networks. The mean mutual information of integration per node of such a global network relies on two factors: the mutual information content of each individual node considered in isolation, and the gross topology of the network. Even if each node, considered in isolation, does not contain *per se* any information, the whole system does. Robustness of the overall network relies upon the values of the products of the catalytic rates constants $k_i$ by the *fractionation factors* $f_i$. If one of the values $k_i f_i$ is small the overall network tends to breakdown at this very place. Such a situation occurs if the ternary complex $E_iA_iB_i$ is strongly stabilized relative to the corresponding transition state $E_iT_i^{\neq}$. Conversely, this implies that an overall network will be robust if the individual mutual information of all the nodes is strongly negative.

The participation of network topology in mean node information is expressed through probability factors that represent a measure of this topological information. Each topological information factor, $p(Y_i)\log_2 p(Y_i)$, expresses how gross network topology affects the mutual information of a given node. It represents the influence of the context on node information, exactly as the information of a word in a sentence depends upon its context.

As for classical Shannon information, mean topological information per node is maximum if the topological probability factors have the same value. This situation is expected to take place if the transition rates connecting the nodes have similar values and if the network is not wide open, i.e., if its input and output rates are small. Conversely, if the rates of entry and exit are different and large, one must expect that, for regular networks, the mean topological information drops. Connectivity varies from node to node for fuzzy-organized networks. If these graphs are nearly closed, one can expect the mean topological information per node to decrease. But if the network is open, the difference of connectivity from node to node may adjust to the same value as the probabilities of occurrence of these nodes. Hence one must expect the mean information per node to increase significantly. If a regular network displays a "fast" transition rate followed by a "slow" one, or conversely, the probability of occurrence of the nodes varies within large limits, the mean mutual information per node will then fall off. Bypassing "slow" steps in highly connected fuzzy-organized networks may bring the probability of occurrence of the nodes to similar values, thus increasing the mean topological information. Hence, contrary to a well-known view, it is not only the degree of connection of nodes that is important, but even more their probability of occurrence. For all these reasons, one may speculate, in a neo-Darwinian perspective, that there exists a functional advantage for a biological network to be of a fuzzy-organized type.

## *References*


1. Jeong, H., Tombor, B., Albert, R., Oltvai, Z.N., and Barabasi, A.L. (2000) The large scale organization of metabolic networks. Nature 407, 651–654.
2. Fell, D.E. and Wagner, A. (2000) The small world of metabolism. Nature Biotech. 18, 1121–1122.
3. Kacser, H. and Burns, J.A. (1979) Molecular democracy. Who shares the control? Biochem. Soc. Trans. 7, 1149–1160.
4. Kacser, H. and Burns, J.A. (1972) The control of flux. Symp. Soc. Exp. Biol. 27, 65–104.
5. Kacser, H. (1983) The control of enzyme systems *in vivo*. Elasticity analysis of the steady state. Biochem. Soc. Trans. 11, 35–40.
6. Heinrich, R. and Rapoport, T.A. (1974) A linear steady state treatment of enzymatic chains. Critique of the crossover theorem and a general procedure to identify interaction site with an effector. Eur. J. Biochem. 42, 97–105.
7. Heinrich, R., Rapoport, S.M., and Rapoport, T.A. (1977) Metabolic regulation and mathematical models. Progress Biophys. Mol. Biol. 32, 1–82.
8. Fell, D.E. and Sauro, H.M. (1985) Metabolic control and its analysis. Additional relationships between elasticities and control coefficients. Eur. J. Biochem. 148, 555–561.
9. Sen, A.K. (1990) Metabolic control analysis. An application of signal flow graphs. Biochem. J. 269, 141–147.
10. Reder, C. (1988) Metabolic control theory: a structural approach. J. Theor. Biol. 135, 175–201.
11. Cornish-Bowden, A. and Cardenas, M.L. (1990) Control of Metabolic Processes. Plenum Press, New York.
12. Giersch, C. (1988a) Control analysis of metabolic networks. Homogeneous functions and the summation theorems for control coefficients. Eur. J. Biochem. 174, 509–513.

13. Giersch, C. (1988b) Control analysis of metabolic networks. 2. Total differentials and general formulation of the connectivity relations. Eur. J. Biochem. 174, 515–519.
14. Ricard, J. (1973) Cinétique et Mécanismes d'Action des Enzymes. Doin, Paris.
15. Cornish-Bowden, A. (1995) Fundamentals of Enzyme Kinetics. Portland Press, London.
16. Wong, J.T. (1975) Kinetics of Enzyme Mechanisms. Academic Press, London.
17. Laidler, K.J. (1969) Theories of Chemical Reaction Rates. McGraw-Hill, New York.
18. Glasstone, S., Laidler, K.J., and Eyring, H. (1941) The Theory of Rate Processes. McGraw-Hill, New York.
19. Kraut, J. (1988) How do the enzymes work? Science 242, 553–559.
20. Laidler, K.J. (1965) Chemical Kinetics. McGraw-Hill, New York.
21. Pauling, L. (1948) Nature of forces between large molecules of biological interest. Nature 161, 707–709.
22. Pauling, L. (1946) Molecular architecture and biological reactions. Chem. Eng. News 24, 1375–1377.
23. Kurz, J.L. (1963) Transition state characterization of catalyzed reactions. J. Amer. Chem. Soc. 85, 987–991.
24. Jencks, W.P. (1975) Binding energy, specificity and enzymic catalysis: the Circe effect. Adv. Enzymol. 43, 219–410.
25. Jencks, W.P. (1969) Catalysis in Chemistry and Enzymology. McGraw-Hill, New York.
26. Fersht, A. (1977) Enzyme Structure and Mechanism. Freeman, Reading and San Francisco.
27. Cha, S. (1968) A simple model for derivation of rate equations for enzyme-catalyzed reactions under rapid equilibrium assumption or combined assumptions of equilibrium and steady state. J. Biol. Chem. 243, 820–825.
28. Albert, R. and Barabasi, A.L. (2002) Statistical mechanics of complex networks. Rev. Mod. Phys. 74, 47–97.
29. Ricard, J. (2004) Reduction, integration and emergence in biochemical networks. Biol. Cell 96, 719–725.

J. Ricard *Emergent Collective Properties, Networks, and Information in Biology*

DOI: 10.1016/S0167-7306(05)40008-3

CHAPTER 8

# Functional connections in multienzyme complexes: information and generalized microscopic reversibility

J. Ricard

In a multi-protein network, one can follow different paths in going from a node to another one. Away from thermodynamic equilibrium, the multiplicity of pathways between two nodes is expressed by a function, the function of connection. Functions of connection which express in mathematical terms the complexity of network topology, vanish if the system is close to thermodynamic equilibrium or if another condition, called generalized microscopic reversibility, holds. This condition, distinct from standard microscopic reversibility, applies to some systems occurring under steady state conditions. It implies that the transition states for substrate, $S^{\neq}$, and catalysis, $X^{\neq}$, stabilize the same conformation of the corresponding enzyme. If this condition is not fulfilled, the information of the overall system is in part defined by the functions of connections. This situation illustrates the importance of nonequilibrium and network topology on the information content of the system.

*Keywords:* function of connection, multienzyme complex, network connections and information, steady state in branched multienzyme networks, steady state in linear multienzyme networks.

Metabolic networks behave as coherent wholes. It is therefore extremely tempting to assume there must exist some form of cooperation between enzyme reactions in order to allow fine tuning of metabolism [1–10]. Moreover it is evident that in order to do so, metabolic sub-networks and cycles have to be coordinated with respect to one another. This coordination is often considered to be mediated through metabolic intermediates such as adenosine triphosphate (ATP), nicotinamide adenine dinucleotide (NAD), nicotinamide adenine dinucleotide phosphate (NADP), etc.... which are nearly ubiquitous in metabolic processes.

The existence of enzyme compartmentalization and spatial organization in the living cell offers new possibilities for understanding how enzyme reactions and metabolic sub-networks could be coordinated. The first idea that springs to mind is to consider that a number of reaction intermediates do not diffuse freely within the cell towards other enzymes that take them up as substrates. Instead they might be channelled within multienzyme complexes from one enzyme active site to another [11–28]. The view that channelling allows fine tuning of cell metabolism is substantiated by the finding that quite often enzymes that are associated as multimolecular complexes catalyze consecutive reactions of cell metabolism [13–21]. However, although the reality of channelling has been demonstrated in a number of cases [29–37], it is often difficult to generalize the validity of this idea to many metabolic processes [22–28].

In a number of cases, however, enzymes that are associated in multimolecular complexes do not catalyze consecutive reactions. This is the case, for instance, of a complex made up of glyceraldehyde phosphate dehydrogenase and phosphoribulokinase of chloroplasts [38–41], of aminoacyl-tRNA synthetases of eukaryotes [42–47], of isoleucyl-tRNA synthetase and threonine deaminase [48], of ornithine carbamoyl transferase and arginase [49] and of many other multimolecular enzyme complexes. Hence if there is a functional advantage in the existence of these multienzyme complexes, it cannot be the channelling from site to site of reaction intermediates.

One can therefore raise the point of a possible connection between different enzyme reactions through a physical association of functionally unrelated enzymes. This raises an even more general question. It is obvious that the functional activity of a multienzyme complex should be quantitatively described by a network and the question that comes to mind is to whether the degree of node connection of this network affects its functioning under steady-state conditions. These two questions will be addressed on very simple models in the present chapter.

## 1. *Network connections and mutual information of integration in multienzyme complexes*

### 1.1. *Linear networks*

We call linear network a graph made up of a linear sequence of connected nodes. Let us consider for instance a bi-enzyme complex (Fig. 1). Each enzyme can bind its substrate ($S_1$ or $S_2$) and one can imagine that either enzyme, $E_1$ or $E_2$, binds its substrate first. But if it is enzyme $E_2$ that has bound substrate $S_2$ first, then the associated enzyme $E_1$ cannot bind substrate $S_2$. Moreover each edge between two nodes displays three types of events: binding of the relevant substrate (either $S_1$ or $S_2$), release of that substrate, conversion of that substrate into a product (Fig. 1). The corresponding graph is made up of a linear sequence of nodes. Two of them ($C_0$ and $C_1$) have a degree of connection of two (i.e., two of them are connected to two other nodes) whereas the two others ($C_2$ and $C_{12}$) have a node connection of one (each is connected with one node only). Moreover the network is assumed to occur under nonequilibrium conditions, usually a steady state. Hence the connection of any node to its nearest neighbor involves three rate constants (associated with the three events already mentioned) and a substrate concentration. For instance, the connection of $C_0$ to $C_1$ involves the rate constant of substrate binding multiplied by the substrate concentration, $k_1[S_1]$, the rate constant, $k_{-1}$, of substrate release from $C_1$, and the catalytic rate constant, $k_{c1}$, of the conversion of substrate $S_1$ into

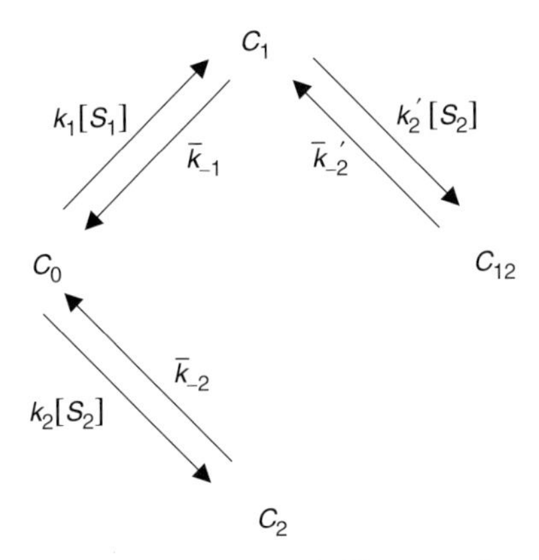


Fig. 1. A bi-enzyme complex that forms a linear network. The system is assumed to be in steady state and each of the apparent rate constants $\overline{k}_{-1}, \overline{k}_2, \overline{k}'_2$ is in fact the sum of two rate constants: a catalytic and a substrate release rate constant. See text.

the corresponding product $P_1$. Long ago [50], it was demonstrated that the two rate constants can be collected and expressed as an apparent rate constant $\overline{k}_{-1}$, i.e.,

$$\overline{k}_{-1} = k_{-1} + k_{c1} \tag{1}$$

The existence of the catalytic constants prevents the system from being in thermodynamic equilibrium. But if the catalytic constants are very small the system can then be close to thermodynamic equilibrium. As already mentioned, this state is called quasi-equilibrium.

If, as already assumed, the system occurs under steady-state conditions, one can define apparent affinity constants of substrates $S_1$ and $S_2$ for $C_0, C_1$, and $C_2$, viz.

$$\begin{aligned} \overline{K}_1 &= \frac{k_1}{k_{-1} + k_{c1}} = \frac{k_1}{\overline{k}_{-1}} \\ \overline{K}_2 &= \frac{k_2}{k_{-2} + k_{c2}} = \frac{k_2}{\overline{k}_{-2}} \\ \overline{K}'_2 &= \frac{k'_2}{k'_{-2} + k'_{c2}} = \frac{k'_2}{\overline{k}'_{-2}} \end{aligned} \tag{2}$$

Under these conditions one can easily demonstrate that

$$\begin{aligned} \frac{[C_1]}{[C_0]} &= \overline{K}_1[S_1] \\ \frac{[C_2]}{[C_0]} &= \overline{K}_2[S_2] \\ \frac{[C_{12}]}{[C_0]} &= \overline{K}_1\overline{K}'_2[S_1][S_2] \end{aligned} \tag{3}$$

If the system were close to thermodynamic conditions, expressions (3) above would still apply except that the apparent affinity constants $\overline{K}_1, \overline{K}_2, \overline{K}'_2$ would have to be replaced by real affinity constants $K_1, K_2, K'_2$ [51].

The probability, $p(S_1)$, that the complex has bound substrate $S_1$ is

$$p(S_1) = \frac{[C_1]/[C_0] + [C_{12}]/[C_0]}{1 + [C_1]/[C_0] + [C_2]/[C_0] + [C_{12}]/[C_0]} \tag{4}$$

and similarly the probability that the protein complex binds $S_1$ given it has already bound $S_2$ is

$$p(S_1/S_2) = \frac{[C_{12}]/[C_0]}{[C_2]/[C_0] + [C_{12}]/[C_0]} \tag{5}$$

Making use of expression (3) leads to

$$p(S_1) = \frac{\overline{K}_1[S_1] + \overline{K}_1\overline{K}'_2[S_1][S_2]}{1 + \overline{K}_1[S_1] + \overline{K}_2[S_2] + \overline{K}_1\overline{K}'_2[S_1][S_2]} \tag{6}$$

and to

$$p(S_1/S_2) = \frac{\overline{K}_1\overline{K}'_2[S_1]}{\overline{K}_2 + \overline{K}_1\overline{K}'_2[S_1]} \tag{7}$$

One can then derive the expression of the ratio $p(S_1/S_2)/p(S_1)$. One finds

$$\frac{p(S_1/S_2)}{p(S_1)} = \frac{\overline{K}'_2 + \overline{K}_1\overline{K}'_2[S_1] + \overline{K}_2\overline{K}'_2[S_2] + \overline{K}_1\overline{K}'^2_2[S_1][S_2]}{\overline{K}_2 + \overline{K}_1\overline{K}'_2[S_1] + \overline{K}_2\overline{K}'_2[S_2] + \overline{K}_1\overline{K}'^2_2[S_1][S_2]} \tag{8}$$

This ratio can be larger or smaller than one depending on the respective values of $\overline{K}'_2$ and $\overline{K}_2$. As already done previously, one can define the mutual information of integration as

$$I(S_1 : S_2) = \log_2 \frac{p(S_1/S_2)}{p(S_1)} \tag{9}$$

When $\overline{K}'_2 > \overline{K}_2$, $I(S_1 : S_2) > 0$ and when $\overline{K}'_2 < \overline{K}_2$, $I(S_1 : S_2) < 0$.

It is possible to associate the latter situation with the emergence, or the enhancement, of a function. Owing to the definition of an apparent affinity constant, one can expect that the smaller the value of $\overline{K}'_2$, the larger the values of the rate constants $k'_2$, and $k'_{c2}$. If this were occurring, it would mean that the binding of substrate $S_1$ to enzyme $E_1$, within the complex, results in an enhancement of the catalytic activity of enzyme $E_2$ towards its own substrate $S_2$. This enhancement can be considered a process somewhat similar to the emergence, or the enhancement, of a function.

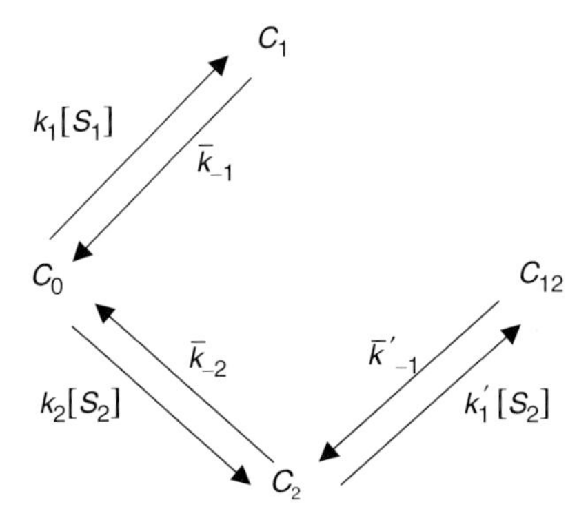


Fig. 2. A different type of node connection for a bi-enzyme network. The system is in steady state. See text.

The same kind of reasoning can be followed on a slightly different model shown in Fig. 2. The number of nodes of this network is the same as that shown in Fig. 1 but their connections are different. Now the ratio $p(S_1/S_2)/p(S_1)$ is expressed as

$$\frac{p(S_1/S_2)}{p(S_1)} = \frac{\overline{K}'_1 + \overline{K}_1\overline{K}'_1[S_1] + \overline{K}'_1\overline{K}_2[S_2] + \overline{K}'^2_1\overline{K}_2[S_1][S_2]}{\overline{K}_1 + \overline{K}_1\overline{K}'_1[S_1] + \overline{K}'_1\overline{K}_2[S_2] + \overline{K}'^2_1\overline{K}_2[S_1][S_2]} \tag{10}$$

and the sign of the corresponding information of integration is defined by the respective values of $\overline{K}'_1$ and $\overline{K}_1$. In particular, if $\overline{K}'_1 < \overline{K}_1$ the corresponding mutual information is negative. As previously mentioned, this suggests that the binding of substrate $S_2$ to the corresponding enzyme $E_2$ within the complex results in the enhancement of the activity of the enzyme $E_1$ towards its own substrate $S_1$. Again, the enhancement of a catalytic activity can be considered the emergence of a function.

### *1.2. Functions of connection*

Let us assume now that the network of either Fig. 1 or 2 be modified as to allow to go from node $C_0$ to node $C_{12}$ by two different routes (Fig. 3). Under steady-state conditions the relative concentrations $[C_1]/[C_0]$, $[C_2]/[C_0]$, and $[C_{12}]/[C_0]$ are expressed as

$$\begin{aligned}
\frac{[C_1]}{[C_0]} &= \overline{K}_1[S_1] + u_1 \\
\frac{[C_2]}{[C_0]} &= \overline{K}_2[S_2] + u_2 \\
\frac{[C_{12}]}{[C_0]} &= \overline{K}_1\overline{K}'_2[S_1][S_2] + u_{12} = \overline{K}'_1\overline{K}_2[S_1][S_2] + u'_{12}
\end{aligned} \tag{11}$$

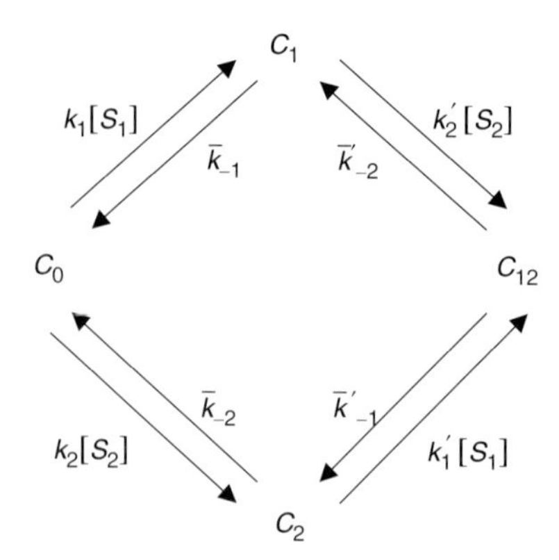


Fig. 3. A simple lattice in which two different routes lead to the same nodes. The system is in steady state and the expressions of the probabilities of occurrence of the nodes involve the functions of connection $u$.

Here the relative steady-state concentrations $[C_1]/[C_0]$, $[C_2]/[C_0]$, and $[C_{12}]/[C_0]$ cannot be expressed in terms of the sole apparent affinity constants $\overline{K}_i$. The functions $u_1$, $u_2$, $u_{12}$, or $u'_{12}$, contribute to the expression of these steady-state concentrations. As we shall see later, these functions express how the additional connection of two nodes, by creating an alternative pathway in the network, alters the mutual information of organization of the system.

The expression of the functions $u$ can be obtained by writing down the steady-state equations of $[C_1]/[C_0]$, $[C_2]/[C_0]$, and $[C_{12}]/[C_0]$. One finds

$$\begin{aligned}
&k_1[S_1] + \bar{k}'_{-2}\frac{[C_{12}]}{[C_0]} - (\bar{k}_{-1} + k'_2[S_2])\frac{[C_1]}{[C_0]} = 0 \\
&k_2[S_2] + \bar{k}'_{-1}\frac{[C_{12}]}{[C_0]} - (\bar{k}_{-2} + k'_1[S_1])\frac{[C_2]}{[C_0]} = 0 \\
&k'_2[S_2]\frac{[C_1]}{[C_0]} + k'_1[S_1]\frac{[C_2]}{[C_0]} - (\bar{k}'_{-2} + \bar{k}'_{-1})\frac{[C_{12}]}{[C_0]} = 0
\end{aligned} \tag{12}$$

In these equations, substituting for the ratios $[C_1]/[C_0]$, $[C_2]/[C_0]$, $[C_{12}]/[C_0]$ the corresponding expressions from (11) leads to

$$\begin{bmatrix} -(\bar{k}_{-1} + k'_2[S_2]) & 0 & \bar{k}'_{-2} \\ 0 & -(\bar{k}_{-2} + k'_1[S_1]) & \bar{k}'_{-1} \\ k'_2[S_2] & k'_1[S_1] & -(\bar{k}'_{-2} + \bar{k}'_{-1}) \end{bmatrix} \begin{bmatrix} u_1 \\ u_2 \\ u_{12} \end{bmatrix}$$

$$= \begin{bmatrix} 0 \\ \bar{k}'_{-1}[S_1][S_2](\overline{K}_1\overline{K}'_2 - \overline{K}'_1\overline{K}_2) \\ -\bar{k}'_{-1}[S_1[S_2](\overline{K}_1\overline{K}'_2 - \overline{K}'_1\overline{K}_2) \end{bmatrix} \tag{13}$$

and to

$$\begin{bmatrix} -(\overline{k}_{-1}+k_2'[S_2]) & 0 & \overline{k}'_{-2} \\ 0 & -(\overline{k}_{-2}+k_1'[S_1]) & \overline{k}'_{-1} \\ k_2'[S_2] & k_1'[S_1] & -(\overline{k}'_{-2}+\overline{k}'_{-1}) \end{bmatrix} \begin{bmatrix} u_1 \\ u_2 \\ u'_{12} \end{bmatrix} = \begin{bmatrix} -\overline{k}'_{-2}[S_1][S_2](\overline{K}_1\overline{K}'_2-\overline{K}'_1\overline{K}_2) \\ 0 \\ \overline{k}'_{-2}[S_1][S_2](\overline{K}_1\overline{K}'_2-\overline{K}'_1\overline{K}_2) \end{bmatrix} \tag{14}$$

The functions $u$, which we call functions of connection, all become identical to zero if

$$\overline{K}_1\overline{K}'_2-\overline{K}'_1\overline{K}_2=0 \tag{15}$$

This expression is the so-called generalized microscopic reversibility condition [51–54]. It is not a true physical principle that should of necessity apply and, for that reason, has to be distinguished from the classical principle of microscopic reversibility which expresses the first principle of thermodynamics. However relation (15) is not purely fortuitous and relies upon physical grounds that will be discussed later. Be that as it may, relation (15) may be fulfilled or not.

The functions of connection can be derived by solving Cramer systems (13) and (14). One finds [55,56]

$$\begin{aligned} u_1 &= -\frac{\overline{k}_{-2}\overline{k}'_{-1}\overline{k}'_{-2}[S_1][S_2](\overline{K}_1\overline{K}'_2-\overline{K}'_1\overline{K}_2)}{\overline{k}_{-1}\overline{k}_{-2}(\overline{k}'_{-1}+\overline{k}'_{-2})+\overline{k}_{-1}\overline{k}'_{-2}k_1'[S_1]+\overline{k}_{-2}\overline{k}'_{-1}k_2'[S_2]} \\ u_2 &= \frac{\overline{k}_{-1}\overline{k}'_{-2}\overline{k}'_{-1}[S_1][S_2](\overline{K}_1\overline{K}'_2-\overline{K}'_1\overline{K}_2)}{\overline{k}_{-1}\overline{k}_{-2}(\overline{k}'_{-1}+\overline{k}'_{-2})+\overline{k}_{-1}\overline{k}'_{-2}k_1'[S_1]+\overline{k}_{-2}\overline{k}'_{-1}k_2'[S_2]} \\ u_{12} &= -\frac{\overline{k}'_{-1}\overline{k}_{-2}[S_1][S_2](\overline{k}_{-1}+k_2'[S_2])(\overline{K}_1\overline{K}'_2-\overline{K}'_1\overline{K}_2)}{\overline{k}_{-1}\overline{k}_{-2}(\overline{k}'_{-1}+\overline{k}'_{-2})+\overline{k}_{-1}\overline{k}'_{-2}k_1'[S_1]+\overline{k}_{-2}\overline{k}'_{-1}k_2'[S_2]} \\ u'_{12} &= \frac{\overline{k}_{-1}\overline{k}'_{-2}[S_1][S_2](\overline{k}_{-2}+k_1'[S_1])(\overline{K}_1\overline{K}'_2-\overline{K}'_1\overline{K}_2)}{\overline{k}_{-1}\overline{k}_{-2}(\overline{k}'_{-1}+\overline{k}'_{-2})+\overline{k}_{-1}\overline{k}'_{-2}k_1'[S_1]+\overline{k}_{-2}\overline{k}'_{-1}k_2'[S_2]} \end{aligned} \tag{16}$$

The probabilities $p(S_1)$ and $p(S_1/S_2)$ are now expressed as

$$\begin{aligned} p(S_1) &= \frac{\overline{K}_1[S_1]+\overline{K}_1\overline{K}'_2[S_1][S_2]+u_1+u_{12}}{1+\overline{K}_1[S_1]+\overline{K}_2[S_2]+\overline{K}_1\overline{K}'_2[S_1][S_2]+u_1+u_2+u_{12}} \\ p(S_1/S_2) &= \frac{\overline{K}_1\overline{K}'_2[S_1][S_2]+u_{12}}{\overline{K}_2[S_2]+\overline{K}_1\overline{K}'_2[S_1][S_2]+u_2+u_{12}} \end{aligned} \tag{17}$$

or as

$$
\begin{aligned}
p(S_1) &= \frac{\overline{K}_1[S_1] + \overline{K}'_2\overline{K}_2[S_1][S_2] + u_1 + u'_{12}}{1 + \overline{K}_1[S_1] + \overline{K}_2[S_2] + \overline{K}'_1\overline{K}_2[S_1][S_2] + u_1 + u_2 + u'_{12}} \\
p(S_1/S_2) &= \frac{\overline{K}'_1\overline{K}_2[S_1][S_2] + u'_{12}}{\overline{K}_2[S_2] + \overline{K}'_1 K_2[S_1][S_2] + u_2 + u'_{12}}
\end{aligned}
\tag{18}
$$

Setting for simplicity

$$
\begin{aligned}
\alpha(S_1, S_2) &= \overline{K}_1[S_1] + \overline{K}_1\overline{K}'_2[S_1][S_2] \\
\beta(S_1, S_2) &= 1 + \overline{K}_1[S_1] + \overline{K}_2[S_2] + \overline{K}_1\overline{K}'_2[S_1][S_2] \\
\gamma(S_1, S_2) &= \overline{K}_1\overline{K}'_2[S_1][S_2] \\
\delta(S_1, S_2) &= \overline{K}_2[S_2] + \overline{K}_1\overline{K}'_2[S_1][S_2]
\end{aligned}
\tag{19}
$$

and

$$
\begin{aligned}
u_\alpha &= u_1 + u_{12} \\
u_\beta &= u_1 + u_2 + u_{12} \\
u_\gamma &= u_{12} \\
u_\delta &= u_2 + u_{12}
\end{aligned}
\tag{20}
$$

expressions (17) can be rewritten as

$$
\begin{aligned}
p(S_1) &= \frac{\alpha(S_1, S_2) + u_\alpha}{\beta(S_1, S_2) + u_\beta} \\
p(S_1/S_2) &= \frac{\gamma(S_1, S_2) + u_\gamma}{\delta(S_1, S_2) + u_\delta}
\end{aligned}
\tag{21}
$$

It is in fact the four functions $u_\alpha$, $u_\beta$, $u_\gamma$, and $u_\delta$ that reveal how the expressions for $p(S_1)$ and $p(S_1/S_2)$ depart from standard generalized microscopic reversibility conditions. If these conditions are fulfilled then expressions (21) become

$$
\begin{aligned}
p(\overline{S}_1) &= \frac{\alpha(S_1, S_2)}{\beta(S_1, S_2)} \\
p(\overline{S}_1/\overline{S}_2) &= \frac{\gamma(S_1, S_2)}{\delta(S_1, S_2)}
\end{aligned}
\tag{22}
$$

Here the overbars are used to emphasize the idea that the system fulfils generalized microscopic reversibility conditions. It is then of interest to determine how the departure from these conditions affects the expression of $I(S_1 : S_2)$. One has

$$\begin{aligned}\log_2 p(S_1) &= \log_2 \frac{\alpha(S_1, S_2) + u_\alpha}{\beta(S_1, S_2) + u_\beta} = \log_2 \frac{\alpha(S_1, S_2)[1 + u_\alpha/\alpha(S_1, S_2)]}{\beta(S_1, S_2)[1 + u_\beta/\beta(S_1, S_2)]} \\ \log_2 p(S_1/S_2) &= \log_2 \frac{\gamma(S_1, S_2) + u_\gamma}{\delta(S_1, S_2) + u_\delta} = \log_2 \frac{\gamma(S_1, S_2)[1 + u_\gamma/\gamma(S_1, S_2)]}{\delta(S_1, S_2)[1 + u_\delta/\delta(S_1, S_2)]}\end{aligned} \tag{23}$$

Setting for simplicity

$$u^\bullet_\alpha = \frac{u_\alpha}{\alpha(S_1, S_2)}, \; u^\bullet_\beta = \frac{u_\beta}{\beta(S_1, S_2)}, \; u^\bullet_\gamma = \frac{u_\gamma}{\gamma(S_1, S_2)}, \; u^\bullet_\delta = \frac{u_\delta}{\delta(S_1, S_2)} \tag{24}$$

expressions (23) become

$$\begin{aligned}\log_2 \; p(S_1) &= \log_2 \; p(\overline{S}_1) + \log_2 \frac{1 + u^\bullet_\alpha}{1 + u^\bullet_\beta} \\ \log_2 \; p(S_1/S_2) &= \log_2 \; p(\overline{S}_1/\overline{S}_2) + \log_2 \frac{1 + u^\bullet_\gamma}{1 + u^\bullet_\delta}\end{aligned} \tag{25}$$

Close to generalized microscopic reversibility conditions, the $u^\bullet$ values are small enough to allow one to expand $\log_2(1 + u^\bullet)$ in power series and keep only the first term, i.e.,

$$\log_2(1 + u^\bullet) = 1.4426u^\bullet - 1.4426\frac{u^{\bullet 2}}{2} + \cdots \tag{26}$$

Hence it follows that

$$\begin{aligned}\log_2 p(S_1) &\approx \log_2 p(\overline{S}_1) + 1.4426u^\bullet_\alpha - 1.4426u^\bullet_\beta \\ \log_2 p(S_1/S_2) &\approx \log_2 p(\overline{S}_1/\overline{S}_2) + 1.4426u^\bullet_\gamma - 1.4426u^\bullet_\delta\end{aligned} \tag{27}$$

Therefore the expression for the mutual information of integration of the system assumes the form

$$I(S_1 : S_2) = I(\overline{S}_1 : \overline{S}_2) + 1.4426[(u^\bullet_\gamma - u^\bullet_\delta) - (u^\bullet_\alpha - u^\bullet_\beta)] \tag{28}$$

Although

$$u_\alpha - u_\beta = u_\gamma - u_\delta = -u_2 \tag{29}$$

it is easy to show that, in general

$$u^\bullet_\alpha - u^\bullet_\beta \neq u^\bullet_\gamma - u^\bullet_\delta \tag{30}$$

As a matter of fact one has

$$
\begin{aligned}
u_{\alpha}^{\bullet} - u_{\beta}^{\bullet} &= \frac{[\beta(S_1, S_2) - \alpha(S_1, S_2)]u_1 + [\beta(S_1, S_2) - \alpha(S_1, S_2)]u_{12} - \alpha(S_1, S_2)u_2}{\alpha(S_1, S_2)\beta(S_1, S_2)} \\
u_{\gamma}^{\bullet} - u_{\delta}^{\bullet} &= \frac{[\delta(S_1, S_2) - \gamma(S_1, S_2)]u_{12} - \gamma(S_1, S_2)u_2}{\gamma(S_1, S_2)\delta(S_1, S_2)}
\end{aligned}
\tag{31}
$$

and it is clear that these expressions are, in general, different. Hence departure from generalized microscopic reversibility conditions can result in a decrease, or increase of the mutual information of integration of the network.

## *2. Generalized microscopic reversibility*

We have already seen that generalized microscopic reversibility implies that

$$
\overline{K}_1\overline{K}_2' = \overline{K}_1'\overline{K}_2 \tag{32}
$$

This relationship is formally identical to standard microscopic reversibility except that in relation (32) the constants $\overline{K}_i$ do not refer to equilibrium constants but to apparent affinity constants of reaction steps that are away from thermodynamic equilibrium. However, as already alluded to, generalized microscopic reversibility is not devoid of physical significance.

Let us consider, for instance, the two rate constants $k_{-1}$ and $k_{c1}$ i.e., the rate constants for substrate release and catalysis, respectively [51]. If we assume, although this is not necessarily the case, that the transition states for substrate release, $S^{\neq}$, and catalysis, $X^{\neq}$, stabilize the same conformation, one has the situation shown in Fig. 4. In this figure, $\Delta G_{-1}^{\neq}$ and $\Delta G_{c1}^{\neq}$ are the corresponding free energies of activation. These energy values are, in general, different from $\Delta G_{-S1}^{\neq *}$ and $\Delta G_{c1}^{\neq *}$ i.e., the free energy of activation of release and catalysis of substrate $S_1$, respectively, upon assuming that the enzyme is free, or naked. Then one has

$$
\begin{aligned}
\Delta G_{-1}^{\neq} &= \Delta G_{-S1}^{\neq *} - U_{\tau} + U_{\gamma} \\
\Delta G_{c1}^{\neq} &= \Delta G_{c1}^{\neq *} - U_{\tau} + U_{\gamma}
\end{aligned}
\tag{33}
$$

In these expressions, $U_{\gamma}$ and $U_{\tau}$ are the stabilization–destabilization energies of the ground and the transition states of enzyme $E_1$ by its association with another enzyme $E_2$. Indeed relationships (33) above are valid only because the two transition states stabilize the same conformation of the enzyme. Identical reasoning can be applied to the other rate constants $k'_{-1}$ and $k'_{c1}$, giving

$$
\begin{aligned}
\Delta G_{-1}^{\neq'} &= \Delta G_{-S1}^{\neq *} - U'_{\tau} + U'_{\gamma} \\
\Delta G_{c1}^{\neq'} &= \Delta G_{c1}^{\neq *} - U'_{\tau} + U'_{\gamma}
\end{aligned}
\tag{34}
$$

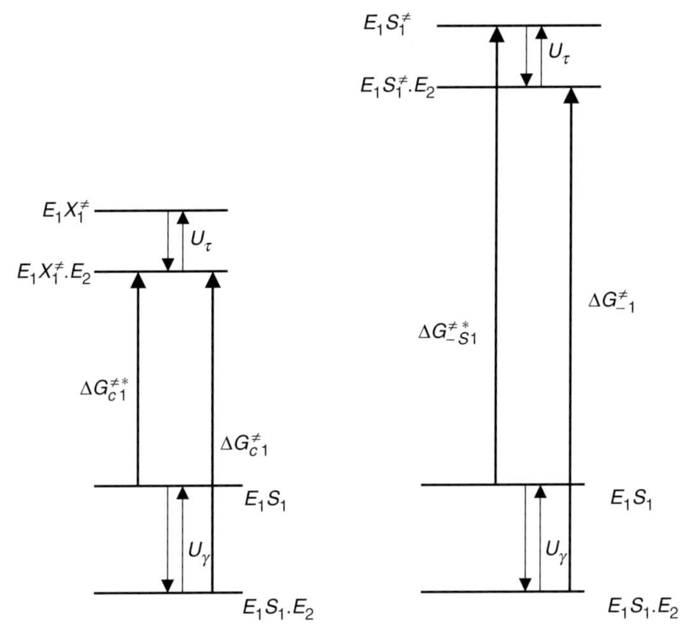


Fig. 4. Thermodynamic significance of generalized microscopic reversibility for a bi-enzyme complex. If enzyme–enzyme interactions stabilize (or destabilize) to the same extent the transition states for substrate release and catalysis, generalized microscopic reversibility should of necessity apply. See text. Adapted from [51].

Comparing Eqs. (33) and (34) leads to

$$\Delta G_{c1}^{\neq} - \Delta G_{-1}^{\neq} = \Delta G_{c1}^{\neq'} - \Delta G_{-1}^{\neq'} = \Delta G_{c1}^{\neq *} - \Delta G_{-S1}^{\neq *} \tag{35}$$

and this is equivalent to

$$\frac{k_{c1}}{k_{-1}} = \frac{k'_{c1}}{k'_{-1}} \tag{36}$$

If this reasoning is applied to the other substrate $S_2$, one has

$$\Delta G_{c2}^{\neq} - \Delta G_{-2}^{\neq} = \Delta G_{c2}^{\neq'} - \Delta G_{-2}^{\neq'} = \Delta G_{c2}^{\neq *} - \Delta G_{-S2}^{\neq *} \tag{37}$$

$\Delta G_{c2}^{\neq}$ and $\Delta G_{c2}^{\neq'}$ are the free energies of activation associated with catalytic rate constants $k_{c2}$ and $k'_{c2}$, $\Delta G_{-2}^{\neq}$ and $\Delta G_{-2}^{\neq'}$ the free energies of activation associated with rate constants $k_{-2}$ and $k'_{-2}$. Last but not least, $\Delta G_{c2}^{\neq *}$ and $\Delta G_{-S2}^{\neq *}$ are the free energies of activation for catalysis and release of substrate $S_2$ when assuming that enzyme $E_2$ is naked. Relationship (37) is equivalent to

$$\frac{k_{c2}}{k_{-2}} = \frac{k'_{c2}}{k'_{-2}} \tag{38}$$

Expressions (36) and (38) imply that relationship (32) is fulfilled. Hence although generalized microscopic reversibility may not apply in a number of cases, there is nevertheless little doubt that it is not fortuitous and possesses precise thermodynamic grounds.

### 3. *Connections in a network displaying generalized microscopic reversibility*

If we consider the model of Fig. 3 in which it is assumed that generalized microscopic reversibility applies, the functions of connections $u$ are all equal to zero, and the ratio $p(S_1/S_2)/p(S_1)$ which is directly related to the mutual information of integration of the system assumes one of the two following forms

$$\frac{p(S_1/S_2)}{p(S_1)} = \frac{\overline{K}'_2 + \overline{K}_1\overline{K}'_2[S_1] + \overline{K}_2\overline{K}'_2[S_2] + \overline{K}_1\overline{K}'^2_2[S_1][S_2]}{\overline{K}_2 + \overline{K}_1\overline{K}'_2[S_1] + \overline{K}_2\overline{K}'_2[S_2] + \overline{K}_1\overline{K}'^2_2[S_1][S_2]} \tag{39}$$

or

$$\frac{p(S_1/S_2)}{p(S_1)} = \frac{\overline{K}'_1 + \overline{K}_1\overline{K}'_1[S_1] + \overline{K}'_1\overline{K}_2[S_2] + \overline{K}'^2_1\overline{K}_2[S_1][S_2]}{\overline{K}_1 + \overline{K}_1\overline{K}'_1[S_1] + \overline{K}'_1\overline{K}_2[S_2] + \overline{K}'^2_1\overline{K}_2[S_1][S_2]} \tag{40}$$

One can notice that these expressions are consistent with Eqs. (8) and (10). This means that when a network occurs in steady state and when generalized microscopic reversibility applies, its mutual information of integration no longer depends on its degree of connection.

### 4. *Mutual information of integration and reaction rate*

We are now in a position to raise the interesting point of whether the mutual information of integration of a multienzyme system is related to the enzyme reaction rate. In the classical view of cell metabolism, enzyme reactions are considered functional entities solely connected through chemical substances that are the product of an enzyme reaction and the substrate of the next one. As previously outlined, however, there is experimental evidence that different enzymes which exist as multimolecular complexes catalyze consecutive reactions in such a way, it has been proposed that the product of an enzyme reaction is channelled within the complex from one enzyme active site to another [14,18]. At least in a number of cases the reality of channelling has been experimentally demonstrated [34].

However, many enzymes in eukaryotic cells exist as multienzyme complexes that do not catalyze consecutive reactions. If there exists a functional advantage to this situation, one may wonder whether this advantage could not be a functional interaction between different chemical reactions. This is precisely what is suggested by the models of bi-enzyme networks that have been discussed previously. In the case of such a complex that catalyzes the conversion of two substrates $S_1$ and $S_2$ into two products $P_1$ and $P_2$, the steady state rate of consumption of substrate $S_1$ is

$$\frac{v}{[C]_T} = \frac{k_{c1}\overline{K}_1[S_1] + k'_{c1}\overline{K}_1\overline{K}'_2[S_1][S_2] + k_{c1}u_1 + k'_{c1}u_{12}}{1 + \overline{K}_1[S_1] + \overline{K}_2[S_2] + \overline{K}_1\overline{K}'_2[S_1][S_2] + u_1 + u_2 + u_{12}} \tag{41}$$

where $[C]_T$ is the concentration of the enzyme complex. Hence one can expect from this equation that the presence of substrate $S_2$ affects the consumption of substrate $S_1$. If we assume, for simplicity, that generalized microscopic reversibility applies to this situation, the reaction rate assumes the much simpler equation

$$\frac{v_1}{[C]_T} = \frac{k_{c1}\overline{K}_1[S_1] + k'_{c1}\overline{K}_1\overline{K}'_2[S_1][S_2]}{1 + \overline{K}_1[S_1] + \overline{K}_2[S_2] + \overline{K}_1\overline{K}'_2[S_1][S_2]} \tag{42}$$

If $\overline{K}'_2 = \overline{K}_2$ (or $\overline{K}'_1 = \overline{K}_1$) and $k'_{c1} = k_{c1}$, expression (42) reduces to

$$\frac{v_1}{[C]_T} = \frac{k_{c1}\overline{K}_1[S_1]}{1 + \overline{K}_1[S_1]} \tag{43}$$

and under these conditions the corresponding network does not possess any mutual information of integration. The corresponding reaction rate $v_1$ is insensitive to the presence of $S_2$.

We have already outlined that when $\overline{K}'_2 > \overline{K}_2$ (or $\overline{K}'_1 > \overline{K}_1$) the network displays positive mutual information of integration. Conversely, if $\overline{K}_2 > \overline{K}'_2$ (or $\overline{K}_1 > \overline{K}'_1$) the corresponding information of the system is negative. In the absence of substrate $S_2$ the corresponding reaction rate, $v_0$, follows Eq. (43). In order to determine whether $S_2$ plays the part of an activator or an inhibitor of the reaction $S_1 \rightarrow P_1$, one can derive the expression for $v_1/v_0$, i.e., the ratio of the steady-state rates in the presence of both $S_1$ and $S_2(v_1)$ and in the absence of $S_2(v_0)$. Under generalized microscopic reversibility conditions one finds [51]

$$\frac{v_1}{v_0} = \frac{k_{c1} + k_{c1}\overline{K}_1[S_1] + k'_{c1}\overline{K}'_2[S_2] + k'_{c1}\overline{K}_1\overline{K}'_2[S_1][S_2]}{k_{c1} + k_{c1}\overline{K}_1[S_1] + k_{c1}\overline{K}_2[S_2] + k_{c1}\overline{K}_1\overline{K}'_2[S_1][S_2]} \tag{44}$$

Hence if, for a definite domain of $[S_2]$, one has

$$(k'_{c1}\overline{K}'_2 - k_{c1}\overline{K}_2) + (k'_{c1} - k_{c1})\overline{K}_1\overline{K}'_2[S_1] > 0 \tag{45}$$

equivalent to

$$(k'_{c1}\overline{K}'_1 - k_{c1}\overline{K}_1) + (k'_{c1} - k_{c1})\overline{K}_1\overline{K}'_1[S_1] > 0 \tag{46}$$

$S_2$ will behave, in that domain, as an activator of the reaction $S_1 \rightarrow P_1$.

Let us consider for instance the case where the network displays the relationships $\overline{K}_2 > \overline{K}'_2$ and $\overline{K}_1 > \overline{K}'_1$. If this situation occurs, one should expect that the binding of $S_2$ on enzyme $E_2$ results in a decrease of the rate constant of $S_1$ binding to enzyme $E_1$, and an increase of the rate constants for substrate release and catalysis. Hence the relation $\overline{K}_1 > \overline{K}'_1$ is equivalent to

$$\frac{k_1}{k_{-1} + k_{c1}} > \frac{\alpha k_1}{\beta(k_{-1} + k_{c1})} \tag{47}$$

with $\alpha < 1$ and $\beta > 1$. Moreover the difference $k'_{c1}\overline{K}'_1 - k_{c1}\overline{K}_1$ which appears in expression (46) is equal to

$$k'_{c1}\overline{K}'_1 - k_{c1}\overline{K}_1 = \frac{\alpha k_{c1} k_1}{k_{-1} + k_{c1}} - \frac{k_{c1} k_1}{k_{-1} + k_{c1}} \tag{48}$$

and is negative whereas $k'_{c1} - k_{c1}$ is positive. Therefore, according to the expression (46), $S_2$ behaves as an activator of the reaction $S_1 \rightarrow P_1$ at high concentration of $S_1$ and as an inhibitor of the same reaction at low $[S_1]$ values. If the symmetrical situation occurs, i.e., if $\overline{K}'_1 > \overline{K}_1$ (and $\overline{K}'_2 > \overline{K}_2$), then $S_2$ behaves as an activator of the reaction $S_1 \rightarrow P_1$ at low concentration of $S_1$ and as an inhibitor at high concentrations of this reagent. One can indeed wonder about the possible functional advantages and the kinetic implications of this situation. In the case where $\overline{K}_1 > \overline{K}'_1$, the reaction rate of the enzyme $E_1$ within the complex displays strong positive cooperativity (or sigmoidicity) when the reaction rate is plotted as a function of $S_1$ (not shown). If enzyme $E_2$ were naked the same reaction velocity would display classical hyperbolic behavior. If, alternatively, $\overline{K}'_1 > \overline{K}_1$ the reaction rate of enzyme $E_1$ within the complex *versus* the corresponding substrate concentration displays negative cooperativity.

Another interesting property of this system is that the response of enzyme reaction $v_1$ to substrate $S_2$ can parallel the response of $v_2$ to substrate $S_2$ [51] (Fig. 5). Hence it appears that the supramolecular edifice can respond as a whole, as a coherent system, to *one* molecular signal.

## 5. *Possible functional advantages of physically associated enzymes*

As already outlined, there is little doubt that many enzyme complexes exist that do not catalyze consecutive enzyme reactions. One may therefore wonder about the possible functional advantages of a physical association of enzymes that belong to different metabolic processes [51]. The first possible advantage is a modulation of the reaction velocity of an enzyme reaction by another protein (Fig. 6). This situation is trivial and need not be discussed any further here. The second possible advantage is subtler. It implies that the activity of the enzyme $E_1$ in the complex is different depending on whether the other enzyme is active or not. Hence, the catalytic activity of enzyme $E_1$ is modulated by a substrate of $E_2$ and *vice versa*. In this perspective,

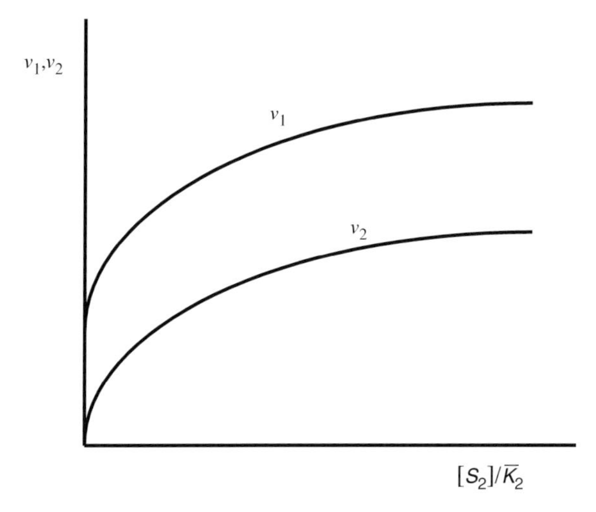


Fig. 5. In a bi-enzyme complex the two enzymes may respond in similar ways to changes of only *one* substrate concentration. Within the complex, the reaction rates, $v_1$ and $v_2$, respond to the same extent to the variation of the substrate concentration $[S_2]$. Adapted from [51].

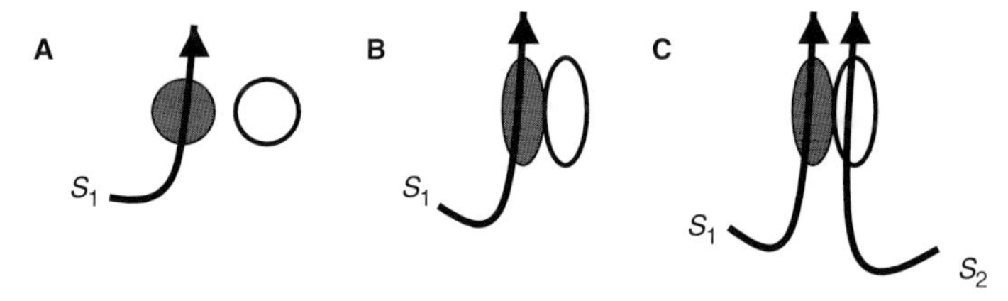


Fig. 6. Different possible types of functional enzyme interactions. (A) The two enzymes do not interfere. (B) The two enzymes form a complex, but one of the proteins is silent because it lacks the substrate. The effect of this protein is to modify the conformation and the activity of its partner. (C) The two enzymes form a complex in the presence of their respective substrate. The substrate $S_2$ acts as a substrate of enzyme $E_2$ and as a modulator of enzyme $E_1$. Alternatively, $S_1$ acts as a substrate of $E_1$ and as a modulator of $E_2$.

a ligand that does not interact with a protein may still modulate the activity of that protein!

If the two enzyme reactions belong to two different metabolic processes, these properties will appear coordinated. An advantage of this situation is that a set of physically associated enzymes, for instance aminoacyl-tRNA synthetases may respond to a sudden change of concentration of one aminoacid only, thus giving rise to a highly coordinated response.

## *References*

1. Kacser, H. and Burns, J.A. (1979) Molecular democracy: who shares the control? Biochem. Soc. Trans. 7, 1149–1160.

2. Kacser, H. and Burns, J.A. (1973) The control of flux. Symp. Soc. Exp. Biol. 27, 65–104.
3. Heinrich, R. and Rapoport, T.A. (1974) A linear steady state treatment of enzymatic chains: general properties, control and effector strength. Eur. J. Biochem. 42, 89–95.
4. Heinrich, R., Rapoport, S.M., and Rapoport, T.A. (1977) Metabolic regulation and mathematical models. Prog. Biophys. Mol. Biol. 32, 1–82.
5. Kacser, H. (1987) Control of metabolism. In: Stumpf P.K. and Conn E.E. (eds.) The Biochemistry of Plants, Academic Press, New York, 11, 39–67.
6. Westerhoff, H.V. and Van Dam, K. (1987) Thermodynamics and Control of Biological Free-Energy Transduction. Elsevier, Amsterdam.
7. Porteous, J.W. (1990) Control analysis: a theory that works. In: Cornish-Bowden, A. and Cardenas, M.L. (eds.) Control of Metabolic Processes, 51–67.
8. Hofmeyer, J.H.S. and Cornish-Bowden, A. (1991) Quantitative assessment of regulation in metabolic systems. Eur. J. Biochem. 200, 223–236.
9. Fell, D.A. (1992) Metabolic control analysis: a survey of its theoretical and experimental development. Biochem. J. 286, 313–330.
10. Fell, D.A. and Sauro, H.M. (1985) Metabolic control and its analysis. Additional relationships between elasticities and control coefficients. Eur. J. Biochem. 148, 555–561.
11. Westerhoff, H.V. and Welch, G.R. (1992) Enzyme organization and the direction of metabolic flow: physicochemical considerations. Curr. Top. Cell Regul. 33, 361–390.
12. Keizer, J. and Smolen, P. (1992) Mechanisms of metabolic transfer between enzymes: diffusional versus direct transfer. Curr. Top. Cell. Regul. 33, 391–405.
13. Ovadi, J. (1991) Physiological significance of metabolite channelling (a review with critical commentaries by many authors). J. Theor. Biol. 152, 1–22.
14. Srere, P.A. and Ovadi, J. (1990) Enzyme-enzyme interactions and their metabolic role. FEBS Letters 268, 360–364.
15. Keleti, T., Ovadi, J., and Batke, J. (1989) Kinetic and physico-chemical analysis of enzyme complexes and their possible role in the control of metabolism. Prog. Biophys. Mol. Biol. 53, 105–152.
16. Keleti, T. and Ovadi, J. (1988) Control of metabolism by dynamic macromolecular interactions. Curr. Top. Cell. Regul. 29, 1–33.
17. Gontéro, B., Cardenas, M., and Ricard, J. (1988) A functional five-enzyme complex of chloroplasts involved in the Calvin cycle. Eur. J. Biochem. 173, 437–443.
18. Srere, P.A. (1987) Complexes of sequential metabolic enzymes. Annu. Rev. Biochem. 56, 89–124.
19. Srivastava, D.K. and Bernhard, S.A. (1986) Enzyme-enzyme interactions and the regulation of metabolic reaction pathways. Curr. Top. Cell. Regul. 28, 1668.
20. Srivastava, D.K. and Bernhard, S.A. (1986) Metabolite transfer via enzyme-enzyme complexes. Science 234, 1081–1086.
21. Srivastava, D.K. and Bernhard, S.A. (1987) Mechanism of transfer of reduced nicotinamide adenine dinucleotide among dehydrogenases. Biochemistry 26, 1240–1246.
22. Chock, P.B. and Gutfreund, H. (1988) Reexamination of the kinetics of the transfer of NADH between its complexes with glycerol-3-phosphate dehydrogenase and with lactate dehydrogenase. Proc. Natl. Acad. Sci. USA 85, 8870–8874.
23. Xu, X., Gutfreund, H.D., Lakatos, S., and Chock. P.B. (1991) Substrate channelling in glycolysis: a phantom phenomenon. Proc. Natl. Acad. Sci. USA 88, 497–501.
24. Gutfreund, H.D. and Chock, P.B. (1991) Substrate channelling among glycolytic enzymes: fact or fiction. J. Theor. Biol. 152, 117–121.
25. Petersson, G. (1991) No convincing evidence is available for metabolite channelling between enzymes forming dynamic complexes. J. Theor. Biol. 152, 65–69.
26. Sainis, J.K., Merriam, K., and Harris, G.C. (1989) The association of D-ribulose-1,5-bisphosphate carboxylase/oxygenase with phosphoribulokinase. Plant Physiol. 89, 368–374.
27. Friedrich, P. (1985) Dynamic compartmentation in soluble multienzyme systems. In: G.R. Welch (ed.) Organized Multienzyme Systems: Catalytic Properties. Academic Press, New York, pp. 141–176.

28. Kurganov, B.I., Sugrobova, N.P., and Mil'man, L.S. (1985) Supramolecular organization of glycolytic enzymes. J. Theor. Biol. 116, 509–526.
29. Siegbahn, N., Mosbach, K., and Welch, G.R. (1985) Models of organized multienzyme systems: use in microenvironmental characterization and in practical application 271–301. In: G.R. Welch (ed.) Organized Multienzyme Systems: Catalytic Properties, Academic Press, New York.
30. Shatalin, K., Lebreton, S., Rault-Leonardon, M., Velot, C., and Srere, P.A. (1999) Electrostatic channelling in a fusion protein of porcine citrate synthase and porcine malate dehydrogenase. Biochemistry 38, 881–889.
31. Pan, P., Whoel, E., and Dunn, M.F. (1997) Protein architecture, dynamics and allostery in tryptophan synthase channelling. Trends Biochem. Sci. 22, 22–27.
32. Hyde, C.C., Ahmed, S.A., Padlan, E.A., Miles, E.W., and Davies, D.R. (1988) Three-dimensianal structure of the tryptophan synthase $\alpha_2\beta_2$ multienzyme complex from Salmonella thyphimurium. J. Biol. Chem. 263, 15857–15871.
33. Rhee, S., Parris, K.D., Ahmed, S.A., Miles, E.W., and Davies, D.R. (1996) Exchange of $K^+$ and $Cs^+$ for $Na^+$ induces local and long-range changes in the three-dimensional structure of the tryptophan synthase $\alpha_2\beta_2$ complex. Biochemistry 35, 4211–4221.
34. Dunn, M.F., Aguilar, V., Brzovic, P., Drewe, W.F., Houben, K.F., Leja, C.A., and Roy, M. (1990) The tryptophan synthase bienzyme complex transfers indole between the $\alpha-$ and $\beta-$ sites via a 25–30 A long tunnel. Biochemistry 29, 8598–8607.
35. Lane, A.N. and Kirschner, K. (1991) Mechanism of the physiological reaction catalyzed by tryptophan synthase from *Escherichia coli*. Biochemistry 30, 479–484.
36. Anderson, K.S., Miles, E.W., and Johnson, K.A. (1991) Serine modulates substrate channelling in tryptophan synthase. J. Biol. Chem. 266, 8020–8033.
37. Houben, K.F. and Dunn, M.F. (1990) Allosteric effects acting over a distance of 20–25 A in the *Escherichia coli* tryptophan synthase increase ligand affinity and cause redistribution of covalent intermediates. Biochemistry 29, 2421–2429.
38. Avilan, L., Gontéro, B., Lebreton, S., and Ricard, J. (1997) Memory and imprinting effects in multi-enzyme complexes. I. Isolation, dissociation and reassociation of a phosphoribulokinase-glyceraldehyde phosphate dehydrogenase from *Chlamydomonas reinhardtii* chloroplasts. Eur. J. Biochem. 246, 78–84.
39. Lebreton, S., Gontéro, B., Avilan, L., and Ricard, J. (1997) Memory and imprinting effects in multi-enzyme complexes. II. Kinetics of the bi-enzyme complex from *Chlamydomonas reinhardtii* and hysteretic activation of chloroplast oxidized phosphoribulokinase. Eur. J. Biochem. 246, 85–91.
40. Lebreton, S., Gontéro, B., Avilan, L., and Ricard, J. (1997) Information transfer in multienzyme complexes. I. Thermodynamics of conformational constraints and memory effects in the bienzyme glyceraldehyde phosphate dehydrogenase-phosphoribulokinase of *Chlamydomonas reinhardtii* chloroplasts. Eur. J. Biochem. 250, 286–295.
41. Avilan, L., Gontéro, B., Lebreton, S., and Ricard, J. (1997) Information transfer in multienzyme complexes. II. The role of Arg 64 of *Chlamydomonas reinhardtii* phosphoribulokinase in the information transfer between glyceraldehyde phosphate dehydrogenase and phosphoribulokinase. Eur. J. Biochem. 250, 296–302.
42. Deutscher, M.P. (1984) The eukaryotic aminoacyl-tRNA synthetase complex: suggestions for its structure and function. J. Cell. Biol. 99, 373–377.
43. Mirande, M., Le Corre, D., and Waller, J.P. (1985) A complex from cultured Chinese hamster ovary cells containing nine aminoacyl-tRNA synthetases. Thermolabile leucyl-tRNA synthetase from the tsH1 mutant cell line is an integral component of this complex. Eur. J. Biochem. 147, 281–289.
44. Harris, C.L. (1987) An aminoacyl-tRNA synthetase complex in *Escherichia coli*. J. Bacteriol. 169, 2718–2723.
45. Cerini, C., Kerjan, P., Astier, M., Cratecos, D., Mirande, M., and Semeriva, M. (1991) A component of the multisynthetase complex is a functional aminoacyl-tRNA synthetase. EMBO J. 10, 4267–4277.

46. Mirande, M. (1991) Aminoacyl-tRNA synthetase family from prokaryotes: structural domains and their implications. Prog. Nucleic Acid Res. Mol. Biol. 40, 95–142.
47. Mirande, M., Lazard, M., Martinez, R., and Latreille, M.T. (1992) Engineering mammalian aspartyl-tRNA synthetase to probe structural features mediating its association with the multisynthetase complex. Eur. J. Biochem. 203, 459–466.
48. Singer, P.A., Levinthal, M., and Williams, L.S. (1984) Synthesis of the isoleucyl- and valyl-tRNA synthetases and isoleucine-valine biosynthetic enzymes in a threonine deaminase regulatory mutant of *Escherichia coli* K 12. J. Mol. Biol. 175, 39–55.
49. Hensley, P. (1988) Ligand binding and multienzyme complex formation between ornithine carbamoyltransferase and arginase from *Saccharomyces cerevisiae*. Curr. Top. Cell. Regul. 29, 35–75.
50. Volkenstein, M.V. and Goldstein, B.N. (1966) A new method for solving the problems of the stationary kinetics of enzymological reactions. Biochim. Biophys. Acta 115, 471–477.
51. Kellershohn, N. and Ricard, J. (1994) Coordination of catalytic activities within enzyme complexes. Eur. J. Biochem. 220, 955–961.
52. Whitehead, E. (1970) The regulation of enzyme activity and allosteric transition. Progr. Biophys. Mol. Biol. 21, 449–456.
53. Whitehead, E. (1976) Simplifications of the derivations and forms of steady-state equations for non-equilibrium random substrate-modifier and allosteric enzyme mechanisms. Biochem. J. 159, 449–456.
54. Ricard, J. (1985) Organized polymeric enzyme systems: catalytic properties 177–240. In: G.R. Welch (ed.) Organized Multienzyme Systems: Catalytic Properties. Academic Press, New York.
55. Laidler, K.J. (1958) The Chemical kinetics of Enzyme Action. Clarendon Press, Oxford.
56. Laidler, K.J. and Bunting, P.S. (1973) The Chemical Kinetics of Enzyme Action (Second Edition). Clarendon Press, Oxford.

J. Ricard *Emergent Collective Properties, Networks and Information in Biology*

DOI: 10.1016/S0167-7306(05)40009-5

CHAPTER 9

# Conformation changes and information flow in protein edifices

J. Ricard

Conformation changes propagate within protein edifices. Hence, one may expect information flow to be associated with the conformational spread. The aim of the present chapter is to study this information flow under thermodynamic equilibrium conditions. This study has been performed on a model protein lattice made up of identical functional protein units. Each unit is made up of two dimeric proteins $A$ and $B$. Protein $A$ binds ligand $X$ and protein $B$ binds ligand $Y$. These binding processes induce conformation changes that may propagate in the protein edifice. The binding constants of $X$ and $Y$ to the functional unit is in fact the product of the intrinsic binding constant of $X$ to $A$ times an energy parameter that expresses how the interactions between $A$ and $B$ alters the ligand binding process. A similar reasoning holds for the binding of $B$ to the functional unit. Hence the probability of occurrence of the nodes of the lattice, as well as their information, is a mathematical expression involving two types of vectors: vectors that describe the successive steps of ligand binding and vectors of conformational constraints that express how these constraints modulate ligand binding. The existence of information in such a lattice is the consequence of conformational flow. It is therefore possible to define an "information landscape" i.e., a surface that shows how local information varies in going from node to node. The landscape, or the surface, displays "ranges of mountains" separated by "valleys" of information. The interesting conclusion of this theoretical study is that the "information landscape" is, to a large extent, the consequence of the propagation of the conformational constraints.

*Keywords:* conformational spread and information landscape, cooperativity, energy contribution of subunit arrangement, integration and emergence in a protein lattice, mutual information of a protein unit in a lattice, Pauling's principle, protein lattice, quaternary constraint energy contribution, conformational transitions in quasi-linear lattices, thermodynamics of induced conformational spread, vectors of conformational constraints.

It is well known that, in multimeric proteins and in more complex protein edifices, an "influence" can propagate from place to place [1,2]. This influence is usually called cooperativity [3–11]. Thus, in oligomeric enzymes bearing several identical active sites, it often happens that these sites are not independent. This means that if one of the sites is occupied by a ligand, a substrate for instance, the ability of the other sites to bind the same ligand will be affected. Hence, the free sites receive an "influence" from the liganded sites. It is this positive, or negative, "influence" that is precisely called cooperativity. One can wonder, however, whether this "influence" is not accompanied by a propagation of information from one place to another of the protein lattice. If the term information is given the broad meaning already referred to, i.e., the ability of a system to associate molecular signals in order to generate a function, one can easily conceive that information flow be associated with conformation spread. If a protein edifice is made up of the association of two different kinds of proteins that can change their conformations, the whole macromolecular edifice associates different molecular signals in its own structure and the protein lattice then possesses information that propagates together with conformation change. In order to understand this situation, which will be the topic of the present chapter, it is mandatory to discuss first the effects of cooperativity on the function of a simple polymeric protein.

## *1. Phenomenological description of equilibrium ligand binding and nonequilibrium catalytic processes*

Let us consider a multimeric protein. We assume that each subunit bears a site that can bind a given ligand. Hence the whole binding process involves $n$ molecules of the same ligand to the $n$-sited protein (Fig. 1). The corresponding equilibrium binding isotherm, $\overline{\nu}$ or $\overline{Y}$, can be defined as

$$\overline{\nu} = n\overline{Y} = \frac{[P_1] + 2[P_2] + \cdots + n[P_n]}{[P_0] + [P_1] + \cdots + [P_n]} \tag{1}$$

One can define two different types of ligand binding constants, namely the macroscopic and microscopic binding constants. Macroscopic constants, $K_i$, are defined as

$$K_i = \frac{[P_{i,T}]}{[P_{i-1,T}][L]} \tag{2}$$

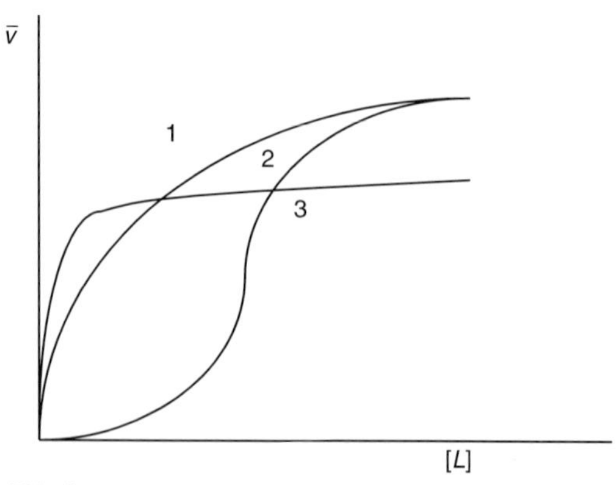


Fig. 1. Different types of binding curves generated by a binding isotherm. Curve 1: Lack of cooperativity. Curve 2: Positive cooperativity. Curve 3: Negative cooperativity.

where $[P_{i,T}]$ and $[P_{i-1,T}]$ represent the concentrations of protein molecules that have bound $i$ and $i-1$ molecules of ligand $L$ on any combination of sites. If the sites are all equivalent, i.e., if there is the same probability to bind the ligand to any site, one can define microscopic constants, $K_i'$, as

$$K_i' = \frac{[P_i]}{[P_{i-1}][L]} \tag{3}$$

where $[P_i]$ and $[P_{i-1}]$ are now the concentrations of $P$ that have bound $i$ and $i-1$ molecules of $L$ on specific combinations of sites. It follows from this definition that

$$[P_{i,T}] = \binom{n}{i}[P_i] \tag{4}$$

and that

$$K_i = \left\{\binom{n}{i}\Big/\binom{n}{i-1}\right\}K_i' = \frac{n-i+1}{i}K_i' \tag{5}$$

With this definition in mind the expression of the binding isotherm becomes

$$\overline{\nu} = n\overline{Y} = \frac{\sum_{i=1}^{n} i\binom{n}{i}K_1'K_2'\ldots K_i'[L]^i}{1+\sum_{i=1}^{n}\binom{n}{i}K_1'K_2'\ldots K_i'[L]^i} \tag{6}$$

This function can generate different types of binding isotherms (Fig. 1).

This reasoning can be extended to a nonequilibrium chemical reaction process carried out by a polymeric enzyme bearing $n$ identical catalytic sites (Fig. 2). One can then define apparent microscopic affinity constants as

$$\overline{K}_i' = \frac{k_i}{k_{-i}+k_i'} \tag{7}$$

where $k_i'$ are the catalytic constants. The steady-state equation is then

$$\frac{v}{[E]_0} = \frac{\sum_{i=1}^{n} ik_i'\binom{n}{i}\overline{K}_1'\overline{K}_2'\ldots\overline{K}_i'[S]^i}{1+\sum_{i=1}^{n}\binom{n}{i}\overline{K}_1'\overline{K}_2'\ldots\overline{K}_i'[S]^i} \tag{8}$$

where $[E_0]$ is the total enzyme concentration and $v$ the steady-state rate. If the binding constants $K_i'$ in Eq. (6), or the apparent binding constants $\overline{K}_i'$ in Eq. (8), all have the same value, the binding isotherm, or the steady-state rate curve, becomes

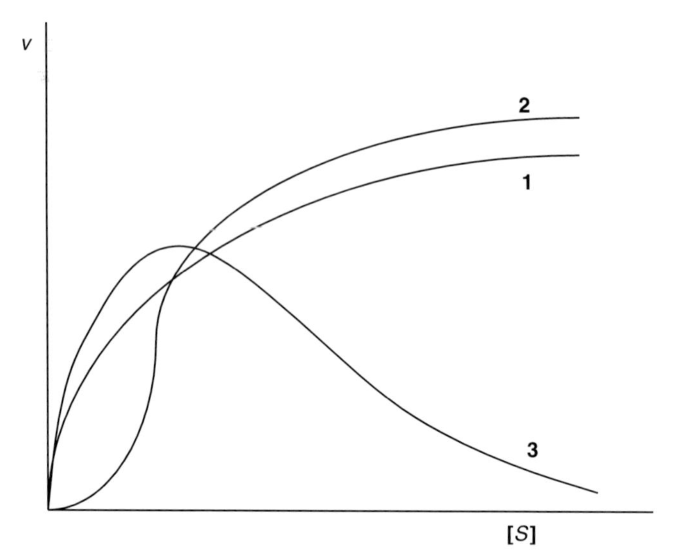


Fig. 2. Inhibition by excess substrate of a reaction rate can be generated by site–site interactions. Inhibition by excess ligand cannot be generated by a binding isotherm. Curve 1: Lack of cooperativity. Curve 2: Positive cooperativity. Curve 3: Inhibition by excess substrate.

hyperbolic. This situation is expected to occur when the binding, or catalytic, sites are all independent. It will be seen later that the network of ligand, or substrate, binding can also be nonlinear and then the situation becomes more complex. Moreover, as already pointed out in the first chapter of this book, Eqs. (6) and (8) can perfectly well generate quite different types of curves (Figs. 1 and 2).

## 2. *Thermodynamic bases of long-range site–site interactions in proteins and enzymes*

### 2.1. *General principles*

In many cases, binding and catalytic sites are not independent. They mutually interact. As already outlined, this means that one site receives an "influence" from another. This process of long-range interaction is called cooperativity. It is important at this stage to know the thermodynamic basis of long-range site–site interaction. The free energy of activation of a kinetic process carried out by a polymeric enzyme, $\Delta G^{\neq}$, can be written as [11–16]

$$\Delta G^{\neq} = \Delta G^{\neq *} + U_{\gamma} - U_{\tau} \tag{9}$$

where $\Delta G^{\neq *}$ is the free energy of activation of the same chemical process carried out by an ideally isolated subunit and where the energy contributions $U_{\gamma}$ and $U_{\tau}$ are the stabilization–destabilization energies exerted by subunit interactions on the ground

($U_\gamma$) and the transition states ($U_\tau$). The difference $U_\gamma - U_\tau$ expresses the mutual interactions between the sites of the polymeric enzyme. If these sites were mutually independent, $U_\gamma = U_\tau$ and $\Delta G^{\neq} = \Delta G^{\neq *}$. In the ground state, the stabilization–destabilization energy, $U_\gamma$, stems from two types of effects. The first is called subunit arrangement energy contribution [14]. It expresses how subunit design controls the rate of conformational transition of the subunit that carries out the chemical reaction. This energy corresponds to the free energy change that accompanies the dissociation of a polymer into its subunits, assuming that no conformation change of the subunits occurs during the dissociation process. The second type of effect is assumed to be exerted through site–site distortion, or intersubunit strain. It implies that subunits exert mutual quaternary constraints that tend to alter the reaction rate of the chemical process. Therefore one has

$$U_\gamma = \sum ({}^\alpha \Delta G) + \sum ({}^\sigma \Delta G) \tag{10}$$

and a similar relation is expected to hold for the expression of $U_\tau$. In expression (10) the first and the second terms of the right-hand side member refer to the two types of effect already mentioned. An illustration of the possible effects of the energy components on the free energy of activation of a chemical process is shown in Fig. 3.

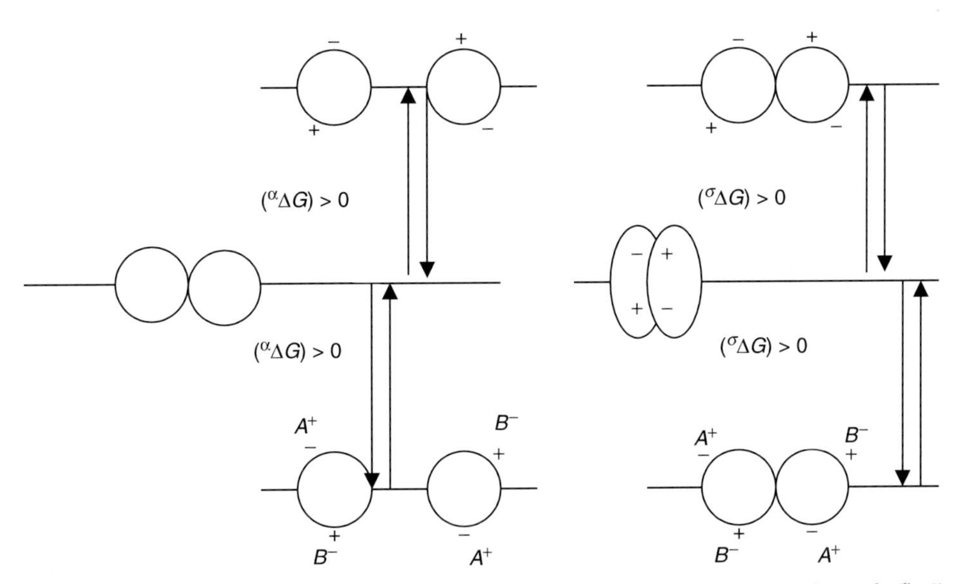


Fig. 3. Energy diagram showing that, in a dimer, ($^\alpha\Delta G$) and ($^\sigma\Delta G$) can be positive or negative. Left: ($^\alpha\Delta G$) is the free energy difference between the dissociated and the aggregated state of the subunits. If the dissociated state is destabilized relative to the aggregated state, then ($^\alpha\Delta G$)>0. If the dissociated state tends to be stabilized relative to the aggregated state, then ($^\alpha\Delta G$)>0. Right: ($^\sigma\Delta G$) is the free energy difference between the unstrained and strained aggregated states. If the unstrained aggregated state is destabilized relative to the strained aggregated state, then ($^\sigma\Delta G$)>0. If alternatively the unstrained state is stabilized relative to the strained state ($^\sigma\Delta G$)>0. Whether ($^\alpha\Delta G$) and ($^\sigma\Delta G$) are positive or negative they correspond to the free energy of *ideal* processes. Adapted from [9].

*2.2. Energy contribution of subunit arrangement*

Let us consider a supramolecular edifice made up of $n$ identical protein subunits that all interact and that can exist in one of two states, $A$ or $B$. If there exists $m$ subunits in the $B$ state and therefore $n-m$ subunits in the $A$ state, the expression of the free energy contribution is

$$\sum(^{\alpha}\Delta G) = \binom{n-m}{2}(^{\alpha}\Delta G_{AA}) + \binom{m}{2}(^{\alpha}\Delta G_{BB}) + m(n-m)(^{\alpha}\Delta G_{AB}) \qquad (11)$$

where $(^{\alpha}\Delta G_{AA})$, $(^{\alpha}\Delta G_{BB})$, and $(^{\alpha}\Delta G_{AB})$ represent the energy required to dissociate unstrained dimers $AA$, $BB$, and $AB$ into their subunits.

In the simplest case, with two enzyme forms $E_{i-1}$ and $E_i$ interconnected, there are three types of rate processes that should be considered: the rate of binding of the substrate to the protein, the rate of release of the substrate from the protein and the apparent rate of catalysis. One has

$$k_i = k_s^* \exp\{-(U_{i-1,\gamma} - U_{i-1,\tau})/RT\} \qquad (12)$$

for substrate binding and

$$\begin{aligned} k_{-i} &= k_{-s}^* \exp\{-(U_{i,\gamma} - U_{i-1,\tau})/RT\} \\ k_i' &= k^* \exp\left\{-(U_{i,\gamma} - U_{i-1,\tau}')/RT\right\} \end{aligned} \qquad (13)$$

for substrate release and catalysis, respectively. The energy parameters $U_\gamma$, $U_\tau$, and $U_\tau'$ have the same significance as in Eq. (9). $k_s^*$, $k_{-s}^*$, and $k^*$ are the intrinsic rate constants for substrate binding, release and catalysis [11]. In the absence of quaternary constraints, Eq. (11) when applied to enzyme form $E_{i-1}$ becomes

$$U_{i-1,\gamma} = \binom{n-i+1}{2}(^{\alpha}\Delta G_{AA}) + \binom{i-1}{2}(^{\alpha}\Delta G_{BB}) + (i-1)(n-i+1)(^{\alpha}\Delta G_{AB}) \qquad (14)$$

for the ground state. A similar relation would hold for the transition state. If relation (11) is applied to enzyme form $E_i$ one has

$$U_{i,\gamma} = \binom{n-i}{2}(^{\alpha}\Delta G_{AA}) + \binom{i}{2}(^{\alpha}\Delta G_{BB}) + i(n-i)(^{\alpha}\Delta G_{AB}) \qquad (15)$$

Moreover the second-order process of product formation has the same stabilization–destabilization as the catalytic process itself. Hence one can define a second-order constant of the catalytic process, $k_i' \overline{K}_i'$, as [11]

$$k_i'\overline{K}_i' = k^*\overline{K}^* \exp\left\{-(U_{i-1,\gamma} - U_{i-1,\tau}')/RT\right\} \qquad (16)$$

and it becomes possible to derive, from the second equation (13) and from Eq. (16), the expression for $\overline{K}'_i$ viz.

$$\overline{K}'_i = \overline{K}^* \exp\{-(U_{i-1,\gamma} - U_{i,\gamma})/RT\} \tag{17}$$

By comparing Eqs. (12), (13) with Eq. (17) it becomes obvious that relationship (17) can be obtained, only because

$$U'_{i-1,\tau} = U_{i-1,\tau} \tag{18}$$

This expression illustrates the validity of Pauling's principle [17,18] which states that an enzyme that has bound a transition state has the same conformation as the free enzyme. This is equivalent to stating that an enzyme should have the same conformation whether it has bound a transition state, or another one. Moreover the quaternary constraints should be relieved when the enzyme reaches the top of the free energy barrier. Since we have been assuming so far there is no quaternary constraint in the ground state, one should have

$$U'_{i-1,\tau} = U_{i-1,\tau} = U_{i-1,\gamma} \tag{19}$$

It is now possible to calculate the difference between the energies $U_{i,\gamma} - U_{i-1,\tau}$ and $U_{i-1,\gamma} - U_{i,\gamma}$. If, as assumed thus far, there is no subunit strain

$$\begin{aligned} U_{i,\gamma} - U_{i-1,\tau} = -(U_{i-1,\gamma} - U_{i,\gamma}) &= -(n-i)({}^{\alpha}\Delta G_{AA}) \\ &+ (i-1)({}^{\alpha}\Delta G_{BB}) + (n-2i+1)({}^{\alpha}\Delta G_{AB}) \end{aligned} \tag{20}$$

Defining the dimensionless coefficients

$$\begin{aligned} \alpha_{AA} &= \exp\{-({}^{\alpha}\Delta G_{AA})/RT\} \\ \alpha_{BB} &= \exp\{-({}^{\alpha}\Delta G_{BB})/RT\} \\ \alpha_{AB} &= \exp\{-({}^{\alpha}\Delta G_{AB})/RT\} \end{aligned} \tag{21}$$

One can now re-express the catalytic and apparent binding constants of a nonequilibrium process as

$$\begin{aligned} k'_i &= k^* \frac{\alpha_{BB}^{i-1}\alpha_{AB}^{n-2i+1}}{\alpha_{AA}^{n-i}} \\ \overline{K}'_i &= \overline{K}^* \frac{\alpha_{AA}^{n-i}}{\alpha_{BB}^{i-1}\alpha_{AB}^{n-2i+1}} \end{aligned} \tag{22}$$

In the case of an equilibrium process, the true microscopic constant, $K'_i$, is equal to

$$K'_i = K^* \frac{\alpha_{AA}^{n-i}}{\alpha_{BB}^{i-1}\alpha_{AB}^{n-2i+1}} \tag{23}$$

The product of apparent, or real, binding constants that appear in steady state, or binding, equations assumes the form

$$
\begin{aligned}
\overline{K}'_1\overline{K}'_2\ldots\overline{K}'_i &= \overline{K}^{\bullet i}\,\frac{\alpha_{AA}^{i(2n-i+1)/2}}{\alpha_{BB}^{i(i-1)/2}\alpha_{AB}^{i(n-i)}} \\
K'_1K'_2\ldots K'_i &= K^{\bullet i}\,\frac{\alpha_{AA}^{i(2n-i-1)/2}}{\alpha_{BB}^{i(i-1)}\alpha_{AB}^{i(n-i)}}
\end{aligned}
\tag{24}
$$

Moreover the product of rate and apparent binding constants that appear in steady-state rate Eq. (8) can be expressed as

$$
k'_i\overline{K}'_1\overline{K}'_2\ldots\overline{K}'_i = k^*\overline{K}^{*i}\,\frac{\alpha_{AA}^{(i-1)(2n-i)/2}}{\alpha_{BB}^{(i-1)(i-2)/2}\alpha_{AB}^{(i-1)(n-i+1)}}
\tag{25}
$$

We are now in a position to derive the so-called "structural" expression of the numerator and denominator of binding and steady-state rate equations. For instance, in the case of a protein tetramer ($n=4$) that does not display any quaternary constraint but whose subunits interact and, under equilibrium conditions, bind ligand $L$, we have the sequence of events shown in Fig. 4. The distribution of the protein molecules over the various liganded and unliganded states is

$$
[E]_T = [E_0] + [E_1] + \cdots + [E_4]
\tag{26}
$$

which is equivalent to

$$
[E]_T = [E_0](1 + 4K'_1[L] + 6K'_1K'_2[L]^2 + 4K'_1K'_2K'_3[L]^3 + K'_1K'_2K'_3K'_4[L]^4)
\tag{27}
$$

and to

$$
[E]_T = [E_0]\left\{1 + 4\frac{\alpha_{AA}^3}{\alpha_{AB}^3}K^*[L] + 6\frac{\alpha_{AA}^5}{\alpha_{AB}^4\alpha_{BB}}K^{*2}[L]^2 + 4\frac{\alpha_{AA}^6}{\alpha_{AB}^3\alpha_{BB}^3}K^{*3}[L]^3 + \frac{\alpha_{AA}^6}{\alpha_{BB}^6}K^{*4}[L]^4\right\}
\tag{28}
$$

It is clear that if the subunits do not interact

$$
\alpha_{AA} = \alpha_{BB} = \alpha_{AB} = 1
\tag{29}
$$

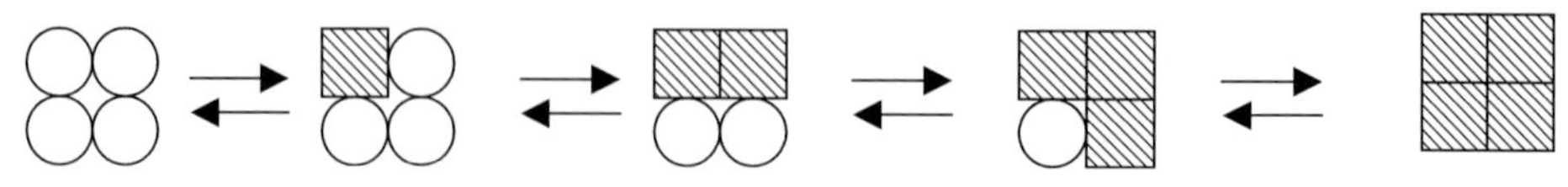

Fig. 4. A linear sequence of conformation changes in a tetrameric protein. The conformation change is induced by the sequential binding of a ligand. No conformational constraint is assumed to occur. See text.

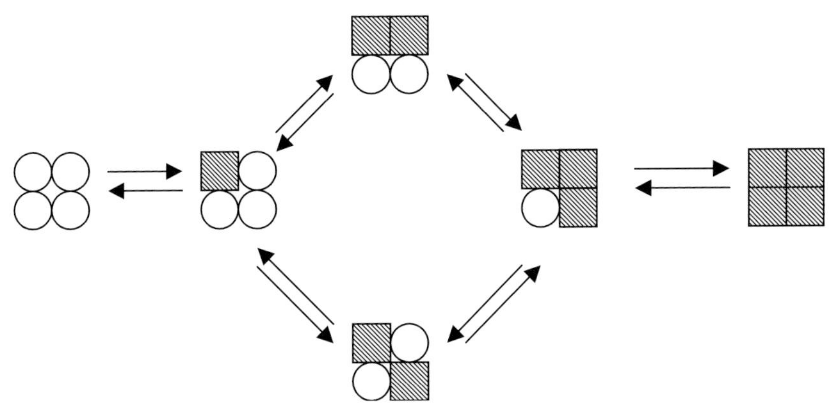

Fig. 5. A branched process of conformation change in a tetrameric protein. The conformation change is induced by the sequential binding of a ligand. No conformational constraint is assumed to occur. See text.

and

$$[E]_T = [E_0](1 + K^*[L])^4 \tag{30}$$

It has been assumed so far that the network of conformation changes was linear. But this is not compulsory. If each subunit interacts with two others, instead of three, which was precisely the case in the previous model, the corresponding network is not linear but branched (Fig. 5). One can show after some algebra that the distribution of the various protein states assumes the following expression

$$[E]_T = [E_0]\left\{1 + \frac{\alpha_{AA}^2}{\alpha_{AB}^2} K^*[L] + \left(\alpha_{AA}\alpha_{BB} + 2\alpha_{AB}^2\right)\frac{\alpha_{AA}^3}{\alpha_{BB}\alpha_{AB}^4} K^{*2}[L]^2 + \frac{\alpha_{AA}^4}{\alpha_{BB}^2\alpha_{AB}^2} K^{*3}[L]^3 + \frac{\alpha_{AA}^4}{\alpha_{BB}^4} K^{*4}[L]^4\right\} \tag{31}$$

and the corresponding equation is seen to be different from expression (28).

### 2.3. *Quaternary constraint energy contribution*

It was postulated in the previous section that no intersubunit strain was taking place. It is therefore of interest to introduce some additional subtleties in the theory in order to take account of the possible existence of quaternary constraints. The effect of intersubunit strain depends on protein design, and this effect may possibly change with the number of protein subunits, or with the degree of saturation of the protein by a ligand. Let us consider, for instance, an unliganded dimer subjected to quaternary constraints. The relief of these constraints involves the breaking of weak bonds between the subunits (indicated by dots in Fig. 6) followed by a spontaneous conformation change of the two subunits. If $({}^{\beta}\Delta G_{AA})$ is the free energy required to

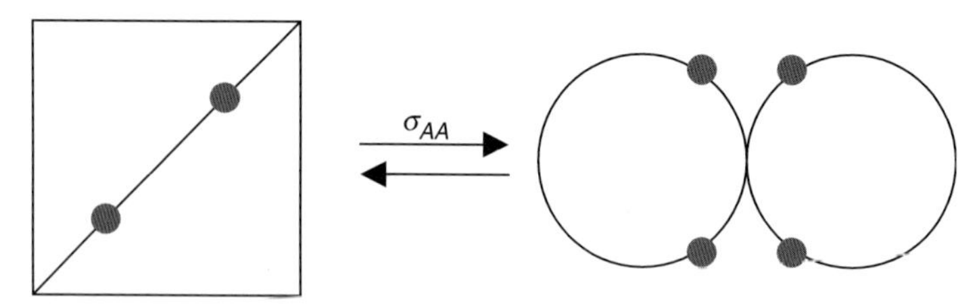


Fig. 6. Intersubunit strain in a dimeric enzyme molecule. On the left side of the figure the enzyme is in strained state and on the right side it has a relieved state. The black dots represent the weak bonds to be broken. Adapted from [14].

break these bonds and $(^{\sigma}\Delta G_A)$ the free energy necessary to strain a relieved subunit to the conformation state it usually possesses within the strain dimer, one must have

$$(^{\sigma}\Delta G_{AA}) = (^{\beta}\Delta G_{AA}) - 2(^{\sigma}\Delta G_A) \tag{32}$$

In this expression, $(^{\sigma}\Delta G_{AA})$ is the free energy of quaternary constraints between conformation $A$ and conformation $A$. Both $(^{\beta}\Delta G_{AA})$ and $(^{\sigma}\Delta G_A)$ are positive (Fig. 3). Hence $(^{\sigma}\Delta G_{AA})$ can be positive or negative (Fig. 3).

One can define both a bond-breaking and an intrasubunit strain coefficient, i.e., $\beta_{AA}$ and $\sigma_A$, and one has

$$\begin{aligned} \beta_{AA} &= \exp\{-(^{\beta}\Delta G_{AA})/RT\} \\ \sigma_A &= \exp\{-(^{\sigma}\Delta G_A)/RT\} \end{aligned} \tag{33}$$

Hence it follows that

$$0 < \beta_{AA} < 1 \quad \text{and} \quad 0 < \sigma_A < 1 \tag{34}$$

The expression for the intersubunit strain coefficient, $\sigma_{AA}$, in the unliganded dimer is then

$$\sigma_{AA} = \beta_{AA}/\sigma_A^2 \tag{35}$$

One will notice that the powers of $\beta_{AA}$ and $\sigma_{\text{A}}$ are different, for the first refers to the number of interactions whereas the second pertains to that of subunits.

One can easily see that the expression of the intersubunit strain coefficient depends upon both subunit number and design. In the case, for instance, of an unliganded trimer (Fig. 7) one as

$$3(^{\sigma}\Delta G_{AA}) = 3(^{\beta}\Delta G_{AA}) - 3(^{\sigma}\Delta G_A) \tag{36}$$

and the corresponding strain coefficient is

$$\sigma_{AA} = \beta_{AA}/\sigma_A \tag{37}$$

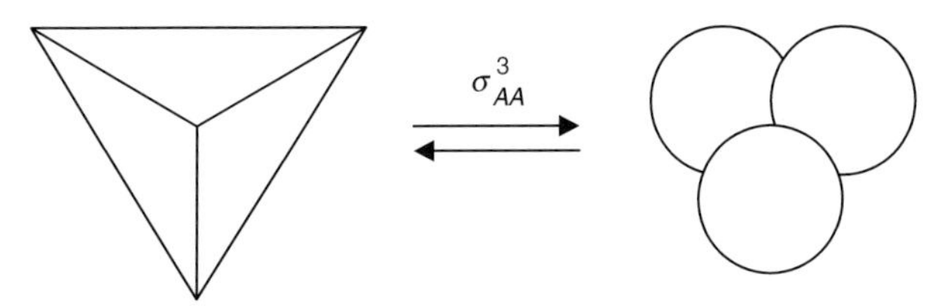


Fig. 7. Intersubunit strain in a trimeric enzyme molecule. On the left side of the figure the enzyme is in a strained state and on the right side it has a relieved state. Adapted from [14].

which is indeed different from its parent coefficient (Eq. (35)). Hence it appears that whereas the bond-breaking coefficient $\beta_{AA}$ is independent of subunit design, the corresponding intersubunit strain coefficient $\sigma_A$ is not. The dependence of the intersubunit strain coefficient on the quaternary constraints within the enzyme implies that, on varying the concentration of the substrate that can be bound to the enzyme active sites, the value of the intersubunit strain coefficient $\sigma_{AA}$ varies. This is illustrated in the Fig. 8. For instance, in a tetrameric enzyme displaying interactions between all the sites, i.e., an enzyme having a "tetrahedral" type of subunit interactions [14], the $\sigma_{AA}$ coefficient will adopt different expressions depending on whether the protein has bound zero, one, or two substrate molecules. If the protein has not bound any substrate, the corresponding strain coefficient, $\sigma_{AA}(0)$, between two unliganded and strained subunits is expressed as

$$\sigma_{AA}(0) = \frac{\beta^5_{AA}}{\sigma^2_A} \tag{38}$$

If the protein has bound one substrate molecule the intersubunit strain coefficient $\sigma_{AA}(1)$ is now

$$\sigma_{AA}(1) = \frac{\beta^3_{AA}\beta^2_{AB}}{\sigma^2_A} \tag{39}$$

But if the protein has bound two substrate molecules, then

$$\sigma_{AA}(2) = \frac{\beta_{AA}\beta^4_{AB}}{\sigma^2_A} \tag{40}$$

When each subunit of a polymeric enzyme interacts with all the others, and when the protein edifice displays quaternary constraints, one can calculate the energies $U_{i-1,\gamma}$ and $U_{i,\gamma}$ of the enzyme in the ground state. One finds

$$\begin{aligned} U_{i-1,\gamma} = &\binom{n-i+1}{2}\{({}^\alpha\Delta G_{AA}) + ({}^\beta\Delta G_{AA})\} + \binom{i-1}{2}\{({}^\alpha\Delta G_{BB}) + ({}^\beta\Delta G_{BB})\} \\ &+ (i-1)(n-i+1)\{({}^\alpha\Delta G_{AB}) + ({}^\beta\Delta G_{AB})\} - (n-i+1)({}^\sigma\Delta G_A) \\ &- (i-1)({}^\sigma\Delta G_B) \end{aligned} \tag{41}$$

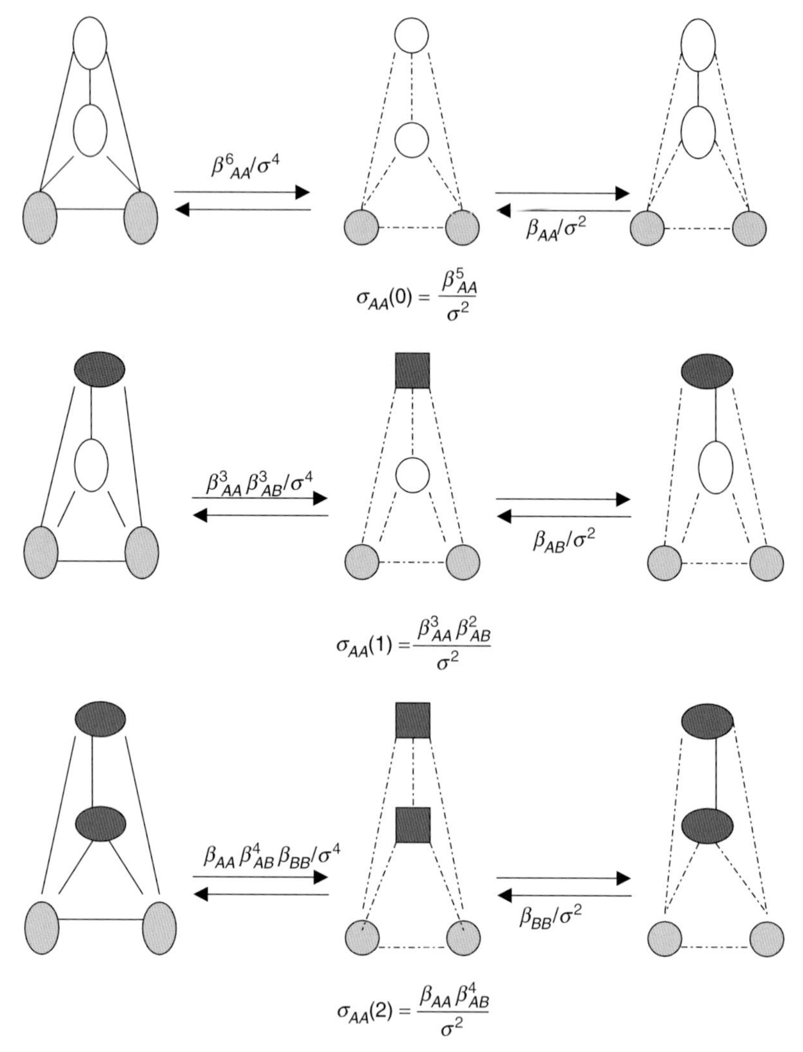


Fig. 8. Strain between two unliganded subunits in a tetramer as a function of the degree of saturation of the protein by the ligand. Top: The tetramer has bound no ligand. Middle: The tetramer has bound one ligand molecule. Bottom: The tetramer has bound two ligand molecules. The dotted lines represent the lack of conformational constraints between two subunits. The "dark" subunits have bound a ligand. The subunits of which we derive the expression of the intersubunit strain are "grey". One can observe that the apparent strain coefficient between the two "grey" unliganded subunits adopts a different value depending on the degree of saturation of the protein by the ligand. Adapted from [14].

and

$$U_{i,\gamma} = \binom{n-i}{2}\left\{({}^{\alpha}\Delta G_{AA}) + ({}^{\beta}\Delta G_{AA})\right\} + \binom{i}{2}\left\{({}^{\alpha}\Delta G_{BB}) + ({}^{\beta}\Delta G_{BB})\right\} + i(n-i)\left\{({}^{\alpha}\Delta G_{AB}) + ({}^{\beta}\Delta G_{AB})\right\} - (n-i)({}^{\sigma}\Delta G_{A}) - i({}^{\sigma}\Delta G_{B}) \tag{42}$$

Moreover, if, as discussed previously, the Pauling principle applies to transition states of polymeric enzymes, i.e., if quaternary constraints are relieved at the top of the energy barriers, the expression of $U_{i-1,\tau}$ is given by Eq. (14). Hence one can calculate $U_{i,\gamma} - U_{i-1,\tau}$ and $U_{i-1,\gamma} - U_{i,\gamma}$. One finds

$$\begin{aligned} U_{i,\gamma} - U_{i-1,\tau} = & -(n-i)({}^{\alpha}\Delta G_{AA}) + (i-1)({}^{\alpha}\Delta G_{BB}) + (n-2i+1)({}^{\alpha}\Delta G_{AB}) \\ & + \binom{n-i}{2}({}^{\beta}\Delta G_{AA}) + \binom{i}{2}({}^{\beta}\Delta G_{BB}) + i(n-i)({}^{\beta}\Delta G_{AB}) \\ & - (n-i)({}^{\sigma}\Delta G_{A}) - i({}^{\sigma}\Delta G_{B}) \end{aligned} \tag{43}$$

and

$$\begin{aligned} U_{i-1,\gamma} - U_{i,\gamma} = & (n-i)\{({}^{\alpha}\Delta G_{AA}) + ({}^{\beta}\Delta G_{AA})\} - (i-1)\{({}^{\alpha}\Delta G_{BB}) + ({}^{\beta}\Delta G_{BB})\} \\ & - (n-2i+1)\{({}^{\alpha}\Delta G_{AB}) + ({}^{\beta}\Delta G_{AB})\} - ({}^{\sigma}\Delta G_{A}) + ({}^{\sigma}\Delta G_{B}) \end{aligned} \tag{44}$$

If one defines the bond-breaking and intramolecular strain coefficients as

$$\begin{aligned} \beta_{AA} &= \exp\{-({}^{\beta}\Delta G_{AA})/RT\} \\ \beta_{BB} &= \exp\{-({}^{\beta}\Delta G_{BB})/RT\} \\ \beta_{AB} &= \exp\{-({}^{\beta}\Delta G_{AB})/RT\} \\ \sigma_{A} &= \exp\{-({}^{\sigma}\Delta G_{A})/RT\} \\ \sigma_{B} &= \exp\{-({}^{\sigma}\Delta G_{B})/RT\} \end{aligned} \tag{45}$$

It then becomes possible to derive the expression for the catalytic constant, $k'_i$, for the apparent substrate binding constant, $\overline{K}'_i$, and for the products $\overline{K}'_1\overline{K}'_2\ldots\overline{K}'_i$ and $k'_i\overline{K}'_1\overline{K}'_2\ldots\overline{K}'_i$. One finds

$$\begin{aligned} k'_i &= k^* \frac{\alpha_{BB}^{i-1}\alpha_{AB}^{n-2i+1}}{\alpha_{AA}^{n-i}} \frac{\beta_{AA}^{(n-i)(n-i-1)/2}\beta_{BB}^{i(i-1)/2}\beta_{AB}^{i(n-i)}}{\sigma_A^{n-i}\sigma_B^{i}} \\ \overline{K}'_i &= \overline{K}^* \frac{\alpha_{AA}^{n-i}}{\alpha_{BB}^{i-1}\alpha_{AB}^{n-2i+1}} \frac{\beta_{AA}^{n-i}}{\beta_{BB}^{i-1}\beta_{AB}^{n-2i+1}} \frac{\sigma_B}{\sigma_A} \\ \overline{K}'_1\overline{K}'_2\ldots\overline{K}'_i &= \overline{K}^{*i} \frac{\alpha_{AA}^{i(2n-i-1)/2}}{\alpha_{BB}^{i(i-1)/2}\alpha_{AB}^{i(n-i)}} \frac{\beta_{AA}^{i(2n-i-1)/2}}{\beta_{BB}^{i(i-1)/2}\beta_{AB}^{i(n-i)}} \frac{\sigma_B^{i}}{\sigma_A^{i}} \\ k'_i\overline{K}'_1\overline{K}'_2\ldots\overline{K}'_i &= k^*\overline{K}^{*i} \frac{\alpha_{AA}^{(i-1)(2n-i)/2}}{\alpha_{BB}^{(i-1)(i-2)/2}\alpha_{AB}^{(i-1)(n-i+1)}} \frac{\beta_{AA}^{n(n-1)/2}}{\sigma_A^{n}} \end{aligned} \tag{46}$$

Thus, for instance, the denominator, $D$, of the reaction rate of a tetrameric enzyme displaying "tetrahedral" site–site interactions is

$$D = 1 + 4\frac{\alpha_{AA}^3}{\alpha_{AB}^3}\frac{\beta_{AA}^3}{\beta_{AB}^3}\frac{\sigma_B}{\sigma_A}\overline{K}^*[S] + 6\frac{\alpha_{AA}^5}{\alpha_{AB}^4\alpha_{BB}}\frac{\beta_{AA}^5}{\beta_{AB}^4\beta_{BB}}\frac{\sigma_B^2}{\sigma_A^2}\overline{K}^{*2}[S]^2$$

$$+ 4\frac{\alpha_{AA}^6}{\alpha_{AB}^3\alpha_{BB}^3}\frac{\beta_{AA}^6}{\beta_{AB}^3\beta_{BB}^3}\frac{\sigma_B^3}{\sigma_A^3}\overline{K}^{*3}[S]^3 + \frac{\alpha_{AA}^6}{\alpha_{BB}^6}\frac{\beta_{AA}^6}{\beta_{BB}^6}\frac{\sigma_B^4}{\sigma_A^4}\overline{K}^{*4}[S]^4 \quad (47)$$

In general, steady-state rate equations of multi-sited enzymes are rather cumbersome and it is sensible to wonder whether some simple axioms could not generate a simplification of these equations.

### 2.4. *Fundamental axioms*

Any model that aims at understanding how the structural changes of a protein can affect its functional properties must incorporate some fundamental axioms about the structural changes induced by the binding of a ligand. These axioms must not be viewed as gratuitous assumptions or as *ad hoc* hypotheses, but as simplifying assumptions that allow one to derive *simple* rate equations and that are, at least approximately, met in nature. The present theory is based on four axioms.

*First axiom.* Current theories of enzyme catalysis postulate that the intramolecular strain of the enzyme is relieved in the transition states [17,18]. For a polymeric protein, this relief of intramolecular strain can only be achieved if the intermolecular strain is itself relieved in the transition states. Hence the first axiom is that quaternary constraints (or intersubunit strain) are relieved when overcoming the energy barriers along the reaction coordinate. This axiom must be met, at least as a first approximation, in order to allow the occurrence of complementarity between the enzyme's active site and the transition states.

*Second axiom.* In the absence of intersubunit strain, the subunits assume either conformation $A$, or conformation $B$. $A$ is the conformation of the unstrained, unliganded, subunit and $B$ the conformation of the liganded one. This postulate, minimizing the number of conformations has already been adopted by Monod *et al.* [3] as well as Koshland *et al.* [4]. This is an economy principle, the so-called "Occam's razor", that is almost always followed in the scientific approach to natural phenomena.

*Third axiom.* The subunit that has bound a transition state has a conformation, $A$, which is identical to that of the unliganded, unstrained subunit. This is in agreement with the result that the unstrained active site of an enzyme is complementary to the transition state of the reaction [17,18] and with the "economy principle" considered above. All these postulates that were implicit in the above reasoning have been used in the mathematical developments discussed above.

*Fourth axiom.* This axiom implies that in the constrained oligomeric enzyme molecules, the potential energy generated by quaternary constraints is evenly stored in all subunits, whether these subunits have bound a ligand or not.

In the equations presented above, quaternary constraints were expressed in terms of bond-breaking energies between subunits and in terms of intrasubunit strain. The second energy parameter could be dropped from the equations if we could express them in terms of intersubunit strain. This cannot be done, however, if the intersubunit strain varies with the degree of saturation of the protein by the ligand. This is not the case if the fourth axiom is an acceptable picture of reality, i.e., if the energies stored in strained liganded and unliganded subunits of the same molecule are about the same. Then $\sigma_A = \sigma_B$ and, for a tetramer displaying "tetrahedral" site–site interactions, one should have

$$\begin{aligned}
\frac{\sigma^6_{AA}(0)}{\sigma^3_{AA}(1)\sigma^3_{AB}(1)} &= \frac{\beta^3_{AA}}{\beta^3_{AB}} = \frac{\sigma^3_{AA}}{\sigma^3_{AB}} \\
\frac{\sigma^6_{AA}(0)}{\sigma_{AA}(2)\sigma^4_{AB}(2)\sigma_{BB}(2)} &= \frac{\beta^5_{AA}}{\beta^4_{AB}\beta_{BB}} = \frac{\sigma^5_{AA}}{\sigma^4_{AB}\sigma_{BB}} \\
\frac{\sigma^6_{AA}(0)}{\sigma^3_{AB}(3)\sigma^3_{BB}(3)} &= \frac{\beta^6_{AA}}{\beta^3_{AB}\beta^3_{BB}} = \frac{\sigma^6_{AA}}{\sigma^3_{AB}\sigma^3_{BB}} \\
\frac{\sigma^6_{AA}(0)}{\sigma^6_{BB}(4)} &= \frac{\beta^6_{AA}}{\beta^6_{BB}} = \frac{\sigma^6_{AA}}{\sigma^6_{BB}}
\end{aligned} \tag{48}$$

An interesting conclusion from these results is that if the intersubunit strain is about the same for each subunit of a polymer, the intersubunit strain of a polymer can be expressed from the strain taking place in a set of dimers. Equation (47) can then be rewritten as

$$\begin{aligned}
D = 1 &+ 4\frac{\alpha^3_{AA}}{\alpha^3_{AB}}\frac{\sigma^3_{AA}}{\sigma^3_{AB}}\overline{K}^*[S] + 6\frac{\alpha^5_{AA}}{\alpha^4_{AB}\alpha_{BB}}\frac{\sigma^5_{AA}}{\sigma^4_{AB}\sigma_{BB}}\overline{K}^{*2}[S]^2 \\
&+ 4\frac{\alpha^6_{AA}}{\alpha^3_{AB}\alpha^3_{BB}}\frac{\sigma^6_{AA}}{\sigma^3_{AB}\sigma^3_{BB}}\overline{K}^{*3}[S]^3 + \frac{\alpha^6_{AA}}{\alpha^6_{BB}}\frac{\sigma^6_{AA}}{\sigma^6_{BB}}\overline{K}^{*4}[S]^4
\end{aligned} \tag{49}$$

where $\sigma_{AA}$, $\sigma_{BB}$, and $\sigma_{AB}$ are strain coefficients of dimers within the polymer molecule.

### 3. *Conformational changes and mutual information of integration in protein lattices*

It was previously outlined that there should exist some relationships between conformation changes and mutual information in protein edifices. Moreover several experimental results suggest that conformation changes propagate within protein edifices. Several authors have presented experimental arguments in favor of such a propagation for chemotaxis [19–24], and ryanodine receptors [25–28], rings of identical proteins [29,30], flagella [31,32], and actin filaments [33]. Let us assume that a protein edifice associates two proteins $A$ and $B$ (Fig. 9). Each of these proteins $A$ and $B$ are made up of two different polypeptide chains. One polypeptide chain of each dimeric protein is associated with its neighbors through isologous, or heterologous, interactions, whereas the other polypeptide chain of each dimer is associated with its neighbors through *both* isologous and heterologous interactions (Fig. 9). The subunits, that are associated with the other polypeptide chains through isologous or heterologous interactions, can bind ligand $X$, or ligand $Y$, respectively. The two other polypeptide chains do not bind anything but play a major role in the specificity of the binding of $X$ and $Y$ to their respective sites.

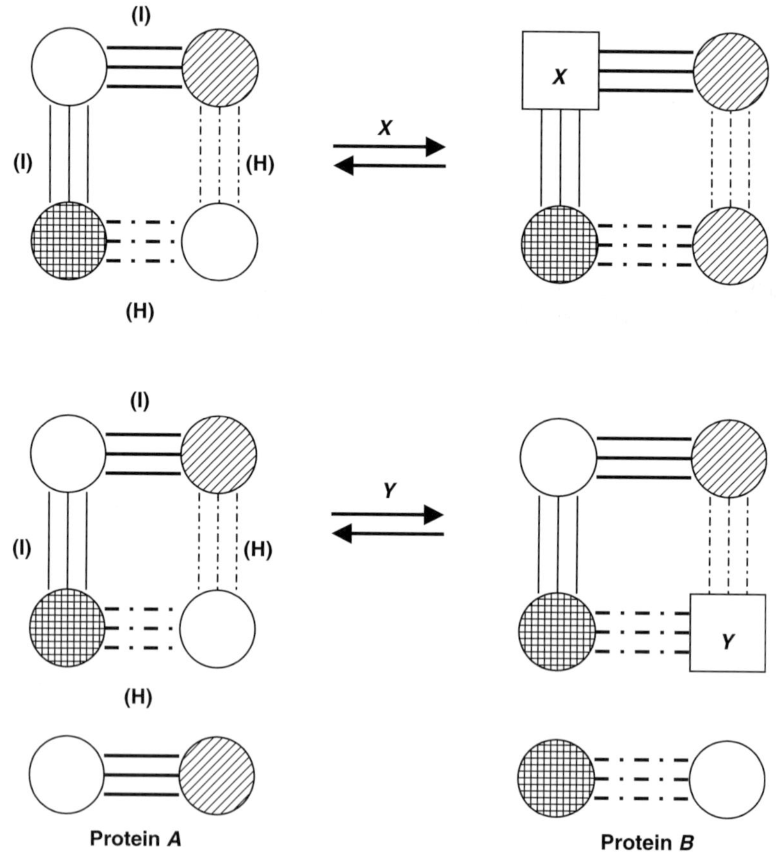


Fig. 9. Simple model of the association of two dimeric proteins as a functional unit. Two dimeric proteins $A$ and $B$ are associated thanks to *both* isologous and heterologous interactions. Isologous interactions (I) are represented by continuous lines and heterologous interactions (H) by dotted lines. Moreover in protein $A$ the two subunits are associated thanks to isologous interactions and in protein $B$ to heterologous interactions. The binding of a ligand $X$ on a subunit of $A$ induces a conformation change of the corresponding subunit. Alternatively another ligand $Y$ binds to a subunit of $B$ and induces a conformation change of that subunit.

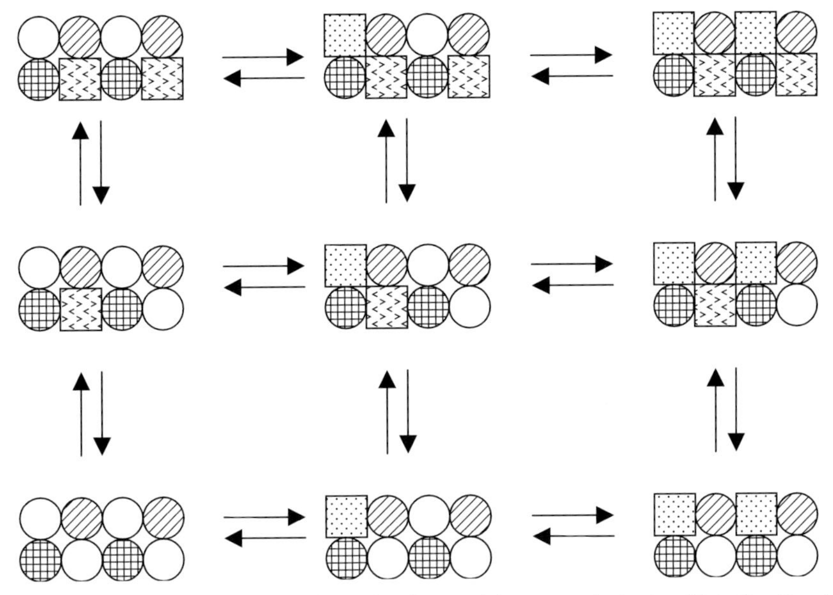

Fig. 10. Simple model of induced conformational spread in a protein lattice. Two functional units undergo induced conformation changes. In fact, as discussed in the previous figure, one subunit of proteins *A* and *B* change their conformation. The subunits that change their conformation are the "white" (or "void") subunits. For the clarity of the figure, this induced conformation change is pictured by a change of both the "shape" and the "color" of the subunit. Hence the binding of ligands *X* and *Y* to the corresponding subunits is not explicitly represented.

The same model can also apply to the situation in which *X* and *Y* do not bind to a subunit but rather at the interface between two subunits. Moreover the binding of *X* or *Y* to its corresponding site induces a conformation change in the two subunits of the same protein, either *A* or *B*. For convenience, we call protein unit the set of four polypeptide chains that change their conformation in a coordinated manner. In the following, we shall consider a protein lattice made up of several, or many, protein units, as shown in Fig. 10, and we shall study the conformational spread such a lattice can display when ligands *X* and *Y* bind to the protein edifice. This situation is actually a model but we believe this model can be applied to many different experimental situations.

### *3.1. Conformation change and mutual information of integration of the elementary protein unit*

Let us consider the elementary protein unit shown in Fig. 9. As already pointed out, the two dimeric proteins of the unit bind ligand *X*, or ligand *Y*, and undergo a conformation change. Each pair of subunits can occur in two conformations, i.e., the liganded and the unliganded ones. Depending on whether the dimeric proteins *A* and *B* have, or have not, bound their ligand they are termed below $X_0$ and $Y_0$ (if they have not bound any ligand), or $X_1$ and $Y_1$ (if they have bound their ligand).

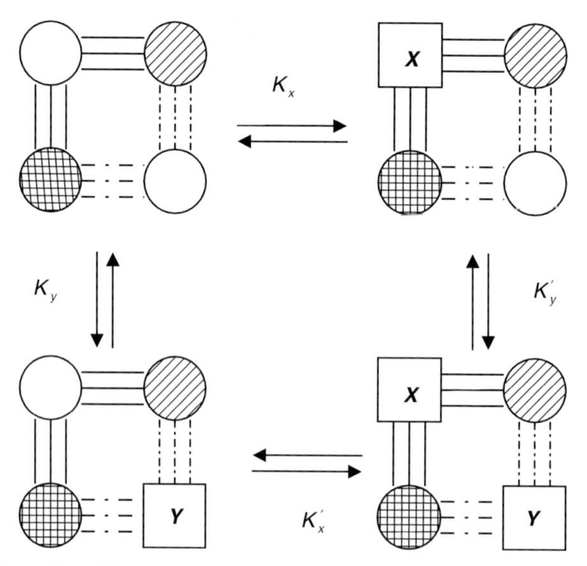


Fig. 11. Mutual information of integration of a functional unit. When the functional unit made up of two dimeric proteins binds ligands $X$ and $Y$ the functional unit can acquire mutual information of integration that expresses a quantitative formulation of its organization. See text.

The questions we raise now are whether such a simple system can possess mutual information of integration, whether this information is related to the conformational spread, and what the main thermodynamic features of this system are. The conformation changes that the protein can undergo are shown in Fig. 11. This situation is similar to other previously discussed (Chapter 6). One can easily calculate the probability that the protein unit has bound $X$ or $Y$, or both $X$ and $Y$ (Fig. 11). One finds

$$\begin{aligned} p(X) &= \frac{K_x[X] + K_x K'_y[X][Y]}{1 + K_x[X] + K_y[Y] + K_x K'_y[X][Y]} \\ p(Y) &= \frac{K_y[Y] + K'_x K_Y[X][Y]}{1 + K_x[X] + K_y[Y] + K'_x K_y[X][Y]} \\ p(X,Y) &= \frac{K_x K'_y[X][Y]}{1 + K_x[X] + K_y[Y] + K_x K'_y[X][Y]} \end{aligned} \tag{50}$$

From these equations one can write

$$\frac{p(X,Y)}{p(X)\,p(Y)} = \frac{K'_y[Y] + K_x K'_y[X][Y] + K_y K'_y[Y]^2 + K_x K'^2_y[X][Y]^2}{K_y[Y] + K'_x K_y[X][Y] + K_y K'_y[Y]^2 + K'_x K_y K'_y[X][Y]^2} \tag{51}$$

As thermodynamics requires that

$$K_x K'_y = K'_x K_y \tag{52}$$

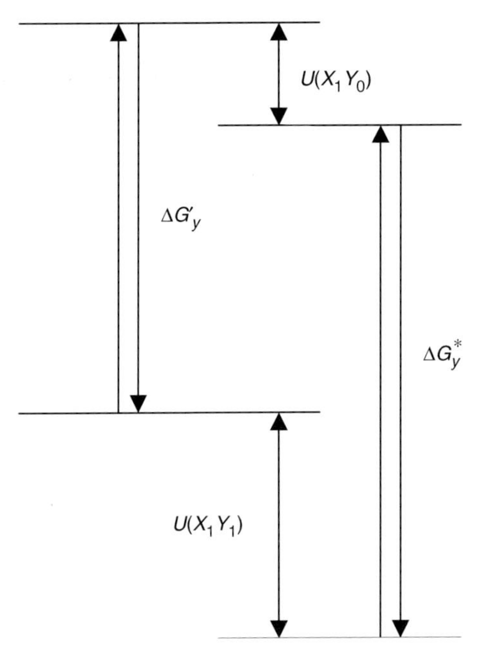


Fig. 12. Relationships between the free energy of ligand binding to the functional protein complex and to the ideally isolated dimeric protein. $\Delta G'_y$ and $\Delta G^*_y$ are the free energies of the binding of ligand $Y$ to the functional protein complex and to the ideally isolated dimeric protein, respectively. $U(X_1 Y_0)$ and $U(X_1 Y_1)$ are the free energies of stabilization–destabilization of the initial and final states brought about by the interaction between proteins $A$ and $B$.

expression (51) will be greater than one if $K'_y > K_y$ and less than one if $K'_y < K_y$. Hence mutual information of integration is positive in the first case and negative in the second. The fact that $K'_y$ is usually different from $K_y$ (and thus $K'_x$ different from $K_x$ as well) implies that there exist conformational interactions between proteins $A$ and $B$. This means that the association of the two proteins can, for instance, stabilize or destabilize both the initial and the final states involved in the process of $Y$ binding to protein $B$, in such a way that the corresponding equilibrium constant $K'_y$ may be larger or smaller than the binding constant of $Y$ to the "ideally isolated" protein $B$. Hence one has

$$K'_y = K^*_y \exp\{-[U(X_1 Y_0) - U(X_1 Y_1)]/RT\} \tag{53}$$

In this expression $K^*_y$ is the so-called intrinsic binding constant of $Y$ to protein $B$ considered "in isolation". $U(X_1 Y_0)$ and $U(X_1 Y_1)$ are the stabilization–destabilization energies of the protein states $X_1 Y_0$ and $X_1 Y_1$. This situation is illustrated by the free energy diagrams of Fig. 12. One can define the free energies associated with the equilibrium binding constants $K_y$ and $K^*_y$ as

$$\begin{aligned} \Delta G'_y &= -RT \ln K'_y \\ \Delta G^*_y &= -RT \ln K^*_y \end{aligned} \tag{54}$$

Moreover one can define energy parameters, $\alpha(X_1 Y_0)$ and $\alpha(X_1 Y_1)$, as

$$\begin{aligned} \alpha(X_1 Y_0) &= \exp\{-U(X_1 Y_0)/RT\} \\ \alpha(X_1 Y_1) &= \exp\{-U(X_1 Y_1)/RT\} \end{aligned} \tag{55}$$

and expression (53) above becomes

$$K'_y = K^*_y \frac{\alpha(X_1 Y_0)}{\alpha(X_1 Y_1)} \tag{56}$$

The ratio of the two $\alpha$'s expresses how the interaction of proteins $A$ and $B$ alters the binding of a ligand ($Y$) to the protein unit. These $\alpha$ coefficients represent an extension to protein edifices of the $\alpha$ and $\sigma$ parameters already referred to in the first part of the present chapter.

The above simple reasoning shows that the mutual information of integration relies upon thermodynamic relationships between the $U$'s or between the $\alpha$'s. As we have seen, positive mutual information of integration of the protein unit implies that $K'_y > K_y$, which is equivalent to

$$\exp\{-[U(X_1 Y_0) - U(X_1 Y_1)]/RT\} > \exp\{-[U(X_0 Y_0) - U(X_0 Y_1)]/RT\} \tag{57}$$

and to

$$U(X_0 Y_0) - U(X_0 Y_1) > U(X_1 Y_0) - U(X_1 Y_1) \tag{58}$$

This relationship can be reexpressed in terms of the $\alpha$'s, viz.

$$\frac{\alpha(X_1 Y_0)}{\alpha(X_1 Y_1)} > \frac{\alpha(X_0 Y_0)}{\alpha(X_0 Y_1)} \tag{59}$$

Alternatively, negative mutual information of integration would imply that

$$\frac{\alpha(X_1 Y_0)}{\alpha(X_1 Y_1)} < \frac{\alpha(X_0 Y_0)}{\alpha(X_0 Y_1)} \tag{60}$$

### *3.2. Ligand binding, conformation changes, and mutual information of integration of protein lattices*

The reasoning that has been followed on a protein unit can be extended to a lattice of protein units.

*3.2.1. A simple protein lattice*

Probably the simplest protein lattice one can think of is shown in Figs. 10 and 13. This lattice is made up of only two protein units. We assume, as previously, that each protein unit can bind two ligands $X$ and $Y$. Under equilibrium conditions the nodes of the lattice are connected through equilibrium binding constants and ligand concentrations. In order to simplify the notation let us write

$$\begin{aligned} K_{i,x}[X] &= ({}^{x}K_{i}) \\ K_{j,y}[Y] &= ({}^{y}K_{j}) \\ K_{x}^{*}[X] &= ({}^{x}K^{*}) \\ K_{y}^{*}[Y] &= ({}^{y}K^{*}) \end{aligned} \tag{61}$$

Under equilibrium conditions one can connect each node to its nearest neighbor. One finds

$$\begin{aligned} ({}^{x}K_{1}) &= ({}^{x}K^{*})\frac{\alpha(X_{0}Y_{0})}{\alpha(X_{1}Y_{0})} & ({}^{x}K_{2}) &= ({}^{x}K^{*})\frac{\alpha(X_{1}Y_{0})}{\alpha(X_{2}Y_{0})} \\ ({}^{x}K_{1}') &= ({}^{x}K^{*})\frac{\alpha(X_{0}Y_{1})}{\alpha(X_{1}Y_{1})} & ({}^{x}K_{2}') &= ({}^{x}K^{*})\frac{\alpha(X_{1}Y_{1})}{\alpha(X_{2}Y_{1})} \\ ({}^{x}K_{1}'') &= ({}^{x}K^{\bullet})\frac{\alpha(X_{0}Y_{2})}{\alpha(X_{1}Y_{2})} & ({}^{x}K_{2}'') &= ({}^{x}K^{\bullet})\frac{\alpha(X_{1}Y_{2})}{\alpha(X_{2}Y_{2})} \end{aligned} \tag{62}$$

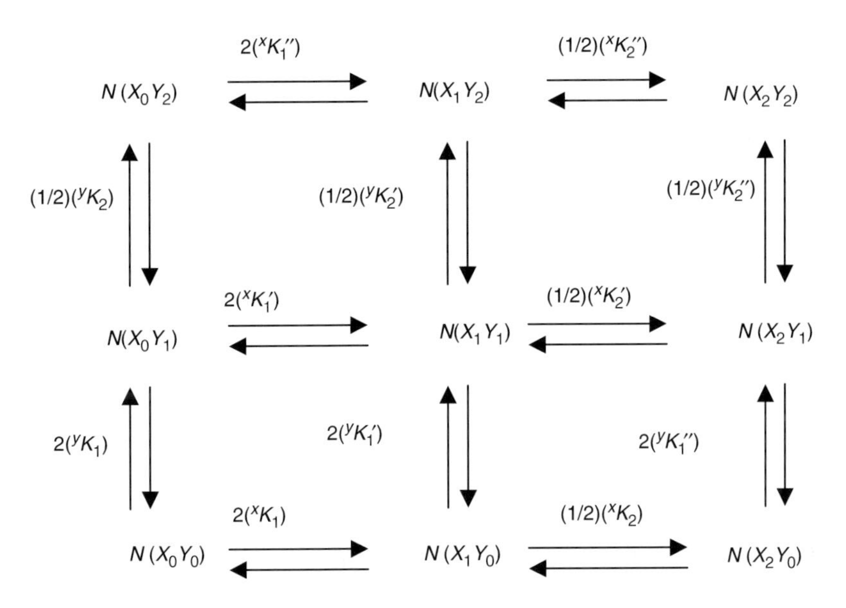


Fig. 13. Formal model of ligand-induced conformational spread in a protein lattice. This model is the formal representation of model of Fig. 10. See text.

and

$$
\begin{aligned}
({}^{y}K_1) &= ({}^{y}K^{*})\frac{\alpha(X_0 Y_0)}{\alpha(X_0 Y_1)} & ({}^{y}K_2) &= ({}^{y}K^{*})\frac{\alpha(X_0 Y_1)}{\alpha(X_0 Y_2)} \\
({}^{y}K'_1) &= ({}^{y}K^{*})\frac{\alpha(X_1 Y_0)}{\alpha(X_1 Y_1)} & ({}^{y}K'_2) &= ({}^{y}K^{*})\frac{\alpha(X_1 Y_1)}{\alpha(X_1 Y_2)} \\
({}^{y}K''_1) &= ({}^{y}K^{\bullet})\frac{\alpha(X_2 Y_0)}{\alpha(X_2 Y_1)} & ({}^{y}K''_2) &= ({}^{y}K^{\bullet})\frac{\alpha(X_2 Y_2)}{\alpha(X_2 Y_1)}
\end{aligned}
\tag{63}
$$

From these relationships one can derive the expressions for the connections between each node, $N(X_i Y_j)$, and the "initial" one, $N(X_0 Y_0)$. One finds

$$
\begin{aligned}
N(X_1 Y_0) &= 2({}^{x}K^{*})\frac{\alpha(X_0 Y_0)}{\alpha(X_1 Y_0)}N(X_0 Y_0) \\
N(X_2 Y_0) &= ({}^{x}K^{\bullet})^2\frac{\alpha(X_0 Y_0)}{\alpha(X_2 Y_0)}N(X_0 Y_0) \\
N(X_0 Y_1) &= 2({}^{y}K^{*})\frac{\alpha(X_0 Y_0)}{\alpha(X_0 Y_1)}N(X_0 Y_0) \\
N(X_1 Y_1) &= 4({}^{y}K^{*})({}^{x}K^{*})\frac{\alpha(X_0 Y_0)}{\alpha(X_1 Y_1)}N(X_0 Y_0) \\
N(X_2 Y_1) &= 2({}^{y}K^{\bullet})({}^{x}K^{\bullet})^2\frac{\alpha(X_0 Y_0)}{\alpha(X_2 Y_1)}N(X_0 Y_0) \\
N(X_0 Y_2) &= ({}^{y}K^{*})^2\frac{\alpha(X_0 Y_0)}{\alpha(X_0 Y_2)}N(X_0 Y_0) \\
N(X_1 Y_2) &= 2({}^{y}K^{*})^2({}^{x}K^{*})\frac{\alpha(X_0 Y_0)}{\alpha(X_1 Y_2)}N(X_0 Y_0) \\
N(X_2 Y_2) &= ({}^{y}K^{*})^2({}^{x}K^{*})^2\frac{\alpha(X_0 Y_0)}{\alpha(X_2 Y_2)}N(X_0 Y_0)
\end{aligned}
\tag{64}
$$

Hence the total "concentration" or "density" of nodes is

$$
\begin{aligned}
\frac{N_T}{N(X_0 Y_0)} = {} & \left\{1 + 2({}^{x}K^{*})\frac{\alpha(X_0 Y_0)}{\alpha(X_1 Y_0)} + ({}^{x}K^{*})^2\frac{\alpha(X_0 Y_0)}{\alpha(X_2 Y_0)}\right\} \\
& + 2({}^{y}K^{*})\left\{\frac{\alpha(X_0 Y_0)}{\alpha(X_0 Y_1)} + 2({}^{x}K^{*})\frac{\alpha(X_0 Y_0)}{\alpha(X_1 Y_1)} + ({}^{x}K^{*})^2\frac{\alpha(X_0 Y_0)}{\alpha(X_2 Y_1)}\right\} \\
& + ({}^{y}K^{*})^2\left\{\frac{\alpha(X_0 Y_0)}{\alpha(X_0 Y_2)} + 2({}^{x}K^{*})\frac{\alpha(X_0 Y_0)}{\alpha(X_1 Y_2)} + ({}^{x}K^{*})^2\frac{\alpha(X_0 Y_0)}{\alpha(X_2 Y_2)}\right\}
\end{aligned}
\tag{65}
$$

*3.2.2. Mutual information of integration and conformational constraints in the lattice*
For reasons that will appear later, it is advantageous to rewrite Eq. (65) in a more compact form. This can be done thanks to matrix notation. Let us define the row vector

$$\Pi_x = \begin{bmatrix} 1 & 2({}^xK^*) & ({}^xK^*)^2 \end{bmatrix} \tag{66}$$

and the three column vectors

$$\Psi(X_i Y_0) = \begin{bmatrix} \dfrac{\alpha(X_0 Y_0)}{\alpha(X_0 Y_0)} \\ \dfrac{\alpha(X_0 Y_0)}{\alpha(X_1 Y_0)} \\ \dfrac{\alpha(X_0 Y_0)}{\alpha(X_2 Y_0)} \end{bmatrix} \quad \Psi(X_i Y_1) = \begin{bmatrix} \dfrac{\alpha(X_0 Y_0)}{\alpha(X_0 Y_1)} \\ \dfrac{\alpha(X_0 Y_0)}{\alpha(X_1 Y_1)} \\ \dfrac{\alpha(X_0 Y_0)}{\alpha(X_2 Y_1)} \end{bmatrix} \quad \Psi(X_i Y_2) = \begin{bmatrix} \dfrac{\alpha(X_0 Y_0)}{\alpha(X_0 Y_2)} \\ \dfrac{\alpha(X_0 Y_0)}{\alpha(X_1 Y_2)} \\ \dfrac{\alpha(X_0 Y_0)}{\alpha(X_2 Y_2)} \end{bmatrix} \tag{67}$$

These expressions are the vectors of conformational constraints, for they express the succession of constraints the system undergoes in going from the initial state $(X_0 Y_0)$ to any final $(X_i Y_j)$ state. With this notation, Eq. (65) becomes

$$\frac{N_T}{N(X_0 Y_0)} = \Pi_x \Psi(X_i Y_0) + 2({}^yK^*)\Pi_x \Psi(X_i Y_1) + ({}^yK^*)^2 \Pi_x \Psi(X_i Y_2) \tag{68}$$

But Eq. (65) can also be rewritten as

$$\frac{N_T}{N(X_0 Y_0)} = \Pi_y \Psi(X_0 Y_j) + 2({}^xK^*)\Pi_y \Psi(X_1 Y_j) + ({}^xK^*)^2 \Pi_y \Psi(X_2 Y_j) \tag{69}$$

where

$$\Pi_y = \begin{bmatrix} 1 & 2({}^yK^*) & ({}^yK^*)^2 \end{bmatrix} \tag{70}$$

and

$$\Psi(X_0 Y_j) = \begin{bmatrix} \dfrac{\alpha(X_0 Y_0)}{\alpha(X_0 Y_0)} \\ \dfrac{\alpha(X_0 Y_0)}{\alpha(X_0 Y_1)} \\ \dfrac{\alpha(X_0 Y_0)}{\alpha(X_0 Y_2)} \end{bmatrix} \quad \Psi(X_1 Y_j) = \begin{bmatrix} \dfrac{\alpha(X_0 Y_0)}{\alpha(X_1 Y_0)} \\ \dfrac{\alpha(X_0 Y_0)}{\alpha(X_1 Y_1)} \\ \dfrac{\alpha(X_0 Y_0)}{\alpha(X_1 Y_2)} \end{bmatrix} \quad \Psi(X_2 Y_j) = \begin{bmatrix} \dfrac{\alpha(X_0 Y_0)}{\alpha(X_2 Y_0)} \\ \dfrac{\alpha(X_0 Y_0)}{\alpha(X_2 Y_1)} \\ \dfrac{\alpha(X_0 Y_0)}{\alpha(X_2 Y_2)} \end{bmatrix} \tag{71}$$

If the ratios of $\alpha$ parameters that appear in the vectors of conformational constraints are all equal to 1 it is then evident that

$$\frac{N_T}{N(X_0 Y_0)} = \{1 + ({}^{y}K^*)\}^2 \{1 + ({}^{x}K^*)\}^2 \tag{72}$$

This reasoning can be extended to the case of a $n \times n$ lattice. For such a lattice one can define the two row vectors

$$\begin{aligned} \Pi_x &= \left[1 \quad n({}^{x}K^*) \quad \{n(n-1)/2\}\,({}^{x}K^*)^2 \ldots ({}^{x}K^*)^n\right] \\ \Pi_y &= \left[1 \quad n({}^{y}K^*) \quad \{n(n-1)/2\}\,({}^{y}K^*)^2 \ldots ({}^{y}K^*)^n\right] \end{aligned} \tag{73}$$

and the column vectors

$$\Psi(X_k Y_j) = \begin{bmatrix} \dfrac{\alpha(X_0 Y_0)}{\alpha(X_k Y_0)} \\ \dfrac{\alpha(X_0 Y_0)}{\alpha(X_k Y_1)} \\ \cdots \\ \dfrac{\alpha(X_0 Y_0)}{\alpha(X_k Y_n)} \end{bmatrix} \qquad \Psi(X_i Y_l) = \begin{bmatrix} \dfrac{\alpha(X_0 Y_0)}{\alpha(X_0 Y_l)} \\ \dfrac{\alpha(X_0 Y_0)}{\alpha(X_1 Y_l)} \\ \cdots \\ \dfrac{\alpha(X_0 Y_0)}{\alpha(X_n Y_l)} \end{bmatrix} \tag{74}$$

where the indices $k$ and $l$ are such that

$$0 \le k \le n \quad \text{and} \quad 0 \le l \le n \tag{75}$$

Equations (68) and (69) then become

$$\begin{aligned} \frac{N_T}{N(X_0 Y_0)} &= \Pi_x \Psi(X_i Y_0) + n({}^{y}K^*)\Pi_x \Psi(X_i Y_1) + \frac{n(n-1)}{2}({}^{y}K^*)^2 \Pi_x \Psi(X_i Y_2) \\ &\quad + \cdots ({}^{x}K^*)^n \Pi_x \Psi(X_i Y_n) \end{aligned} \tag{76}$$

and

$$\begin{aligned} \frac{N_T}{N(X_0 Y_0)} &= \Pi_y \Psi(X_0 Y_j) + n({}^{x}K^*)\Pi_y \Psi(X_1 Y_j) + \frac{n(n-1)}{2}({}^{x}K^*)^2 \Pi_y \Psi(X_2 Y_j) \\ &\quad + \cdots ({}^{x}K^*)^n \Pi_y \Psi(X_n Y_j) \end{aligned} \tag{77}$$

Coming back to the simple model of Fig. 13, one can also express the "density" of nodes in state $(X_1)$, $N(X_1)$, and in state $(X_2)$, $N(X_2)$, as

$$\begin{aligned} \frac{N(X_1)}{N(X_0 Y_0)} &= 2({}^{x}K^*)\Pi_y \Psi(X_1 Y_j) \\ \frac{N(X_2)}{N(X_0 Y_0)} &= ({}^{x}K^*)^2 \Pi_y \Psi(X_2 Y_j) \end{aligned} \tag{78}$$

Similarly one has

$$\frac{N(Y_1)}{N(X_0 Y_0)} = 2({}^y K^*)\Pi_x \Psi(X_i Y_1)$$
$$\frac{N(Y_2)}{N(X_0 Y_0)} = ({}^y K^*)^2 \Pi_x \Psi(X_i Y_2) \tag{79}$$

One can now derive the expression for the probabilities of occurrence of $X_1$, $X_2$, $Y_1$, and $Y_2$. One finds

$$p(X_1) = \frac{2({}^x K^*)\Pi_y \Psi(X_1 Y_j)}{\Pi_y \Psi(X_0 Y_j) + 2({}^x K^*)\Pi_y \Psi(X_1 Y_j) + ({}^x K^*)^2 \Pi_y \Psi(X_2 Y_j)}$$
$$p(X_2) = \frac{({}^x K^*)^2 \Pi_y \Psi(X_2 Y_j)}{\Pi_y \Psi(X_0 Y_j) + 2({}^x K^*)\Pi_y \Psi(X_1 Y_j) + ({}^x K^*)^2 \Pi_y \Psi(X_2 Y_j)} \tag{80}$$

and

$$p(Y_1) = \frac{2({}^y K^*)\Pi_x \Psi(X_i Y_1)}{\Pi_x \Psi(X_i Y_0) + 2({}^y K^*)\Pi_x \Psi(X_i Y_1) + ({}^y K^*)^2 \Pi_x \Psi(X_i Y_2)}$$
$$p(Y_2) = \frac{({}^y K^*)^2 \Pi_x \Psi(X_i Y_2)}{\Pi_x \Psi(X_i Y_0) + 2({}^y K^*)\Pi_x \Psi(X_i Y_1) + ({}^y K^*)^2 \Pi_x \Psi(X_i Y_2)} \tag{81}$$

It becomes obvious from these expressions that $p(X_1)$, $p(X_2)$, $p(Y_1)$, and $p(Y_2)$ depend on both $X$ and $Y$. As already pointed out (Chapter 5), this implies that the protein lattice possesses mutual information of integration. If, however, all the ratios of $\alpha$ parameters that appear in the vectors of conformational constraints are all equal to 1, then the expressions of $p(X_1)$, $p(X_2)$, $p(Y_1)$, and $p(Y_2)$ reduce to

$$p(X_1) = \frac{2({}^x K^*)}{\{1 + ({}^x K^*)\}^2}$$
$$p(X_2) = \frac{({}^x K^*)^2}{\{1 + ({}^x K^*)\}^2}$$
$$p(Y_1) = \frac{2({}^y K^*)}{\{1 + ({}^y K^*)\}^2}$$
$$p(Y_2) = \frac{({}^y K^*)^2}{\{1 + ({}^y K^*)\}^2} \tag{82}$$

It then appears that under these specific conditions $p(X_1)$ and $p(X_2)$ depend solely on $X$ whereas $p(Y_1)$ and $p(Y_2)$ depend on $Y$ but not on $X$. As we have already pointed out (Chapter 5), this situation prevents any mutual information to be in the lattice. The important conclusion that can be derived from these results is that mutual information of integration of a lattice stems from the variations of conformational constraints between protein molecules of the lattice. This conclusion is quite general and applies to any kind of square lattice.

*3.2.3. Integration and emergence in a protein lattice*

The nodes of the lattice that bear both $X$ and $Y$ carry information of integration. In order to determine whether node information is positive, negative, or nil, one has to derive the expression for the joint probabilities $p(X_1 Y_1)$, $p(X_1 Y_2)$, $p(X_2 Y_1)$, and $p(X_2 Y_2)$. One finds

$$
\begin{aligned}
p(X_1 Y_1) &= \frac{4({}^{y}K^{*})({}^{x}K^{*})\alpha(X_0 Y_0)/\alpha(X_1 Y_1)}{\Pi_y \Psi(X_0 Y_j) + 2({}^{x}K^{*})\Pi_y \Psi(X_1 Y_j) + ({}^{x}K^{*})^2 \Pi_y \Psi(X_2 Y_j)} \\
p(X_1 Y_2) &= \frac{2({}^{y}K^{*})^2({}^{x}K^{*})\alpha(X_0 Y_0)/\alpha(X_1 Y_2)}{\Pi_y \Psi(X_0 Y_j) + 2({}^{x}K^{*})\Pi_y \Psi(X_1 Y_j) + ({}^{x}K^{*})^2 \Pi_y \Psi(X_2 Y_j)} \\
p(X_2 Y_1) &= \frac{2({}^{y}K^{*})({}^{x}K^{*})^2\alpha(X_0 Y_0)/\alpha(X_2 Y_1)}{\Pi_x \Psi(X_i Y_0) + 2({}^{y}K^{*})\Pi_x \Psi(X_i Y_1) + ({}^{y}K^{*})^2 \Pi_x \Psi(X_i Y_2)} \\
p(X_2 Y_2) &= \frac{({}^{y}K^{\bullet})^2({}^{x}K^{\bullet})^2\alpha(X_0 Y_0)/\alpha(X_2 Y_2)}{\Pi_x \Psi(X_i Y_0) + 2({}^{y}K^{*})\Pi_x \Psi(X_i Y_1) + ({}^{y}K^{*})^2 \Pi_x \Psi(X_i Y_2)}
\end{aligned}
\tag{83}
$$

From Eqs. (80), (81), and (83) one can find out the general conditions that define the sign of $I(X_1 : Y_1)$, $I(X_1 : Y_2)$, $I(X_2 : Y_1)$, and $I(X_2 : Y_2)$. The mutual information of integration of these nodes will be positive if

$$
\begin{aligned}
&\frac{\alpha(X_0 Y_0)}{\alpha(X_1 Y_1)}\left[\Pi_y \Psi(X_0 Y_j) + 2({}^{x}K^{*})\Pi_y \Psi(X_1 Y_j) + ({}^{x}K^{*})^2 \Pi_y \Psi(X_2 Y_j)\right] \\
&\quad > \Pi_y \Psi(X_1 Y_j)\Pi_x \Psi(X_i Y_1) \\
&\frac{\alpha(X_0 Y_0)}{\alpha(X_1 Y_2)}\left[\Pi_y \Psi(X_0 Y_j) + 2({}^{x}K^{*})\Pi_y \Psi(X_1 Y_j) + ({}^{x}K^{*})^2 \Pi_y \Psi(X_2 Y_j)\right] \\
&\quad > \Pi_y \Psi(X_1 Y_j)\Pi_x \Psi(X_i Y_2) \\
&\frac{\alpha(X_0 Y_0)}{\alpha(X_2 Y_1)}\left[\Pi_x \Psi(X_i Y_0) + 2({}^{y}K^{*})\Pi_x \Psi(X_i Y_1) + ({}^{y}K^{*})^2 \Pi_x \Psi(X_i Y_2)\right] \\
&\quad > \Pi_y \Psi(X_2 Y_j)\Pi_x \Psi(X_i Y_1) \\
&\frac{\alpha(X_0 Y_0)}{\alpha(X_2 Y_2)}\left[\Pi_x \Psi(X_i Y_0) + 2({}^{y}K^{*})\Pi_x \Psi(X_i Y_1) + ({}^{y}K^{*})^2 \Pi_x \Psi(X_i Y_2)\right] \\
&\quad > \Pi_y \Psi(X_2 Y_j)\Pi_x \Psi(X_i Y_2)
\end{aligned}
\tag{84}
$$

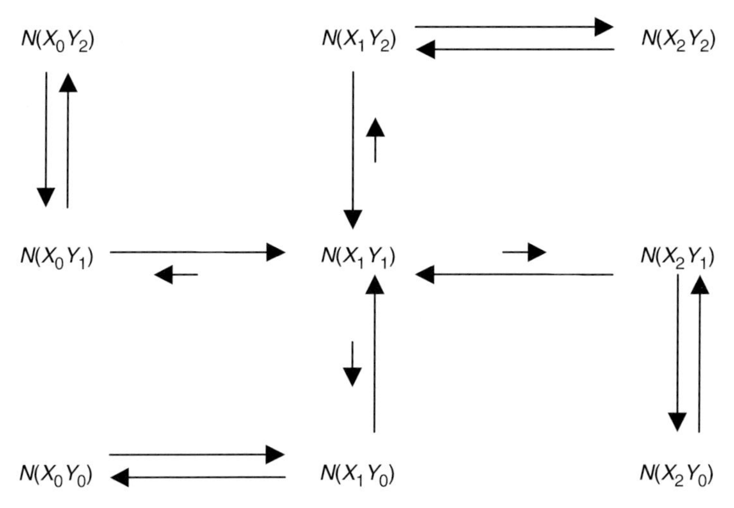


Fig. 14. Reduction of the formal model of Fig. 13 that generates expressions (85). A "short" arrow associated with a "long" one indicates that the corresponding equilibrium is shifted in the direction of the "long" arrow. See text.

and nil or negative otherwise. In the case of the model lattice of Fig. 13, one can show after some algebra that the first expression (84) is fulfilled, i.e., $I(X_1 : Y_1) > 0$, if

$$\begin{aligned} &\frac{\alpha(X_0 Y_1)}{\alpha(X_1 Y_1)} > \frac{\alpha(X_0 Y_0)}{\alpha(X_1 Y_0)} \qquad \frac{\alpha(X_1 Y_0)}{\alpha(X_1 Y_1)} > \frac{\alpha(X_2 Y_0)}{\alpha(X_2 Y_1)} \\ &\frac{\alpha(X_1 Y_2)}{\alpha(X_1 Y_1)} > \frac{\alpha(X_0 Y_2)}{\alpha(X_0 Y_1)} \qquad \frac{\alpha(X_1 Y_2)}{\alpha(X_1 Y_1)} > \frac{\alpha(X_2 Y_2)}{\alpha(X_2 Y_1)} \end{aligned} \tag{85}$$

As shown in Fig. 14, these conditions, which are in fact sufficient conditions, all imply that the probability of occurrence of the node $N(X_1 Y_1)$ (or the "concentration" of the $X_1 Y_1$ species) is as large as possible.

Similarly one can expect that $I(X_1 : Y_2) > 0$ if

$$\begin{aligned} &\frac{\alpha(X_1 Y_0)}{\alpha(X_1 Y_2)} > \frac{\alpha(X_0 Y_0)}{\alpha(X_0 Y_2)} \qquad \frac{\alpha(X_1 Y_0)}{\alpha(X_1 Y_2)} > \frac{\alpha(X_2 Y_0)}{\alpha(X_2 Y_2)} \\ &\frac{\alpha(X_0 Y_2)}{\alpha(X_1 Y_2)} > \frac{\alpha(X_0 Y_1)}{\alpha(X_1 Y_1)} \qquad \frac{\alpha(X_2 Y_2)}{\alpha(X_1 Y_2)} > \frac{\alpha(X_2 Y_1)}{\alpha(X_1 Y_1)} \end{aligned} \tag{86}$$

As shown in Fig. 16, these inequalities imply that the steps leading to the node $N(X_1 Y_2)$ are faster than the processes that leave this node. Moreover comparison of Eqs. (85) and (86) and inspection of Figs. 14 and 15 show that the mathematical conditions that generate $I(X_1 : Y_1) > 0$ and $I(X_1 : Y_2) > 0$ are antagonistic, i.e., a high value of $I(X_1 : Y_1)$ results in a small value of $I(X_1 : Y_2)$, and conversely.

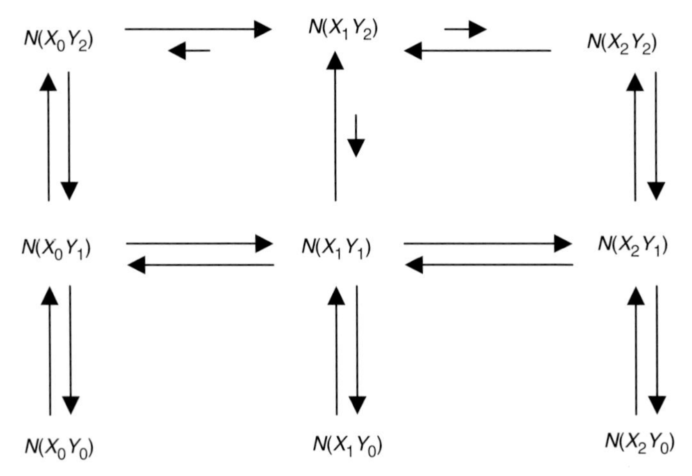


Fig. 15. Reduction of the formal model of Fig. 13 that generates expressions (86). A "short" arrow associated with a "long" one indicates that the corresponding equilibrium is shifted in the direction of the "long" arrow. See text.

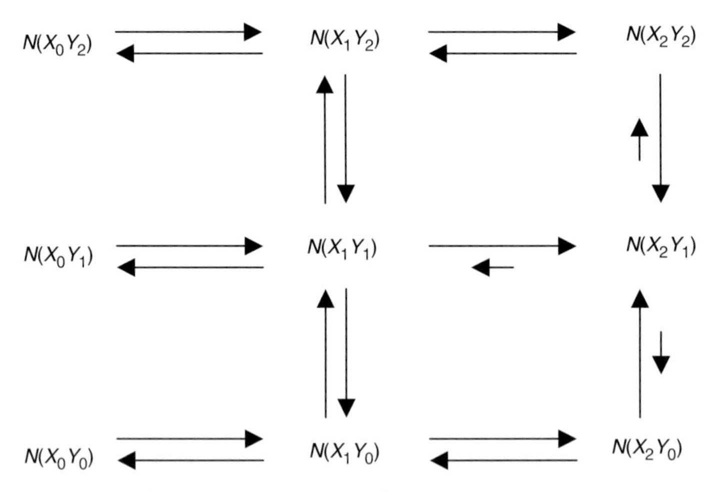


Fig. 16. Reduction of the formal model of Fig. 13 that generates expressions (87). A "short" arrow associated with a "long" one indicates that the corresponding equilibrium is shifted in the direction of the "long" arrow. See text.

In the same way, the conditions that generate $I(X_2 : Y_1)>0$ are

$$\begin{array}{ll} \frac{\alpha(X_0 Y_1)}{\alpha(X_2 Y_1)} > \frac{\alpha(X_0 Y_0)}{\alpha(X_2 Y_0)} & \frac{\alpha(X_0 Y_1)}{\alpha(X_2 Y_1)} > \frac{\alpha(X_0 Y_2)}{\alpha(X_2 Y_2)} \\ \frac{\alpha(X_1 Y_1)}{\alpha(X_2 Y_1)} > \frac{\alpha(X_1 Y_0)}{\alpha(X_2 Y_0)} & \frac{\alpha(X_1 Y_1)}{\alpha(X_2 Y_1)} > \frac{\alpha(X_1 Y_2)}{\alpha(X_2 Y_2)} \end{array} \tag{87}$$

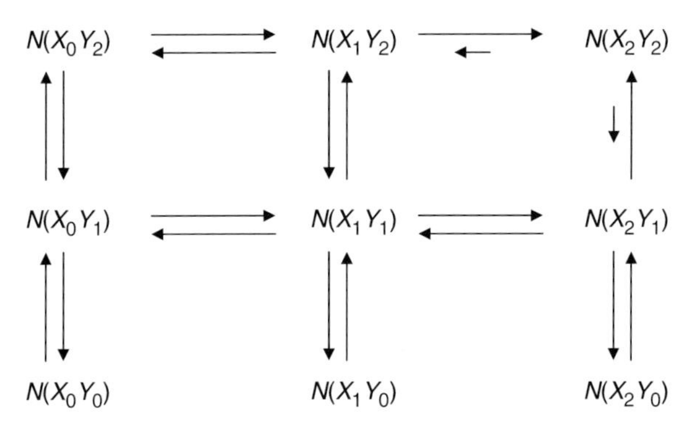


Fig. 17. Reduction of the formal model of Fig. 13 that generates expressions (88). A "short" arrow associated with a "long" one indicates that the corresponding equilibrium is shifted in the direction of the "long" arrow. See text.

As previously, some of these relationships are incompatible with several expressions in (85). Relationships (85) imply that the probability of occurrence of node $N(X_2 Y_1)$ is high, leading in turn to $I(X_2 : Y_1) > 0$ (Fig. 16).

The last possibility is obtained if

$$\frac{\alpha(X_2 Y_0)}{\alpha(X_2 Y_2)} > \frac{\alpha(X_0 Y_0)}{\alpha(X_0 Y_2)} \qquad \frac{\alpha(X_2 Y_0)}{\alpha(X_2 Y_2)} > \frac{\alpha(X_1 Y_0)}{\alpha(X_1 Y_2)}$$
$$\frac{\alpha(X_0 Y_2)}{\alpha(X_2 Y_2)} > \frac{\alpha(X_0 Y_1)}{\alpha(X_2 Y_1)} \qquad \frac{\alpha(X_2 Y_1)}{\alpha(X_2 Y_2)} > \frac{\alpha(X_1 Y_1)}{\alpha(X_1 Y_2)} \tag{88}$$

Again, some of these relationships are incompatible with inequalities (86) and (87), but quite compatible with relations (85) (Fig. 17). More generally speaking, directly connected nodes such as $N(X_2 Y_2)$ and $N(X_1 Y_2)$, $N(X_2 Y_2)$ and $N(X_2 Y_1)$, $N(X_1 Y_1)$ and $N(X_1 Y_2)$, $N(X_1 Y_1)$ and $N(X_2 Y_1)$, cannot both have a high probability of occurrence. Conversely, indirectly connected and "symmetrical" nodes, such as $N(X_1 Y_1)$ and $N(X_2 Y_2)$, can.

We have seen previously (Chapters 5–8) that probably the most interesting situation is that of emergence which implies that $I(X_i : Y_j) < 0$. We have already shown that this situation implies that the corresponding nodes possess a high energy level. As it is impossible that all the nodes of a lattice bearing both $X$ and $Y$ possess a high or, conversely, a low mutual information of integration, one can expect from relations (85)–(88) the existence of a rugged "information landscape" associated with both positive and negative values of $I(X_i : Y_j)$. The information landscape can be obtained by plotting the value of $-I(X_i : Y_j)$ as a function of $X_i$ and $Y_j$ (Fig. 18). It consists of ranges of "mountains" separated by "valleys". All the nodes located in "mountains" have a low probability of occurrence owing to high values of rate constants originating from these nodes and to low values of rate constants leading to these nodes. The opposite situation should hold for the nodes located in the "valleys".

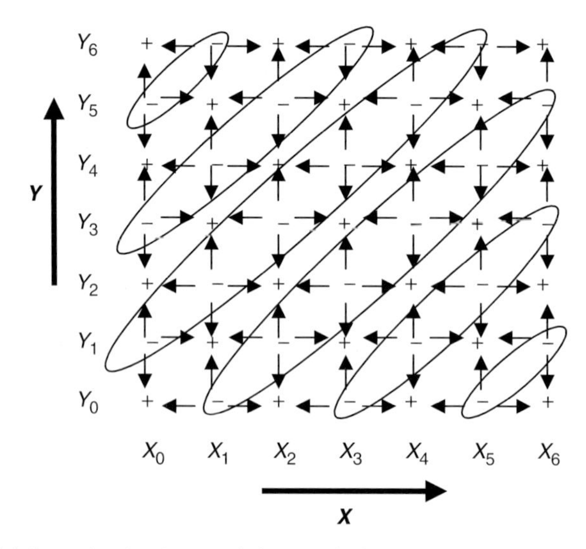


Fig. 18. A rugged information landscape of the protein lattice. The landscape of the protein lattice displays a succession of mountains and valleys. Ranges of mountains are regions of the information landscape that display emergence of information, i.e., negative values of $I(X:Y)$. They are represented in the figure by ellipses containing negative signs, i.e., the sign of $I(X:Y)$. Three or four arrows leave each of these negative signs indicating the preferential direction of the evolution of the system. These arrows lead to positive signs that can be considered attractors and which are located in the valleys. Hence valleys are regions of integration of the system.

The parallel "ranges of mountains" are regions of the "information landscape" that display emergence whereas the valleys display integration effects. This kind of rugged "information landscape" is not the only type to be expected. If all the backward and forward rate constants connecting the nodes have rather similar values the "landscape" of the protein lattice will be fairly smooth and will not display this succession of "mountains" and "valleys". The condition that any node $N(X_i Y_j)$ be located in a "mountain" of the "information landscape" is

$$\frac{\alpha(X_{i-1}Y_j)}{\alpha(X_i Y_j)} < \frac{\alpha(X_i Y_j)}{\alpha(X_{i+1}Y_j)} > \frac{\alpha(X_{i+1}Y_j)}{\alpha(X_{i+2}Y_j)} \qquad (i, j \in N) \tag{89}$$

This relation implies that for any fixed odd value of $j$ one has in the case of Fig. 18

$$\frac{\alpha(X_0 Y_j)}{\alpha(X_1 Y_j)} > \frac{\alpha(X_1 Y_j)}{\alpha(X_2 Y_j)} < \frac{\alpha(X_2 Y_j)}{\alpha(X_3 Y_j)} > \frac{\alpha(X_3 Y_j)}{\alpha(X_4 Y_j)} < \frac{\alpha(X_4 Y_j)}{\alpha(X_5 Y_j)} > \cdots \tag{90}$$

Conversely, for any fixed even value of $j$ one has for the same figure

$$\frac{\alpha(X_0 Y_j)}{\alpha(X_1 Y_j)} < \frac{\alpha(X_1 Y_j)}{\alpha(X_2 Y_j)} > \frac{\alpha(X_2 Y_j)}{\alpha(X_3 Y_j)} < \frac{\alpha(X_3 Y_j)}{\alpha(X_4 Y_j)} > \frac{\alpha(X_4 Y_j)}{\alpha(X_5 Y_j)} < \cdots \tag{91}$$

Relations (89)–(91) are defined along the $X_i$ axis. Similar relationships could be derived along the $Y_j$ axis. The interesting general idea of these theoretical considerations is that the information surface of a protein lattice can display emergent effects only if it possesses "mountains" separated by "valleys". One cannot expect emergence to come out of a smooth surface.

### *3.3. Conformational changes in quasi-linear lattices*

We have considered so far a square lattice. However, many protein edifices such as actin filaments, tubulin, etc.... can be considered, at least to a first approximation, quasi-linear protein lattices. We have also assumed, in the previous theoretical model, that conformational spread occurred as the consequence of the binding of two ligands $X$ and $Y$ to the protein edifice in a sort of an induced-fit process. But conformation change can also take place as a spontaneous process that propagates along the polymer [1].

#### *3.3.1. The basic unit of conformation change*

We are now assuming, as already done previously, that the quasi-linear lattice is made up of identical functional units each comprising two different dimeric proteins (Fig. 19). One of the dimeric proteins can take either the $A_0$ or $A_1$ conformation and therefore undergoes some kind of allosteric transition. Likewise the other dimeric protein can be considered allosteric for it exists in either of the two conformations $B_0$ or $B_1$.

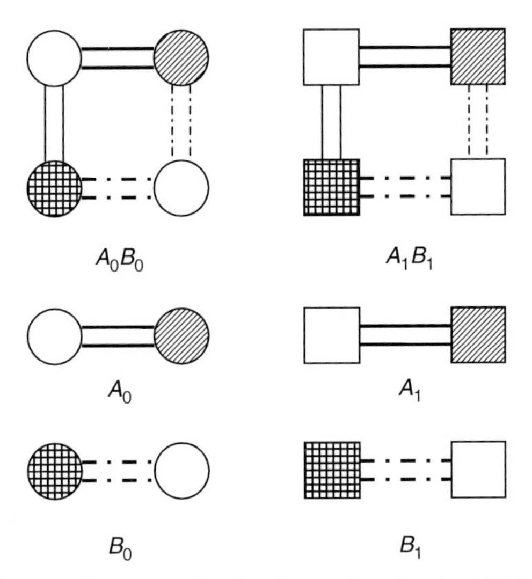


Fig. 19. Conformational states of two proteins that form a functional unit. Here it is assumed that the dimeric proteins $A$ and $B$ can spontaneously (in the absence of any ligand) undergo a conformational transition. $A_0, A_1, B_0, B_1$ are these conformational states. As previously, the functional unit is made up of the association of the two proteins. The states $A_0B_0$ and $A_1B_1$ are shown in the figure. The continuous lines indicates isologous interactions whereas dotted lines are heterologous interactions between subunits.

If $K_a^*$ and $K_b^*$ are the equilibrium constants of the conformational transitions of the two ideally isolated dimeric proteins, $K_a$ and $K_b$ the equilibrium constants of the transconformation of the functional unit, one has (Fig. 20)

$$
\begin{aligned}
K_a &= K_a^* \frac{\alpha(A_0 B_0)}{\alpha(A_1 B_0)} \\
K_b &= K_b^* \frac{\alpha(A_0 B_0)}{\alpha(A_0 B_1)}
\end{aligned}
\tag{92}
$$

As discussed previously, the $\alpha$ coefficients derive from the energy functions $U(A_0B_0)$, $U(A_1B_0)$, and $U(A_0B_1)$ through the relationships

$$
\begin{aligned}
\alpha(A_0B_0) &= \exp\{-U(A_0B_0)/RT\} \\
\alpha(A_1B_0) &= \exp\{-U(A_1B_0)/RT\} \\
\alpha(A_0B_1) &= \exp\{-U(A_0B_1)/RT\}
\end{aligned}
\tag{93}
$$

In fact, the $\alpha$ coefficients express how the structure of the functional unit facilitates or hinders the conformational transitions of the unit.

### 3.3.2. *Thermodynamics of spontaneous conformational transitions in a simple quasi-linear protein lattice*

Let us consider the ideally simple quasi-linear protein lattice made up of two functional units (Fig. 21). In this model we assume that conformation change starts at a fixed defined functional unit and propagates to the other only when the two dimeric proteins have shifted from states $A_0$ and $B_0$ to states $A_1$ and $B_1$. Hence the model of conformational spread can be depicted, as shown in Figs. 21 and 22. One will notice that the assumption made above implies that the topology of the network is different from that considered so far and this in turn implies that the statistical factors assume values different from those in the previous model (Fig. 13).

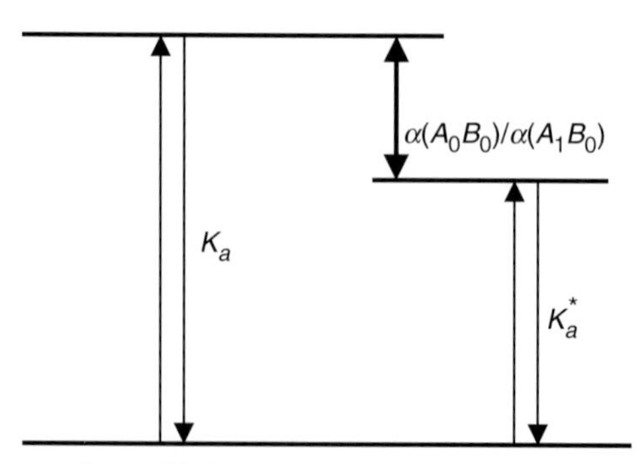


Fig. 20. Relationship between the equilibrium constants of a conformation change for the same protein, naked or inserted in the functional unit. $K_a^*$ is the equilibrium constant of the conformation change for the naked protein, $K_a$ the equilibrium constant of the same process for the same protein inserted in the functional unit. See text.

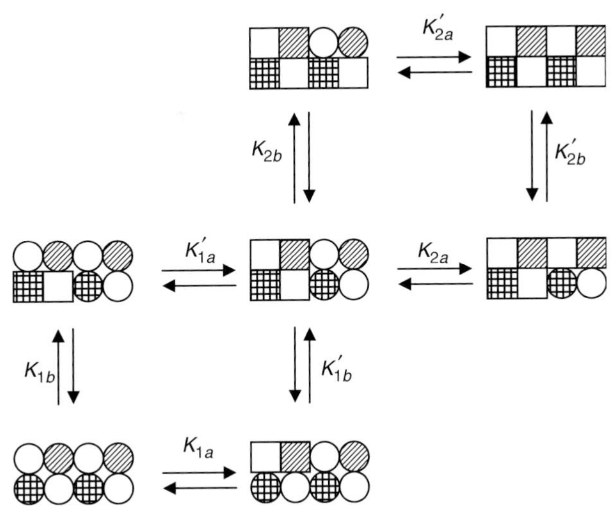


Fig. 21. A simple model for a conformational spread in a quasi-linear lattice. The lattice is made up of two functional units. Each unit consists of two dimeric proteins. The system is defined as quasi-linear because the proteins of a unit undergo a conformational transition only when the proteins of the adjacent unit have already changed their conformation. Then, contrary to the previous square lattice (Fig. 10), the statistical factors associated with equilibrium constants are all equal to one.

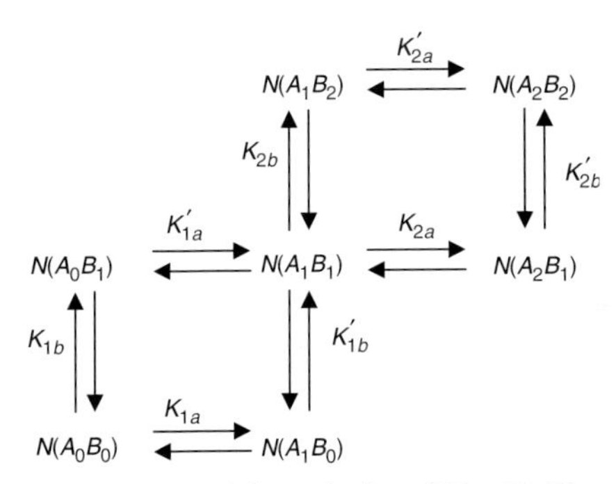


Fig. 22. A formal representation of the quasi-linear lattice of Fig. 21. The states $A_0B_2$ and $A_2B_0$ do not exist. See text.

The equilibrium constants involved in this network takes the form

$$
\begin{aligned}
K_{1a} &= K_a^* \frac{\alpha(A_0B_0)}{\alpha(A_1B_0)} \qquad & K'_{1a} &= K_a^* \frac{\alpha(A_0B_1)}{\alpha(A_1B_1)} \\
K_{2a} &= K_a^* \frac{\alpha(A_1B_1)}{\alpha(A_2B_1)} \qquad & K'_{2a} &= K_a^* \frac{\alpha(A_1B_2)}{\alpha(A_2B_2)} \\
K_{1b} &= K_b^* \frac{\alpha(A_0B_0)}{\alpha(A_0B_1)} \qquad & K'_{1b} &= K_b^* \frac{\alpha(A_1B_0)}{\alpha(A_1B_1)} \\
K_{2b} &= K_b^* \frac{\alpha(A_1B_1)}{\alpha(A_1B_2)} \qquad & K'_{2b} &= K_b^* \frac{\alpha(A_1B_1)}{\alpha(A_1B_2)}
\end{aligned}
\tag{94}
$$

As previously, the node "concentrations" or "densities" can be expressed as

$$
\begin{aligned}
N(A_1B_0) &= K_a^* \frac{\alpha(A_0B_0)}{\alpha(A_1B_0)} N(A_0B_0) \\
N(A_0B_1) &= K_b^* \frac{\alpha(A_0B_0)}{\alpha(A_0B_1)} N(A_0B_0) \\
N(A_1B_1) &= K_a^* K_b^* \frac{\alpha(A_0B_0)}{\alpha(A_1B_1)} N(A_0B_0) \\
N(A_1B_2) &= K_a^* K_b^{*2} \frac{\alpha(A_0B_0)}{\alpha(A_1B_2)} N(A_0B_0) \\
N(A_2B_1) &= K_a^{*2} K_b^* \frac{\alpha(A_0B_0)}{\alpha(A_2B_1)} N(A_0B_0) \\
N(A_2B_2) &= K_a^{*2} K_b^{*2} \frac{\alpha(A_0B_0)}{\alpha(A_2B_2)} N(A_0B_0)
\end{aligned} \tag{95}
$$

and the total "density" or "concentration" of nodes can be expressed as

$$
\begin{aligned}
\frac{N_T}{N(A_0B_0)} = 1 &+ K_a^* \frac{\alpha(A_0B_0)}{\alpha(A_1B_0)} + K_b^* \frac{\alpha(A_0B_0)}{\alpha(A_0B_1)} + K_a^* K_b^* \frac{\alpha(A_0B_0)}{\alpha(A_1B_1)} + K_a^{*2} K_b^* \frac{\alpha(A_0B_0)}{\alpha(A_2B_1)} \\
&+ K_a^* K_b^{*2} \frac{\alpha(A_0B_0)}{\alpha(A_1B_2)} + K_a^{*2} K_b^{*2} \frac{\alpha(A_0B_0)}{\alpha(A_2B_2)}
\end{aligned} \tag{96}
$$

If all the $\alpha$'s are equal to 1, or if their ratios are equal to 1, Eq. (96) above reduces to

$$
\frac{N_T}{N(A_0B_0)} = 1 + (1 + K_a^* K_b^*)(K_a^* + K_b^* + K_a^* K_b^*) \tag{97}
$$

*3.3.3. Mutual information of integration and conformation changes*

For reasons that have been already referred to, it is advantageous to express Eq. (96) in a slightly different form. Defining, as previously, the row vector

$$
\Pi_r = \begin{bmatrix} K_a^* & K_b^* & K_a^* K_b^* \end{bmatrix} \tag{98}
$$

and the column vectors

$$
\Psi_1(AB) = \begin{bmatrix} \dfrac{\alpha(A_0B_0)}{\alpha(A_1B_0)} \\[2ex] \dfrac{\alpha(A_0B_0)}{\alpha(A_0B_1)} \\[2ex] \dfrac{\alpha(A_0B_0)}{\alpha(A_1B_1)} \end{bmatrix} \qquad \Psi_2(AB) = \begin{bmatrix} \dfrac{\alpha(A_0B_0)}{\alpha(A_2B_1)} \\[2ex] \dfrac{\alpha(A_0B_0)}{\alpha(A_1B_2)} \\[2ex] \dfrac{\alpha(A_0B_0)}{\alpha(A_2B_2)} \end{bmatrix} \tag{99}
$$

Equation (96) can be rewritten in more condensed form as

$$\frac{N_T}{N(A_0B_0)} = 1 + \Pi_r\Psi_1(AB) + K_a^*K_b^*\Pi_r\Psi_2(AB) \tag{100}$$

One can check that, if the ratios of the $\alpha$'s in the column vectors are all equal to one, Eq. (100) reduces to expression (97). The elements of the row vector $\Pi_r$ describe the intrinsic steps of the conformational changes of $A_0$ into $A_1$ and $B_0$ into $B_1$ upon disregarding the influence of the overall organization of the quasi-linear lattice, i.e., on the assumption that there is no conformational constraint between the proteins that constitute the lattice. Conversely, the two vectors $\Psi_1(AB)$ and $\Psi_2(AB)$ express how the general organization of the quasi-linear network and the conformational constraints between proteins affect the conformational spread within the protein edifice so as to generate two *different* cycles I and II in the model of Fig. 21. Put in other words, this means that the vector $\Pi_r$ expresses the intrinsic properties of a "functional unit" whereas vectors $\Psi_1(AB)$ and $\Psi_2(AB)$ show how the global organization of the supramolecular edifice affects these intrinsic properties. Expressions (99) and (100) can be easily extended to the situation of a quasi-linear network of any length. One then has

$$\frac{N_T}{N(A_0B_0)} = 1 + \Pi_r\Psi_1(AB) + K_a^*K_b^*\Pi_r\Psi_2(AB) + \cdots + K_a^{*i-1}K_b^{*i-1}\Pi_r\Psi_i(AB) + \cdots \tag{101}$$

with

$$\Psi_i(AB) = \begin{bmatrix} \dfrac{\alpha(A_0B_0)}{\alpha(A_iB_{i-1})} \\ \dfrac{\alpha(A_0B_0)}{\alpha(A_{i-1}B_i)} \\ \dfrac{\alpha(A_0B_0)}{\alpha(A_iB_i)} \end{bmatrix} \tag{102}$$

One can define two functions $N(A_1)$ and $N(A_2)$ as

$$\begin{aligned} \frac{N(A_1)}{N(A_0B_0)} &= \frac{N(A_1B_0) + N(A_1B_1) + N(A_1B_2)}{N(A_0B_0)} \\ \frac{N(A_2)}{N(A_0B_0)} &= \frac{N(A_2B_1) + N(A_2B_2)}{N(A_0B_0)} \end{aligned} \tag{103}$$

As previously mentioned, it is useful to express these functions in matrix notation. One can define the two row vectors

$$\begin{aligned} \Pi(A_1) &= \begin{bmatrix} 1 & K_b^* & K_b^{*2} \end{bmatrix} \\ \Pi(A_2) &= \begin{bmatrix} 1 & K_b^* \end{bmatrix} \end{aligned} \tag{104}$$

and the two column vectors

$$\Psi(A_1B) = \begin{bmatrix} \dfrac{\alpha(A_0B_0)}{\alpha(A_1B_0)} \\[2ex] \dfrac{\alpha(A_0B_0)}{\alpha(A_1B_1)} \\[2ex] \dfrac{\alpha(A_0B_0)}{\alpha(A_1B_2)} \end{bmatrix} \qquad \Psi(A_2B) = \begin{bmatrix} \dfrac{\alpha(A_0B_0)}{\alpha(A_2B_1)} \\[2ex] \dfrac{\alpha(A_0B_0)}{\alpha(A_2B_2)} \end{bmatrix} \tag{105}$$

Therefore Eqs. (103) assume the following form

$$\begin{aligned} \frac{N(A_1)}{N(A_0B_0)} &= K_a^* \Pi(A_1)\Psi(A_1B) \\ \frac{N(A_2)}{N(A_0B_0)} &= K_a^{*2} K_b^* \Pi(A_2)\Psi(A_2B) \end{aligned} \tag{106}$$

Taking advantage of Eqs. (100) and (106) one can easily obtain the probabilities of occurrence of the nodes associated with the states $A_1$ or $A_2$. One finds

$$\begin{aligned} p(A_1) &= \frac{K_a^* \Pi(A_1)\Psi(A_1B)}{1 + \Pi_r \Psi_1(AB) + K_a^* K_b^* \Pi_r \Psi_2(AB)} \\ p(A_2) &= \frac{K_a^{*2} K_b^* \Pi(A_2)\Psi(A_2B)}{1 + \Pi_r \Psi_1(AB) + K_a^* K_b^* \Pi_r \Psi_2(AB)} \end{aligned} \tag{107}$$

In the same way, one can obtain the expressions for $N(B_1)/N(A_0B_0)$, $N(B_2)/N(A_0B_0)$, $p(B_1)$, and $p(B_2)$. Setting the two row vectors

$$\begin{aligned} \Pi(B_1) &= \begin{bmatrix} 1 & K_a^* & K_a^{*2} \end{bmatrix} \\ \Pi(B_2) &= \begin{bmatrix} 1 & K_a^* \end{bmatrix} \end{aligned} \tag{108}$$

and the two column vectors

$$\Psi(AB_1) = \begin{bmatrix} \dfrac{\alpha(A_0B_0)}{\alpha(A_0B_1)} \\[2ex] \dfrac{\alpha(A_0B_0)}{\alpha(A_1B_1)} \\[2ex] \dfrac{\alpha(A_0B_0)}{\alpha(A_2B_1)} \end{bmatrix} \qquad \Psi(AB_2) = \begin{bmatrix} \dfrac{\alpha(A_0B_0)}{\alpha(A_1B_2)} \\[2ex] \dfrac{\alpha(A_0B_0)}{\alpha(A_2B_2)} \end{bmatrix} \tag{109}$$

one finds

$$\frac{N(B_1)}{N(A_0B_0)} = K_b^*\Pi(B_1)\Psi(AB_1)$$
$$\frac{N(B_2)}{N(A_0B_0)} = K_a^*K_b^{*2}\Pi(B_2)\Psi(AB_2) \tag{110}$$

and therefore

$$p(B_1) = \frac{K_b^*\Pi(B_1)\Psi(AB_1)}{1 + \Pi_r\Psi_1(AB) + K_a^*K_b^*\Pi_r\Psi_2(AB)}$$
$$p(B_2) = \frac{K_a^*K_b^{*2}\Pi(B_2)\Psi(AB_2)}{1 + \Pi_r\Psi_1(AB) + K_a^*K_b^*\Pi_r\Psi_2(AB)} \tag{111}$$

The vectors $\Pi(A_1)$ and $\Pi(A_2)$ describe in a quantitative manner the succession of steps involved in the conversion of the protein states that carry either one or two $A$ conformations. Similarly, $\Pi(B_1)$ and $\Pi(B_2)$ offer a quantitative description of the succession of states bearing either one or two $B$ conformations (Fig. 23). In either situation, it is postulated that the organization of the protein edifice, i.e., the physical constraints that exist between the elements of the multi-molecular structure, does not alter the partition of the various states.

The vectors $\Psi(A_1B)$ and $\Psi(A_2B)$ express how the conformational constraints between the proteins of the supramolecular edifice alter the partition of the protein states bearing either one or two $A$ conformations. Similarly, the vectors $\Psi(AB_1)$ and $\Psi(AB_2)$ give a quantitative description of how conformational constraints in the edifice change the partition of the protein states bearing either one or two $B$ conformations. Figure 23 illustrates the physical significance of $\Pi_r$, $\Pi(A_1)$, $\Pi(A_2)$, $\Psi_1(AB)$, $\Psi(A_1B)$ and $\Psi(A_2B)$.

One can also define joint probabilities $p(A_1B_1)$, $p(A_1B_2)$, $p(A_2B_1)$, and $p(A_2B_2)$ in the usual way. One finds

$$p(A_1B_1) = \frac{K_a^*K_b^*\alpha(A_0B_0)/\alpha(A_1B_1)}{1 + \Pi_r\Psi_1(AB) + K_a^*K_b^*\Pi_r\Psi_2(AB)}$$
$$p(A_1B_2) = \frac{K_a^*K_b^{*2}\alpha(A_0B_0)/\alpha(A_1B_2)}{1 + \Pi_r\Psi_1(AB) + K_a^*K_b^*\Pi_r\Psi_2(AB)}$$
$$p(A_2B_1) = \frac{K_a^{*2}K_b^*\alpha(A_0B_0)/\alpha(A_2B_1)}{1 + \Pi_r\Psi_1(AB) + K_a^*K_b^*\Pi_r\Psi_2(AB)} \tag{112}$$
$$p(A_2B_2) = \frac{K_a^{*2}K_b^{*2}\alpha(A_0B_0)/\alpha(A_2B_2)}{1 + \Pi_r\Psi_1(AB) + K_a^*K_b^*\Pi_r\Psi_2(AB)}$$

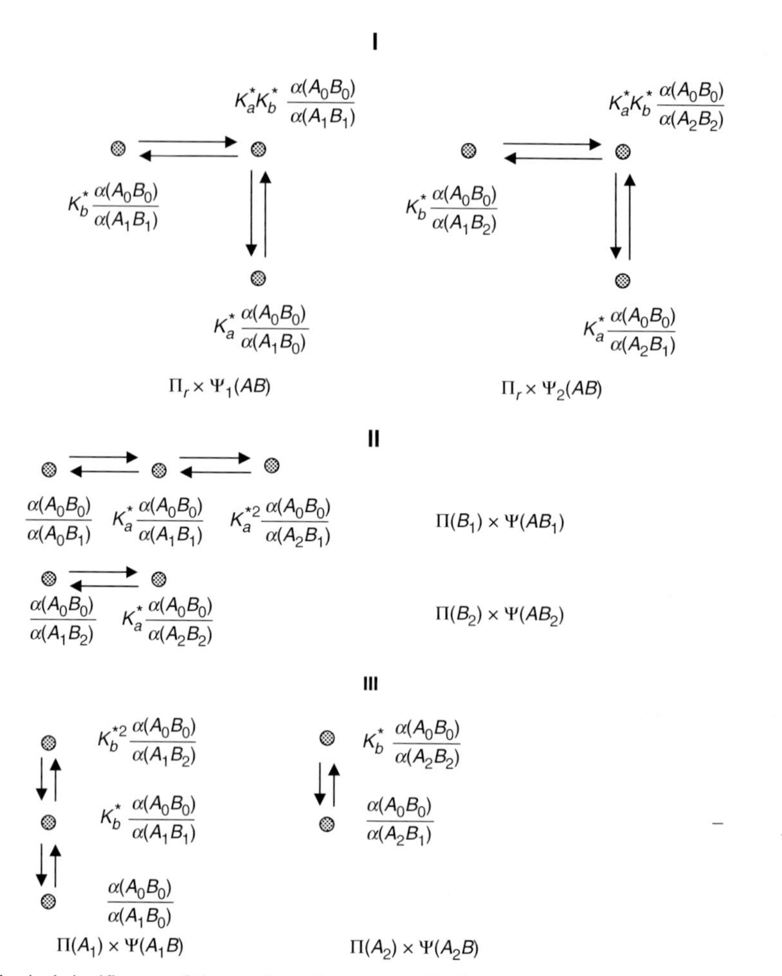


Fig. 23. Physical significance of the product of the row and column vectors involved in the model system of Fig. 13. I: $\Pi_r x\Psi_1(AB)$ and $\Pi_r x\Psi_2(AB)$. II: $\Pi(B_1)x\Psi(AB_1)$ and $\Pi(B_2)x\Psi(AB_2)$. III: $\Pi(A_1)x\Psi(A_1B)$ and $\Pi(A_2)x\Psi(A_2B)$. See text.

The signs of $I(A_1 : B_1)$, $I(A_1 : B_2)$, $I(A_2 : B_1)$, and $I(A_2 : B_2)$ will be positive if

$$
\begin{aligned}
&\frac{\alpha(A_0B_0)}{\alpha(A_1B_1)}\{1+\Pi_r\Psi_1(AB)+K_a^*K_b^*\Pi_r\Psi_2(AB)\} > \Pi(A_1)\Psi(A_1B)\Pi(B_1)\Psi(AB_1)\\
&\frac{\alpha(A_0B_0)}{\alpha(A_1B_2)}\{1+\Pi_r\Psi_1(AB)+K_a^*K_b^*\Pi_r\Psi_2(AB)\} > K_a^*\Pi(A_1)\Psi(A_1B)\Pi(B_2)\Psi(AB_2)\\
&\frac{\alpha(A_0B_0)}{\alpha(A_2B_1)}\{1+\Pi_r\Psi_1(AB)+K_a^*K_b^*\Pi_r\Psi_2(AB)\} > K_b^*\Pi(A_2)\Psi(A_2B)\Pi(B_1)\Psi(AB_1)\\
&\frac{\alpha(A_0B_0)}{\alpha(A_2B_2)}\{1+\Pi_r\Psi_1(AB)+K_a^*K_b^*\Pi_r\Psi_2(AB)\} > K_a^*K_b^*\Pi(A_2)\Psi(A_2B)\Pi(B_2)\Psi(AB_2)
\end{aligned}
\tag{113}
$$

and nil or negative otherwise. When these expressions are written explicitly, it is seen that the signs of $I(A_1 : B_1)$, $I(A_1 : B_2)$, $I(A_2 : B_1)$, and $I(A_2 : B_2)$ depend, to a significant extent, on the sign of the differences

$$\frac{\alpha(A_0B_1)}{\alpha(A_1B_1)} - \frac{\alpha(A_0B_0)}{\alpha(A_1B_0)} \quad \text{and} \quad \frac{\alpha(A_2B_1)}{\alpha(A_1B_1)} - \frac{\alpha(A_2B_2)}{\alpha(A_1B_2)} \tag{114}$$

The left-hand side expression refers to the first functional unit, whereas the right-hand side one refers to the second. Inspection of Fig. 22 shows that the signs of $I(A_1 : B_1)$ and $I(A_2 : B_2)$ are the same and opposite to the signs of $I(A_1 : B_2)$ and $I(A_2 : B_1)$.

## *4. Conformational spread and information landscape*

Classical theories of allostery [3,4] are based on the view that the spontaneous, or ligand-induced, conformation change of a subunit can propagate to the other subunit of an oligomeric protein. On the basis of indirect, but conclusive, evidence it has been suggested that the same idea can be applied to much more complex supramolecular edifices [1]. Aggregates of chemotaxis and ryanodine receptors, flagella and cilia apparently display conformational spread [19–28,31,32]. In the perspective of the two models considered above (Section 3) the conformational constraints that exist between the two proteins in interaction depend, in general, upon the whole structure of the supramolecular edifice. This situation applies whether conformation changes are spontaneous or induced by a ligand. These conformational constraints, in general, imply that the information landscape is rugged, i.e., that the information content of the nodes varies according to a certain organization. Computing the mean information of integration for such a surface would be meaningless since one could not distinguish, for instance, a smooth surface whose mean information of integration is equal to zero from a rugged surface displaying ranges of mountains separated by valleys and still possessing the same mean information equal to zero.

The problem of the conservation, or lack of conservation, of symmetry within oligomeric enzymes that undergo conformational transitions has been a matter of dispute in the past. Today there is little doubt that proteins can undergo either substrate-induced conformation changes, or spontaneous transconformations that keep the symmetry of the molecule unaltered. There are examples of both situations in the literature. Hence one can raise the question of whether the symmetry of a node is associated with a positive or a negative value of its mutual information of integration. Here symmetry means that a node is associated with both $i$ molecules of $X$ and $i$ molecules of $Y$ and is therefore denoted $N(X_iY_i)$ in the $n \times n$ lattice previously studied. Symmetry may also mean that a node of the quasi-linear network bears one protein in conformation $A_i$ and the other protein in conformation $B_i$. Such a node is denoted $N(A_iB_i)$ in the model of the quasi-linear network. Conversely, the unsymmetrical nodes are denoted $N(X_iY_j)$ in the first model and $N(A_iB_j)$

(with $i \neq j$) in the second. The results, which have already been presented, imply that if a symmetrical node, $N(X_iY_i)$ or $N(A_iB_i)$, is stabilized relative to an unsymmetrical one, $N(X_iY_j)$ or $N(A_iB_j)$, it should not display emergence, i.e., its mutual information of integration should be positive. The emergent nodes should then be the unsymmetrical ones. The conclusion should indeed be opposite if the unsymmetrical nodes are more stable than the symmetrical ones. As a general rule it is the stabilization that defines emergence.

## *References*


1. Duke, T.A.J., Le Novère, N., and Bray, D. (2001) Conformational spread in a ring of proteins: a stochastic approach to allostery. J. Mol. Biol. 308, 541–553.
2. Bray, D. and Duke, T. (2004) Conformational spread: the propagation of allosteric states in large multiprotein complexes. Annu. Rev. Biophys. Biomol. Struct. 33, 53–73.
3. Monod, J., Wyman, J., and Changeux, J.P. (1965) On the nature of allosteric transitions: a plausible model. J. Mol. Biol. 12, 88–118.
4. Koshland, D.E., Néméthy, G.D., and Filmer, D. (1966) Comparison of experimental binding data and theoretical models in protein containing subunits. Biochemistry 5, 365–385.
5. Levitzki, A. (1978) Quantitative Aspects of Allosteric Mechanisms. Springer-Verlag, Berlin.
6. Cantor, C.R. and Schimmel, P.R. (1980) Biophysical Chemistry. Part III. The Behavior of Biological Macromolecules. Freeman, San Francisco.
7. Wyman, J. (1964) Linked functions and reciprocal effects in haemoglobin: a second look. Adv. Prot. Chem. 19, 223–286.
8. Wyman, J. (1967) Allosteric linkage. J. Amer. Chem. Soc. 89, 2202–2218.
9. Ricard, J. and Cornish-Bowden, A. (1987) Co-operative and allosteric enzymes: 20 years on. Eur. J. Biochem. 166, 255–272.
10. Ricard, J. (1985) Organized polymeric enzyme systems: catalytic properties, In: Organized Multienzyme Systems. Catalytic Properties, Academic Press, New York, pp. 177–240.
11. Ricard, J. (1999) Biological Complexity and the Dynamics of Life Processes. Elsevier, Amsterdam.
12. Ricard, J. and Noat, G. (1984) Subunit interactions in enzyme transition states. Antagonism between substrate binding and reaction rate. J. Theor. Biol. 111, 737–753.
13. Ricard, J. and Noat, G. (1985) Subunit coupling and kinetic co-operativity of polymeric enzymes. Amplification, attenuation and inversion effects. J. Theor. Biol. 117, 633–649.
14. Ricard, J., Giudici-Orticoni, M.T., and Buc, J. (1990) Thermodynamics of information transfer between subunits in oligomeric enzymes and kinetic cooperativity. 1. Thermodynamics of subunit interactions, partition functions and enzyme reaction rate. Eur. J. Biochem. 194, 463–473.
15. Giudici-Orticoni, M.T., Buc, J., and Ricard, J. (1990) Thermodynamics of information transfer between subunits in oligomeric enzymes and kinetic cooperativity. 2. Thermodynamics of kinetic cooperativity. Eur. J. Biochem. 194, 475–481.
16. Giudici-Orticoni, M.T., Buc, J., Bidaud, M., and Ricard, J. (1990) Thermodynamics of information transfer between subunits in oligomeric enzymes and kinetic cooperativity. 3. Information transfer between the subunits of chloroplast fructose bisphosphatase. Eur. J. Biochem. 194, 483–490.
17. Pauling, L. (1946) Molecular architecture and biological reactions. Chem. Eng. News 24, 1375–1377.
18. Pauling, L. (1948) Nature of forces between large molecules of biological interest. Nature 161, 707–709.
19. Bray, D. (2002) Bacterial chemotaxis and the question of gain. Proc. Natl. Acad. Sci. USA 99, 7–9.
20. Bray, D., Levin, M.D., and Morton-Firth, C.J. (1998) Receptor clustering as a cellular mechanism to control sensitivity. Nature 393, 85–88.

21. Bren, A. and Eisenbach, M. (2000) How signals are heard during bacterial chemotaxis: protein–protein interactions in sensory signal propagation. J. Bacteriol. 182-6865–6873.
22. Francis, N.R., Levin, M.N., Shaikh, T.R., Melanson, L.A., Stock, J.B., and De Rosier, D.J. (2002) Subunit organization in a soluble complex of Tar, CheW, and CheA by electron microscopy. J. Biol. Chem. 277, 36755–36759.
23. Gestwicki, J.R. and Kiessling, L.L. (2001) Interreceptor communication through arrays of bacterial chemoreceptors. Nature 415, 81–84.
24. Kim, S.H.,Wang, W., and Kim, K.K. (2002) Dynamic and clustering model of bacterial chemotaxis receptors: structural basis for signaling and high sensitivity. Proc. Natl. Acad. Sci. USA 99, 11611–11615.
25. Keleshian, A.M., Edeson, R.O., Liu, G.D., and Madsen, B.W. (2000) Evidence for cooperativity between nicotinic acetylcholine receptors in patch clamp records. Biophys. J. 78, 1–12.
26. Liu, F., Wan, Q., Pristupa, Z.B., Yu, X.M., Wang, Y.T., and Niznik, H.B. (2000) Direct protein–protein coupling enables cross-talk between dopamine D5 and gamma-aminobutyric acid A receptors. Nature 403, 274–280.
27. Marx, S.O., Ondrias, K., and Marks, A.R. (1998) Coupled gating between individual skeletal muscle $Ca^{++}$ release channels (ryanodine receptors). Science 281, 818–821.
28. Yeramian, E., Trautmann, A., and Claverie, P. (1986) Acetylcholine receptors are not functionally independent. Biophys. J. 50, 253–263.
29. Cluzel, P., Surette, M., and Leibler, S. (2000) An ultrasensitive bacterial motor revelead by monitoring signaling proteins in single cells. Science 287, 1652–1657.
30. Goodsell, D.S. and Olson, A.J. (2000) Structural symmetry and protein function. Annu. Rev. Biophys. Biomol. Struct. 29, 105–153.
31. Macnab, R.M. and Ornston, M.K. (1977) Normal-to-curly flagellar transitions and their role in bacterial tumbling. Stabilization of an alternative quaternary structure by mechanical force. J. Mol. Biol. 112, 1–30.
32. Cibert, C. (2003) Entropy and information in flagellar axoneme cybernetics: a radial spokes integrative function. Cell Motil. Cytoskeleton 54, 296–316.
33. Oster, G. and Wang, H. (2000) Reverse engineering of a protein: the mechanochemistry of ATP synthase. Biochim. Biophys. Acta 1458, 482–510.

J. Ricard *Emergent Collective Properties, Networks, and Information in Biology*

DOI: 10.1016/S0167-7306(05)40010-1

CHAPTER 10

# Gene networks

J. Ricard

Proteins are not the only molecular elements to be mutually connected, genes also form networks. Gene networks can be studied through theoretical and experimental approaches. The present chapter is devoted to these two kinds of approaches. The same gene, under the same experimental conditions, can be active or not and probably exists under different dynamic states. A great simplification of the mathematical description of gene activity is to assume they follow a Boolean logic, i.e., to assume they can be either ON or OFF, and never more or less active. Genetic engineering has allowed one to construct artificial gene circuits, study their activity and model the results obtained. In this perspective it has become possible, for instance, to understand the role of positive and negative feedback loops and the mechanism of periodic oscillations of engineered gene networks.

*Keywords:* bacterial operons, $\beta$-galactosidase, binary logic and gene networks, cis-acting sequences, differential activity of a gene, engineered gene circuits, feedback loops, induction, jacobian matrix, logic of gene circuits, low affinity sites for repressor protein, negative control, negative feedback, oscillations of gene networks, permease, positive control, positive feedback, promoter, repressor, transacetylase, transacting products.

A gene is not active independently of its neighbors, i.e., a group of contiguous genes can be repressed, derepressed, or activated in a coordinated manner. Such groups of genes constitute networks that are also called operons by molecular biologists.

## *1. An overview of the archetype of gene networks: the bacterial operons*

*E. coli* can grow on lactose as a sole source of carbon. An enzyme, $\beta$-galactosidase, hydrolyzes lactose to glucose and galactose. In fact, the presence of lactose in the growth medium induces the synthesis of new molecules of $\beta$-galactosidase. Two other proteins, a membrane protein called galactoside permease and a thiogalactosidase transacetylase, are also synthesized together with $\beta$-galactosidase. Moreover, as soon as $\beta$-galactosidase appears, the two other proteins are synthesized in constant proportions whatever be the experimental conditions.

*1.1. The operon as a coordinated unit of gene expression*

The basic concepts of how regulation is achieved in bacteria relies upon the seminal work of Jacob and Monod [1–3]. They proposed that two types of DNA sequences exist in bacteria: sequences coding for trans-acting products, and cis-acting sequences. Any gene product, usually a protein, that recognizes its target is defined as trans-acting. Cis-acting sequences are DNA sequences that are active as such in the chromosome [4]. Among the trans-acting sequences is a so-called regulator gene which is expressed as mRNA itself translated as a protein. In the Jacob–Monod model, this protein can be identified to the repressor that binds to a specific region of the DNA called operator (Fig. 1). This process prevents RNA polymerase from binding to its specific site called promoter hence impeding the transcription of a battery of structural genes, viz. those that, in *E. coli*, code for β-galactosidase, permease, and transacetylase. In bacteria, the promoter and operator can be

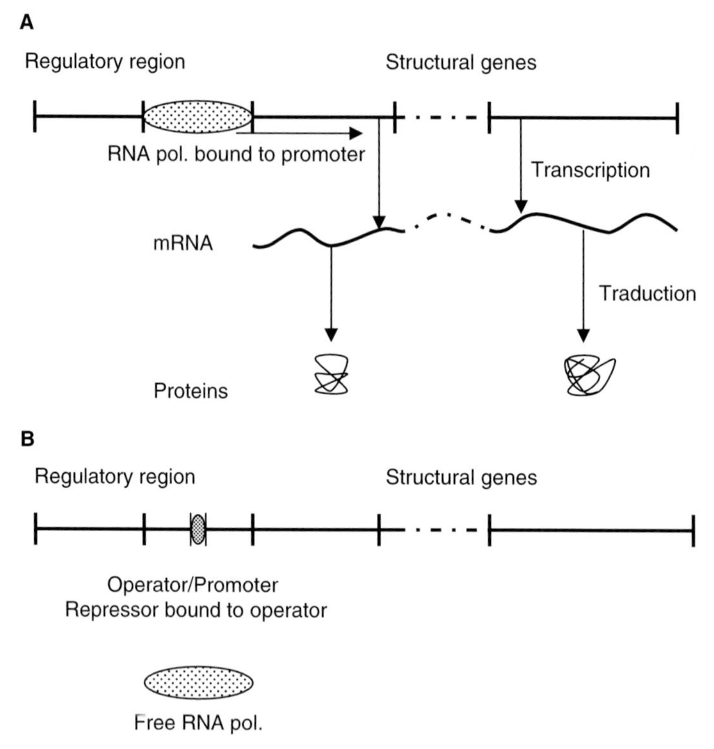


Fig. 1. Typical organization of an operon in bacteria. (A) RNA polymerase can bind to the promoter leading to the transcription and traduction of the structural genes. The precise localization of the operator relative to the promoter can change depending on the biological system. They are arranged so that occupancy of the operator by the repressor prevents RNA polymerase from reaching the promoter. (B) Repressor is bound to the operator thus impeding transcription of the structural genes by RNA polymerase. See text.

arranged one with respect to the other in different manners, but so that the occupancy of the operator by the repressor prevents RNA polymerase from binding to the promoter.

Hence the regulatory region of the bacterial chromosome containing trans-acting sequences acts in a negative manner on the cis-acting sequences governing the structural genes. In *E. coli*, this means that in the absence of the repressor, the three genes *lac*Z, *lac*Y, and *lac*A coding for $\beta$-galactosidase, permease, and trasacetylase are normally transcribed as a unique mRNA which is then translated into polypeptide chains. In the case of $\beta$-galactosidase, four identical polypeptide chains assemble to form the enzyme. In the case of the permease, the polypeptide is integrated in the bacterial membrane whereas two polypeptide chains of transacetylase associate to form the active dimeric enzyme.

The control exerted by trans-acting sequences does not always correspond to a repression but can also involve an activation of RNA-polymerase that cannot initiate its action at the promoter in the absence of transcription factors (Fig. 2). This process of activation exists in bacteria and in eukaryotes. In the latter case, however, a promoter controls one structural gene only. For bacteria, it is the regulation unit which is called operon.

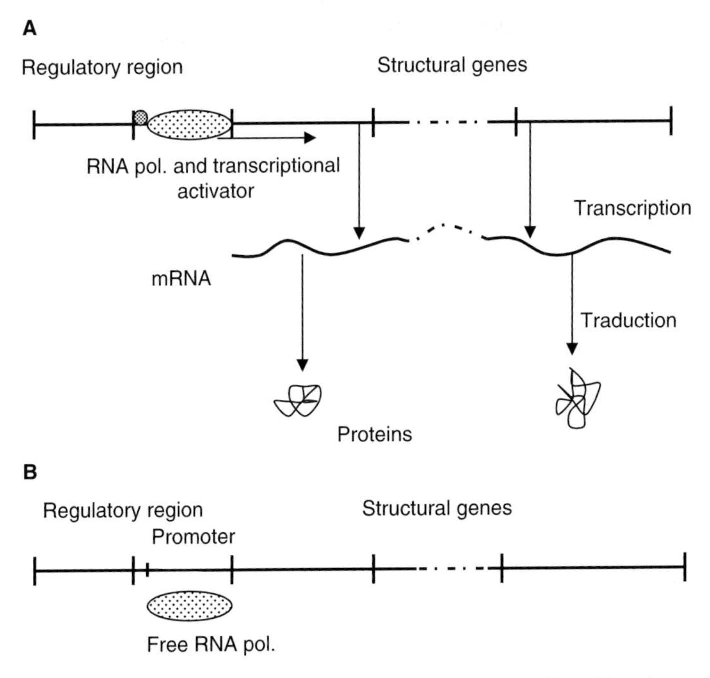


Fig. 2. Activation of RNA polymerase by a transcriptional activator. (A) RNA polymerase requires a protein (the "black" circle in the figure) which plays the part of a transcriptional activator, to initiate its action. (B) In the absence of the transcriptional activator RNA polymerase cannot bind to the promoter and is inactive. See text.

### 1.2. Repressor and induction

The general principles presented in the previous section allow one to understand the logic that exists behind the process of induction. As bacteria live under very different environmental conditions, it would be wasteful for them to possess the chemical machinery required for the consumption of chemicals, for instance lactose, that may, or may not, exist in the external medium. No doubt it is a functional advantage for the cell to synthesize this machinery only if the bacterial cell comes into contact with lactose. When this happens, the *lac*mRNA is then synthesized and translated into β-galactosidase. This mRNA, which contains the information required for the synthesis of the three different proteins already referred to, is unstable and if lactose is removed, it returns back to its initial low level. The enzyme β-galactosidase, however, is much stabler and remains as such in the bacterial cell after removal of the inducer for much longer time periods (Fig. 3).

The basic idea which was put forward in the seminal work of Jacob and Monod [1–3] was that the inducer binds to the repressor thus inducing, or stabilizing, a new conformation of the protein that can no longer bind to the operator. Under these conditions, β-galactosidase, permease, and transacetylase can be synthesized (Fig. 4). This ingenious idea was initially supported by a number of genetic experiments.

The repressor protein has been isolated and purified from a strain of *E. coli* [5–7]. It is a tetrameric protein having subunits of 37 kDal. As expected for a repressor, this protein binds inducers and DNA. Certain molecules, such as isopropylthiogalactoside (IPTG), which are not substrates of β-galactosidase specifically bind to the

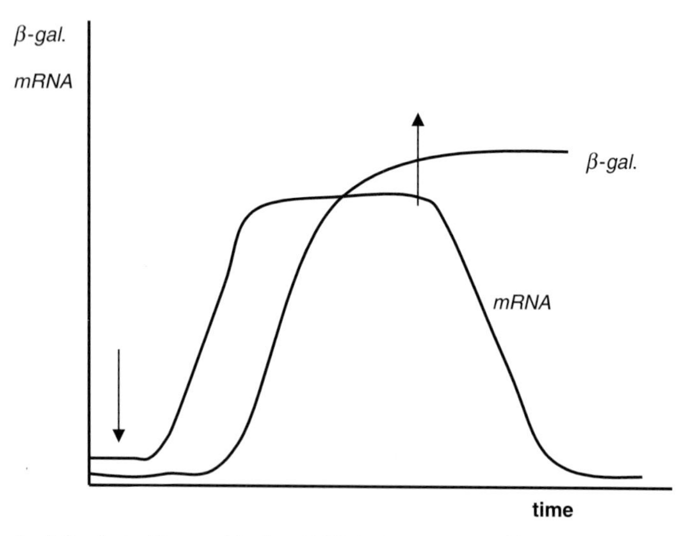


Fig. 3. Synthesis of β-galactosidase and its *lac*mRNA in the presence of lactose. When lactose is added to a culture of *E. coli* (first arrow on the left side of the figure) *lac*mRNA is synthesized after a delay. Upon removal of lactose (second arrow on the right side of the figure), *lac*mRNA is swiftly subjected to degradation. β-galactosidase is synthesized in the cells after the appearance of *lac*mRNA and is not degraded after *lac*mRNA has been destroyed.

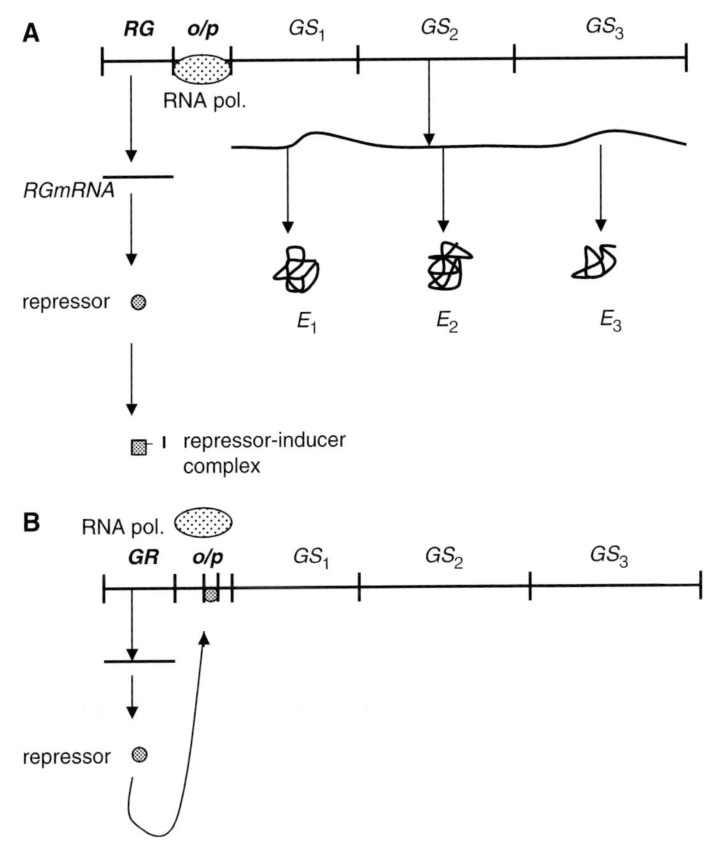


Fig. 4. Action of an inducer on gene expression. (A) An inducer, I, can bind to the repressor thus preventing its interaction with the operator. RNA polymerase can then bind to the promoter and initiate transcription. $GS_1, GS_2, \ldots$ are the structural genes, $E_1, E_2, \ldots$ are the enzymes, $o/p$ is the operator–promoter region of the DNA, and RG is the regulatory region of the DNA and *RGmRNA* its transcription product. (B) As the inducer is absent, the repressor binds to the operator thus impeding transcription. See text.

repressor and play the part of an inducer of $\beta$-galactosidase synthesis. Moreover, as expected from the Jacob-Monod model, binding of IPTG to the repressor brings about its partial dissociation from the operator. When this protein is bound to the DNA, part of the nucleic acid becomes resistant to pancreatic deoxyribonuclease. In fact all the DNA is hydrolyzed except the part that is bound to the repressor. It then becomes possible to determine the sequence of the operator (Fig. 5). This structure is symmetrical about a two-fold axis. Symmetry of the DNA sequence in the operator region reflects the symmetry of the repressor protein which, as already mentioned, is made up of four identical subunits.

The problem of the affinity of the repressor for inducer and for DNA is not a simple one. In fact, in addition to the high-affinity sites for the repressor, there are on DNA many non-specific low-affinity sites for this protein. The affinity of the

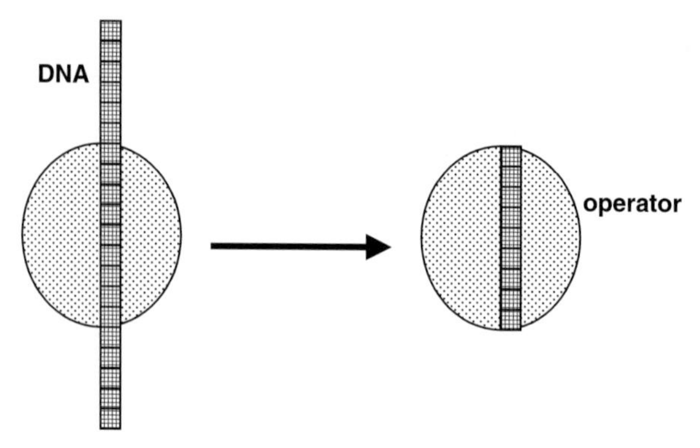


Fig. 5. The complex repressor–operator. The repressor ( the "gray" circle in the figure ) binds to DNA. If the complex is treated with pancreatic ribonuclease, only the operator region bound to the operator is protected against hydrolysis. See text.

repressor for the *lac* operator is about $2 \times 10^{13}\,M^{-1}$. As there are about ten molecules of repressor per cell of *E. coli*, one can conclude that nearly all the repressor molecules are bound to DNA. Moreover the affinity constant of the inducer IPTG for the repressor is about $10^{6}\,M^{-1}$. Hence, from a purely physical viewpoint, it is very unlikely that there is an equilibrium between free and operator-bound repressor that could be shifted by the binding of inducer to the free repressor. In fact, the inducer IPTG binds to the operator-bound repressor as well as to the repressor molecules bound to low-affinity sites of DNA. From this binding process, part of the repressor bound to the operator is released whereas the repressor molecules bound to low-affinity sites do not dissociate from DNA (Fig. 6)

In fact, there are so many low-affinity sites that, whether or not inducer is present, the repressor is always in a bound state. In the presence of inducer, the operator is free and the repressor molecules are bound exclusively to low-affinity sites whereas in the absence of inducer, the repressor is bound to both the operator and the low-affinity sites. The role played by the repressor in induction has also been established through elegant genetic experiments. Thus, for instance, mutants exist that bear an altered operator on their chromosome. Hence the repressor cannot bind to the operator and the three proteins $\beta$-galactosidase, permease, and transacetylase are always present in the cell. A similar situation is observed if the structure of the repressor is altered through a mutation of the corresponding gene.

### *1.3. Positive versus negative control*

Most of the ideas briefly discussed in the previous sections were concerned with negative control. This means that a regulatory gene is expressed as a mRNA which is translated as a repressor protein. The activity of a battery of genes depends on the partitioning of repressor between the high-affinity site, called operator, and many low-affinity sites. The binding of an inducer to the repressor modifies this

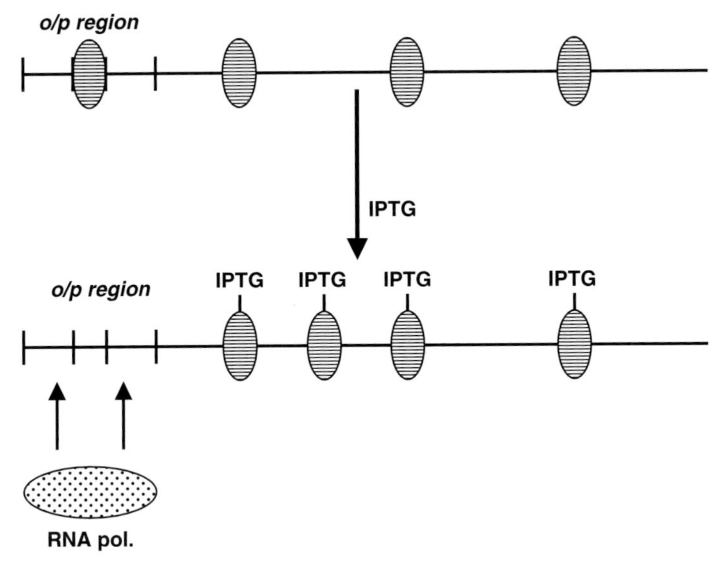


Fig. 6. Multiplicity of sites on DNA able to bind the repressor. Top: The repressor molecules ("gray" ellipses in the figure) bind to the operator as well as to other low-affinity sites on the DNA. Bottom: In the presence of IPTG, the IPTG–repressor complex dissociates from the operator that becomes free. RNA polymerase is then able to bind to the promoter. A number of molecules of the IPTG–repressor complex are then bound to non-specific low-affinity sites on the DNA.

partitioning. Whereas the inducer-bound repressor molecule leaves the operator, which then becomes free, the repressor remains bound to non-specific low-affinity sites (Fig. 6). In other cases, the repressor may be inactive and become active only in the presence of a co-repressor. In either case, a specific region of the DNA, called promoter, has to be free of repressor in order to bind RNA-polymerase.

In bacterial strains, there is also a positive control. The regulatory gene is expressed as an inactive protein that becomes active in the presence of an inducer. The binding of the active protein to the promoter allows the association of RNA-polymerase with the DNA. Repression can take place in a positive control if the association of a ligand, called co-repressor, with the activator inactivates this protein. It therefore appears that positive and negative controls are symmetrical situations that lead to similar consequences.

## 2. *The role of positive and negative feedbacks in the expression of gene networks*

### 2.1. *Multiple dynamic states and differential activity of a gene*

Any gene involved in the process of transcription receives an "input of nucleotides" that become ligated and, as a consequence, an RNA is released as an "output". Hence the transcription process occurs under dynamic conditions. One can easily conceive that if there are multiple states accessible to a gene, these states have different functional activities. Such a situation was suggested long ago by Waddington [8] and Delbruck [9] to be quite likely. Some classical experiments by Novick and Weiner [10] as well as by Cohn and Horibata [11–13] suggest it is precisely so. The genes required for the utilization of lactose by *E. coli* are expressed only if an analog of lactose, which takes the part of an inducer, is present. If the inducer concentrations are moderate, the *lac* genes involved in lactose consumption are either ON or OFF. As outlined by Thomas [14–22], the main point in these experiments is that, for identical experimental conditions (same concentration of inducer), the same gene can be ON or OFF. Hence two different phenotypes can exist for the same genotype and identical experimental conditions. As a matter of fact, one of the genes of the lactose operon codes for the permease involved in the transfer of inducer across the bacterial membrane. The moderate concentrations of inducer used in these experiments, however, are not sufficient to allow permease synthesis. Hence if a cell does not contain any permease, it will be unable to synthesize this protein and no entry of the inducer into the cell will be allowed. Alternatively, if a cell already contains a small number of permease molecules it will be able to concentrate the inducer from the external medium and will accordingly synthesize more permease molecules. The question, which was put forward many years ago by Delbruck [9], is to know whether this difference of behavior could not be explained by the existence of at least two different states of the same gene. In a series of papers, Thomas and his colleagues have convincingly shown [14–22] that the differential activity of a gene is indeed due to the existence of multiple states of this gene.

### *2.2. Formal expression of multiple dynamic states of a gene*

Let us consider a linear differential equation

$$\frac{dx_i}{dt} = f_i(x_1, x_2, \ldots, x_n) \tag{1}$$

the elements of the corresponding jacobian matrix are defined as $a_{ij} = \partial f_i / \partial x_j$. They express how the variables $x_j$ influence the time derivative of the variables $x_i$. We can therefore express this situation by stating that, if $a_{ij} \neq 0$, variables $x_j$ act on variables $x_i$, and we write

$$x_j \xrightarrow{a_{ij}} x_i \tag{2}$$

In this formal representation [21], the arrow does not mean that $x_j$ is being transformed into $x_i$, but that $x_j$ acts, positively or negatively, on $x_i$. Let us consider a sequence of four non-zero terms $a_{21}, a_{32}, a_{43}, a_{14}$ corresponding to

the four-element cyclic circuit

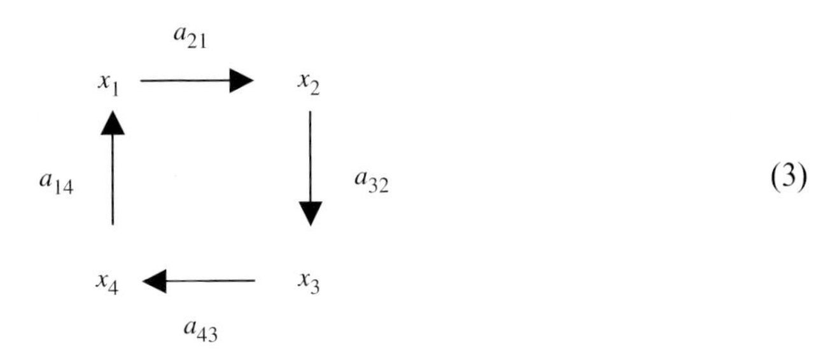


(3)

The corresponding jacobian matrix $J$ is

$$J = \begin{bmatrix} \cdot & \cdot & \cdot & a_{14} \\ a_{21} & \cdot & \cdot & \cdot \\ \cdot & a_{32} & \cdot & \cdot \\ \cdot & \cdot & a_{43} & \cdot \end{bmatrix} \tag{4}$$

In such a cyclic system we state that the elements of the jacobian matrix form a feedback circuit or, as we shall see later, a full-circuit [21]. The elements $a_{ij}$ can be positive or negative. If the number of negative terms is $q$ the sign of a feedback circuit is $(-1)^q$. This means that if the number of negative terms is even (including zero), the circuit is positive. If conversely the number is odd, the circuit is negative. As it appears from the above reasoning, a circuit is defined by a series of terms in such a way that the sequences of $i$ and $j$ (rows and columns of the jacobian matrix) are circular permutations of each other. Multiple dynamic states are generated by positive feedback circuits only. As shown by Thomas [21], $m$ isolated positive circuits generate $3^m$ dynamic states of which $2^m$ are stable. Thus for instance, 8 genes located in eight different positive circuits generate $2^8 = 256$ dynamic states and therefore up to 256 different cell types.

A jacobian matrix may possibly describe the union of disjoint circuits. For instance, the terms $a_{21}$, $a_{32}$, $a_{13}$, and $a_{44}$ describe two disjoint circuits

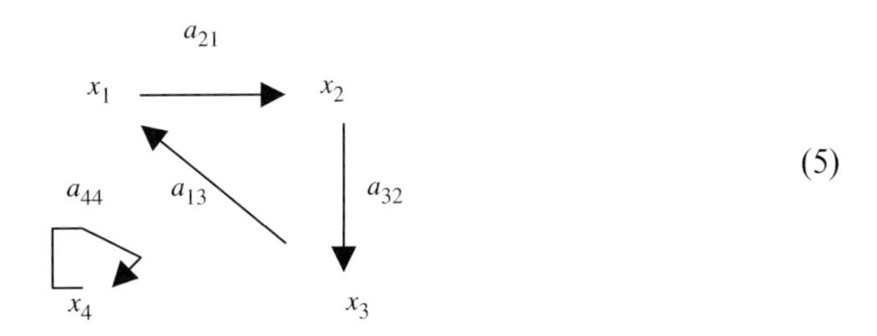


(5)

So far we have only considered linear differential equations. In fact, the equations can also be nonlinear. Then one or several terms of the jacobian matrix can be function of a variable (or of several variables) of the system. Owing to the presence of one or several variables in the matrix, the same circuit can adopt different signs depending on the location in the phase space. From the jacobian matrix, one can derive the corresponding characteristic equation whose solutions are the eigenvalues of the matrix.

### 2.3. *Full-circuits*

In Thomas terminology [21], a full-circuit is a circuit involving all the variables, or a set of disjoint circuits involving all the variables. Thus, for instance, the three matrices

$$\begin{bmatrix} \cdot & a_{12} & \cdot \\ \cdot & \cdot & a_{23} \\ a_{31} & \cdot & \cdot \end{bmatrix} \begin{bmatrix} \cdot & a_{12} & \cdot \\ a_{21} & \cdot & \cdot \\ \cdot & \cdot & a_{33} \end{bmatrix} \begin{bmatrix} a_{11} & \cdot & \cdot \\ \cdot & a_{22} & \cdot \\ \cdot & \cdot & a_{33} \end{bmatrix} \tag{6}$$

are full-circuits for they involve the three variables $x_1$, $x_2$, $x_3$, i.e.,

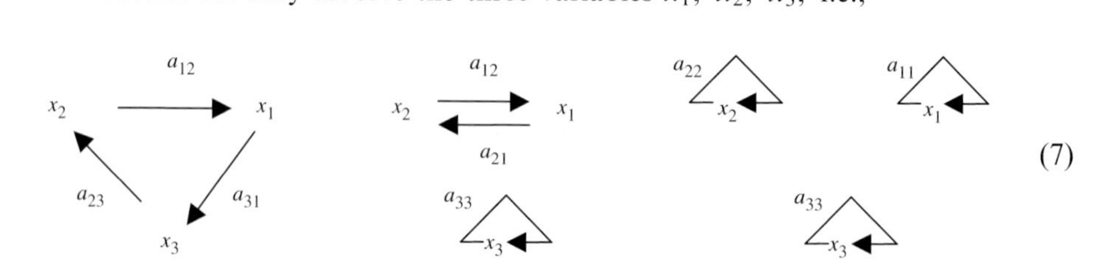


(7)

In order to obtain from a jacobian matrix all the possible full-circuits, one can simply write down the analytic form of the determinant of the matrix. This determinant is the sum of the products of $a_{ij}$, each of these products being a full-circuit of the system. Thus, in the case of the matrix

$$\begin{bmatrix} a_{11} & a_{12} & a_{13} \\ a_{21} & a_{22} & a_{23} \\ a_{31} & a_{32} & a_{33} \end{bmatrix}$$

the determinant

$$\Delta = a_{11}a_{22}a_{33} + a_{12}a_{23}a_{31} + a_{13}a_{32}a_{21} - a_{11}a_{23}a_{32} - a_{22}a_{31}a_{13} - a_{33}a_{12}a_{21} \tag{8}$$

can be expressed as the sum of six terms that are full-circuits. If the corresponding jacobian matrix is not a function of one (or several) variable(s), a full-circuit can specify a dynamic state whose sign and eigenvalues are determined by the signs of the component circuits. If the full-circuit is made up of the union of disjoint circuits, the eigenvalues of the corresponding system are obtained from the union of the eigenvalues corresponding to these disjoint circuits.

In fact, the dynamics of a system relies on the respective values of both the trace, $T_j$, and the determinant, $\Delta_j$, of the corresponding jacobian matrix and on the expression of the discriminant, $T_j^2 - 4\Delta_j$, of the characteristic equation as well [23–26]. Let us consider for instance the circuit

$$x_1 \underset{a_{12}}{\overset{a_{21}}{\rightleftarrows}} x_2 \tag{9}$$

the corresponding jacobian matrix is then

$$J = \begin{bmatrix} \cdot & a_{21} \\ a_{12} & \cdot \end{bmatrix} \tag{10}$$

Its trace $T_j$ and determinant $\Delta_j$ as well as the discriminant of the characteristic equation $T_j^2 - 4\Delta_j$ are

$$T_j = 0, \quad \Delta_j = -a_{12}a_{21}, \quad T_j^2 - 4\Delta_j = 4a_{12}a_{21} \tag{11}$$

If $a_{12}$ and $a_{21}$ are both positive or negative, it follows that

$$\Delta_j < 0 \quad \text{and} \quad T_j^2 - 4\Delta_j > 0 \tag{12}$$

and the system displays a saddle point. If $a_{12}$ and $a_{21}$ are opposite in sign, then

$$\Delta_j > 0 \quad \text{and} \quad T_j^2 - 4\Delta_j < 0 \tag{13}$$

and the eigenvalues are complex. The dynamic system then displays periodic oscillations [23–26]. These conclusions can be expressed schematically as in [21]

$$\begin{matrix} \begin{bmatrix} \cdot & - \\ - & \cdot \end{bmatrix} \text{ or } \begin{bmatrix} \cdot & + \\ + & \cdot \end{bmatrix} \Rightarrow \text{saddle point} \\ \begin{bmatrix} \cdot & + \\ - & \cdot \end{bmatrix} \text{ or } \begin{bmatrix} \cdot & - \\ + & \cdot \end{bmatrix} \Rightarrow \text{focus} \end{matrix} \tag{14}$$

Now, if we consider the union of two 1-circuits, i.e.,

$$a_{11} \circlearrowleft x_1 \qquad a_{22} \circlearrowleft x_2 \tag{15}$$

the jacobian matrix of the system is

$$J = \begin{bmatrix} a_{11} & \cdot \\ \cdot & a_{22} \end{bmatrix} \tag{16}$$

The trace and the determinant of the jacobian matrix as well as the discriminant of the characteristic equation are

$$T_j = a_{11} + a_{22}, \quad \Delta_j = a_{11}a_{22}, \quad T_j^2 - 4\Delta_j = (a_{11} - a_{22})^2 \tag{17}$$

If $a_{11}$ and $a_{22}$ are both negative

$$T_j < 0, \quad \Delta_j > 0, \quad T_j^2 - 4\Delta_j > 0 \tag{18}$$

the system displays a stable node. If, conversely, $a_{11}$ and $a_{22}$ are both positive

$$T_j > 0, \quad \Delta_j < 0, \quad T_j^2 - 4\Delta_j > 0 \tag{19}$$

and the system displays an unstable node. But if $a_{11}$ and $a_{22}$ are opposite in sign, $\Delta_j < 0$ and the dynamic system has a saddle point. As previously, these conclusions can be expressed schematically as in [21]

$$\begin{array}{l} \begin{bmatrix} - & \cdot \\ \cdot & - \end{bmatrix} \Rightarrow \text{stable node} \\ \begin{bmatrix} - & \cdot \\ \cdot & + \end{bmatrix} \text{or} \begin{bmatrix} + & \cdot \\ \cdot & - \end{bmatrix} \Rightarrow \text{saddle point} \\ \begin{bmatrix} + & \cdot \\ \cdot & + \end{bmatrix} \Rightarrow \text{unstable node} \end{array} \tag{20}$$

So far, we have considered linear systems but the concept of circuit also applies to nonlinear systems. Let us consider the system

$$\begin{aligned} \frac{dx}{dt} &= f_x(x, y, z) = y \\ \frac{dy}{dt} &= f_y(x, y, z) = z \\ \frac{dz}{dt} &= f_z(x, y, z) = 2x^3 \end{aligned} \tag{21}$$

of which the last equation is nonlinear. The jacobian matrix of the system is

$$J = \begin{bmatrix} \frac{\partial f_x}{\partial x} & \frac{\partial f_x}{\partial y} & \frac{\partial f_x}{\partial z} \\ \frac{\partial f_y}{\partial x} & \frac{\partial f_y}{\partial y} & \frac{\partial f_y}{\partial z} \\ \frac{\partial f_z}{\partial x} & \frac{\partial f_z}{\partial y} & \frac{\partial f_z}{\partial z} \end{bmatrix} = \begin{bmatrix} 0 & 1 & 0 \\ 0 & 0 & 1 \\ 6x^2 & 0 & 0 \end{bmatrix} \tag{22}$$

In Thomas schematic formulation, the system is a 3-circuit

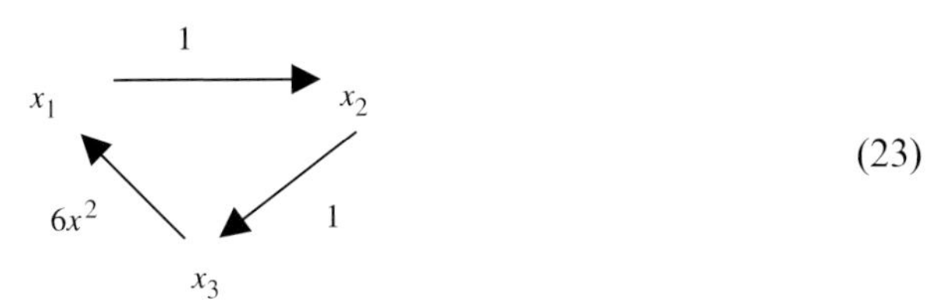


(23)

Although the system is nonlinear, it has one type of dynamic state only (i.e., a focus with a real positive root and a pair of conjugate complex roots) in the phase space. This is the kind of situation we have already met with linear systems. In fact, nonlinearity does not bring anything new to the system for the term $6x^2$ does not change its sign whether $x$ is positive or negative. This situation is far from being general, however. Let us consider a different system [21]

$$\begin{aligned} \frac{dx}{dt} &= f_x(x, y) = 2y \\ \frac{dy}{dt} &= f_y(x, y) = 3x^2 - 1 \end{aligned} \tag{24}$$

The corresponding jacobian matrix is

$$J = \begin{bmatrix} \frac{\partial f_x}{\partial x} & \frac{\partial f_x}{\partial y} \\ \frac{\partial f_y}{\partial x} & \frac{\partial f_y}{\partial y} \end{bmatrix} = \begin{bmatrix} 0 & 2 \\ 6x & 0 \end{bmatrix} \tag{25}$$

Phase space can then be split into two regions. The circuit is positive for $x > 0$ and negative for $x < 0$. In the first case, the system has a saddle point and in the second case, it possesses a focus. Hence owing to the existence of a positive 2-circuit and to the nonlinear character of the system, its dynamics displays two different states that are dependent on the sign of $x$. This conclusion is general. A necessary condition for the system to display two different dynamic states is both to have a positive circuit and to be nonlinear.

Another more complex situation is offered by the apparently simple system [21]

$$\begin{aligned} \frac{dx}{dt} &= x - x^3 \\ \frac{dy}{dt} &= y - y^3 \end{aligned} \tag{26}$$

whose corresponding jacobian matrix is

$$J = \begin{bmatrix} 1 - 3x^2 & 0 \\ 0 & 1 - 3y^2 \end{bmatrix} \tag{27}$$

The trace, $T_j$, and the determinant, $\Delta_j$, of this matrix are

$$T_j = 2 - 3(x^2 + y^2)$$
$$\Delta_j = (1 - 3x^2)(1 - 3y^2) \tag{28}$$

The sign of the determinant $\Delta_j$ is the same as that of

$$\Delta_j^* = \left(\frac{1}{\sqrt{3}} - |x|\right)\left(\frac{1}{\sqrt{3}} - |y|\right) \tag{29}$$

The phase plane corresponding to system (26) is shown in Fig. 7.

The four curves shown in this figure are representative of the equation

$$x^2 + y^2 = \frac{2}{3} \tag{30}$$

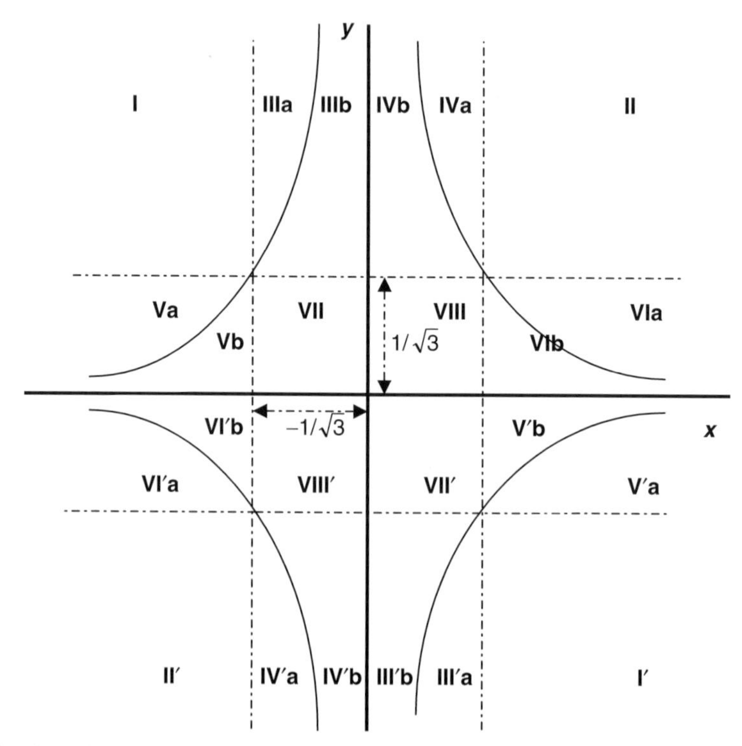


Fig. 7. The phase plane of the system (26).The phase plane of the system (26) is divided in different regions by four curves representative of Eq. (30). In these regions, the trace, $T_j$, the determinant, $\Delta_j$, of the jacobian matrix of the system, and the discriminant, $T_j^2 - 4\Delta_j$, can take different values that define the dynamics of the system. Adapted from [21].

which is equivalent to $T_j = 0$. This implies that if

$$x^2 + y^2 < \frac{2}{3} \tag{31}$$

$T_j > 0$, and if

$$x^2 + y^2 > \frac{2}{3} \tag{32}$$

then $T_j < 0$. Moreover, one can notice that whatever be the signs of $x$ and $y$, one has always

$$T_j^2 \geq 4\Delta_j \tag{33}$$

Hence the nonlinear system (26) cannot display a focus. Comparison of the signs of $\Delta_j^*$ and $T_j$ in the various regions of the phase plane allows one to conclude that in regions I, I′, II, II′ the system has a stable node. In regions III, III′, IV, IV′, V, V′, VI, VI′ it has a saddle point and in regions VII, VII′, VIII, VIII′ it possesses an unstable node (Table 1).

The same conclusions can indeed be obtained through Thomas schematic rules. Owing to the structure of the jacobian matrix of the system, the nonlinear differential equations can be viewed as the union of two one-circuits. In the regions I, I′, II, II′ the jacobian matrix of the union of the two 1-circuits has the following form

$$\begin{bmatrix} - & \cdot \\ \cdot & - \end{bmatrix} \tag{34}$$

and the system has a stable node. In the regions III, III′, IV, IV′, V, V′, VI, VI′ the jacobian matrix assumes either of the two following forms

$$\begin{bmatrix} - & \cdot \\ \cdot & + \end{bmatrix} \text{ or } \begin{bmatrix} + & \cdot \\ \cdot & - \end{bmatrix} \tag{35}$$

Table 1

Different types of dynamic behavior of nonlinear system (26). See text

| | |
|---|---|
| I, I′ | stable node |
| II, II′ | stable node |
| III, III′ | saddle point |
| IV, IV′ | saddle point |
| V, V′ | saddle point |
| VI, VI′ | saddle point |
| VII, VII′ | unstable node |
| VIII, VIII′ | unstable node |

and the system has a saddle point. Last, in the regions VII, VII′, VIII, VIII′ the jacobian matrix is of the following form

$$\begin{bmatrix} + & \cdot \\ \cdot & + \end{bmatrix} \tag{36}$$

and the system has an unstable node.

## 3. *Gene networks and the principles of binary logic*

Some years ago, Thomas attempted to propose a logical description of the activity of gene networks [17,22]. The basic idea of this description is that a gene can only be ON or OFF. In this perspective, it is excluded that a gene can be more or less active. It can only be fully active (ON), or totally inactive (OFF). In the same spirit, it is assumed that the gene products are either present or absent. Hence, when a gene is switched ON, the corresponding boolean variable $X$ changes from 0 to 1. Conversely, a change of $X$ from 1 to 0 implies that the gene has been turned OFF. Similarly, the appearance of a gene product in the cell is expressed by a variable $x$ that takes the successive values 0 and 1. The disappearance of the product is associated with a discrete change of $x$ that adopts the successive values 1 and 0.

Let us consider a gene which is initially inactive and is then switched ON, the corresponding logical variable $X$ varies from 0 to 1. As a result of gene derepression, or activation, the corresponding protein appears in the medium and the logical variable $x$ adopts the successive values 0 and 1. However, this change takes place after a delay $d_x(On)$ which corresponds to the time required for protein synthesis. When, after a while, the same gene is turned OFF the value of $X$ switches from 1 to 0. Similarly, the protein, which is not a stable entity, can be broken down and this is again modelled by a change of the value of $x$ that drops from 1 to 0. The disappearance of the protein occurs after the gene has been turned OFF. Hence there exists another delay, $d_x(Off)$, which is usually different from $d_x(On)$ for the chemical process involved in protein synthesis and degradation can be different. This situation introduces asynchrony in gene expression (Fig. 8).

Let us consider the positive 3-circuit [22]

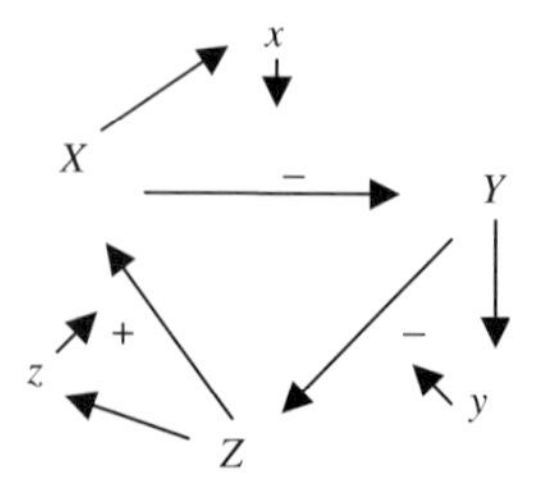

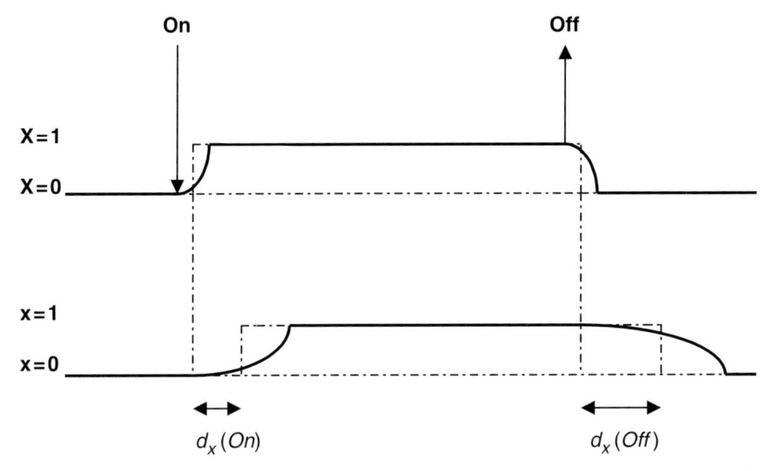


Fig. 8. Boolean representation of switching ON and OFF a gene and the synthesis of a protein. Top: The logical variable $X$ pertaining to the gene varies from 0 to 1 and from 1 to 0 when the gene is switched ON, then OFF. In fact, derepression (or activation) and repression of a gene is not instantaneous but, for simplicity, one assumes they can both be described, as a first approximation, by a step-function. The two steps are represented by dotted lines in the figure. Bottom: The logical variable $x$ pertaining to the protein also varies from 0 to 1 and from 1 to 0 when the gene is turned ON, then OFF. Although the appearance and disappearance of a protein is not instantaneous, it is approximated by a step-function. The two steps are represented by dotted lines. This curve displays a delay for the switching ON, $d_x(On)$, and the switching OFF, $d_x(Off)$. There is no *a priori* reason to believe that these two delays have identical lengths. Adapted from [22].

In this ideal scheme a product, $x$, of gene $X$ inhibits gene $Y$ and a product, $y$, of gene $Y$ inhibits gene $Z$. Conversely a product, $z$, of gene $Z$ activates gene $X$. One can conclude from this situation that if the three products $x$, $y$, $z$ are absent, genes $Y$ and $Z$ will be ON and gene $X$ will be OFF. Under these conditions, products $y$ and $z$ will soon be synthesized. This situation illustrates the interest of Thomas's approach which puts the emphasis on the potential lack of synchrony between the states of genes and of proteins. Similarly, if $x$ and $y$ are absent and $z$ present, the three genes $X$, $Y$, $Z$ will be ON and the products $x$ and $y$ are in the process of being synthesized. Last, if $y$ is present while $x$ and $z$ are absent, genes $X$ and $Z$ will be OFF but $Y$ will be ON. According to Thomas's procedure [22] one can derive the correspondence between proteins and genes for a positive 3-circuit (Table 2).

For a circuit involving three genes and three proteins there are eight different states in the binary variables ($x$, $y$, $z$ and $X$, $Y$, $Z$). Except for two states, i.e., 010 and 101, which are identical in protein and gene languages, all the other states written in their corresponding language are different (Table 2). This situation expresses the lack of synchrony of the system. The two states 010 and 101 correspond in fact to stable states whereas all the others are unstable. The first state 010 implies that $y$ is present, $x$ and $z$ absent. This means that genes $X$ and $Z$ are OFF whereas $Y$ is ON, or has recently been ON. Similarly the second state 101 implies that $x$ and $z$ are present, $y$ absent and again this means that $X$ and $Z$ are ON, or have recently been ON, whereas $Y$ is OFF.

Table 2

Correspondence between proteins and genes (State Table) for a positive 3-circuit. See text. Adapted from [22].

| $x\ y\ z$ | $X\ Y\ Z$ |
|---|---|
| *0 0 0* | *0 1 1* |
| *0 0 1* | *1 1 1* |
| *0 1 0* | *0 1 0* |
| *0 1 1* | *1 1 0* |
| *1 0 0* | *0 0 1* |
| *1 0 1* | *1 0 1* |
| *1 1 0* | *0 0 0* |
| *1 1 1* | *1 0 0* |

Let us consider now the negative 3-circuit [22]

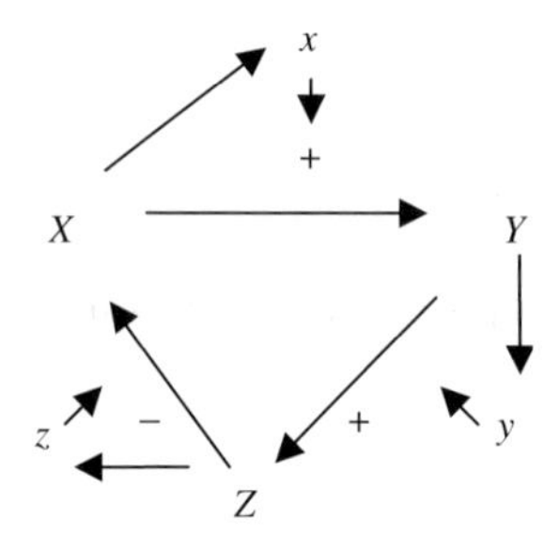


As shown in this circuit, products $x$ and $y$ of genes $X$ and $Y$ activate genes $Y$ and $Z$. Conversely a product, $z$, of gene $Z$ inhibits gene $X$. This means for instance that if the three products $x$, $y$, $z$ are present, genes $Y$ and $Z$ will be ON while gene $X$ will be OFF. Alternatively, if the three products are absent, genes $Y$ and $Z$ are OFF while gene $X$ is ON. As previously, one can set up a State Table (Table 3).

Table 3

Correspondence between proteins and genes (State Table) for a negative 3-circuit. See text. Adapted from [22].

| $x\ y\ z$ | $X\ Y\ Z$ |
|---|---|
| *0 0 0* | *1 0 0* |
| *0 0 1* | *0 0 0* |
| *0 1 0* | *1 0 1* |
| *0 1 1* | *0 0 1* |
| *1 0 0* | *1 1 0* |
| *1 0 1* | *0 1 0* |
| *1 1 0* | *1 1 1* |
| *1 1 1* | *0 1 1* |

In binary alphabet, there are eight possible states of proteins and genes (Table 3). As previously, the results of Table 3 show a lack of synchrony. Thus, for instance, when $x$, $y$, $z$ are absent, this situation is expressed by the state 000. Under these conditions, gene $X$ is ON. This means that the corresponding state for $X$, $Y$, $Z$ is 100. In this table, no state for $x$, $y$, $z$ is identical to the corresponding state for $X$, $Y$, $Z$. Hence no state can be considered stable. In fact, such a situation implies the existence of a six-state cycle viz.

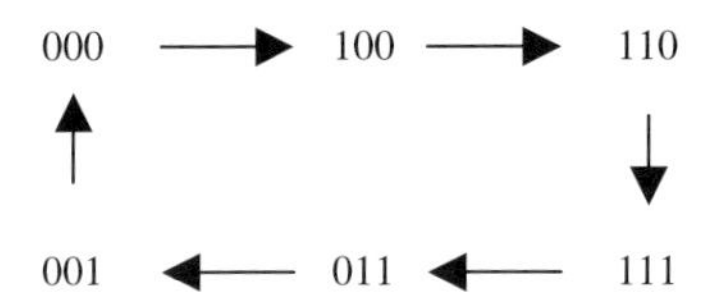


## 4. *Engineered gene circuits*

A very promising approach to the study of gene networks is to construct artificial gene circuits through genetic engineering, insert them in plasmids and study their expression in *E. coli* cells. The guideline for these studies is that gene networks follow the same rules as electric circuits. Experimental results that can be obtained with this approach are usually explained in the light of simple empirical mathematical models.

### 4.1. *The role of feedback loops in gene circuits*

Most studies devoted to the role of feedback loops in the behavior of gene networks were in fact aimed at studying whether or not a negative feedback loop manages to maintain an artificial gene circuit close to a steady state despite unavoidable perturbations. Recent experiments performed in this line [27] use a promoter (the PLtet01 promoter) that controls the *tetR-egfp* gene responsible for the production of tetracycline repressor (TetR-EGFP). The autoregulatory system makes use of the TetR repressor linked to the green fluorescent protein (EGFP). This artificial system displays a negative feedback (Fig. 12) because TetR represses transcription of the PLtet01 operator. If, however, the same promoter controls the modified *tetRY42a-egfp* gene, the structure of the corresponding repressor is altered (Tet-RY42A-EGFP), the negative feedback cannot take place, and the system becomes unregulated (Fig. 9).

A direct experimental test of the stability of the steady states can be obtained from the frequency distribution of the fluorescence intensities of cells labelled with the green fluorescent protein (GFP). These frequency distributions can be obtained by fluorescent microscopy on a statistical sample of labelled cells. One can observe, with the same number of cells in the two samples, that the coefficient of variation (or the standard deviation) is much smaller for the negatively regulated than for the unregulated cells (Fig. 10). As a consequence, the frequency of the modal value of

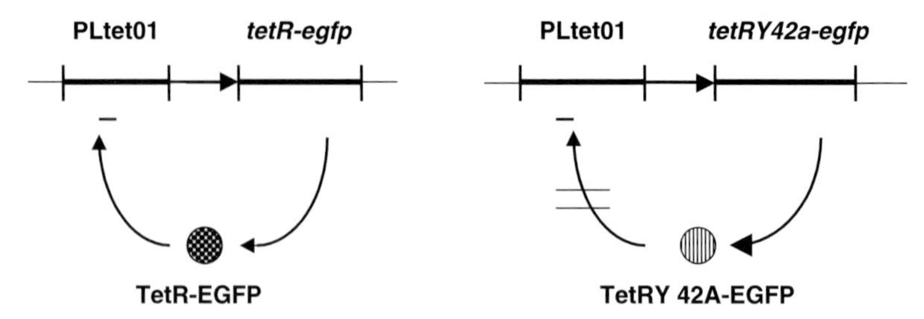


Fig. 9. Negative feedback loop in gene networks. Left: The PLtet01 promoter controls the expression of the *tetR-egfp* gene. In this artificial system, the corresponding protein, TetR-EGFP, represses transcription initiated from PLtet01, thus leading to a negative feedback loop. Right: If *tetR-egfp* is replaced by *tetRY42A-egfp*, the product of expression of this gene, TetRY 42A-EGFP, does not repress the transcription initiated from PLtet01. The corresponding system does not possess any feedback loop. Adapted from [27].

the distribution is much higher for the negatively regulated by feedback, than for the unregulated cells (Fig. 10). As the degree of stability of the system is inversely proportional to the width of the frequency distribution, one can conclude that negative feedback stabilizes the system.

There exists in biology metabolic pathways that display positive feedback control. This is the case, for instance, for the so-called MAP kinase cascade involved in *Xenopus* oocyte maturation [28–31]. As already outlined, a positive feedback control system can display multiple dynamic states. This situation was experimentally discovered in a synthetic network that associates a tetracycline-responsive transactivator (rtTA) and its promoter (tetreg). Moreover, rtTA activates the promoter (Fig. 11). The whole system was implemented in yeast *Saccharomyces cerevisiae*. As previously, the transactivator can be linked with the green protein GFP, thus making it possible to detect the steady state levels by fluorescence microscopy [30,31]. Under certain experimental conditions, the frequency distributions are bimodal, thus suggesting that two classes of cells are present which express different amounts of the protein in the overall cell population (Fig. 11).

The partitioning of the cell population in two classes is not permanent and depends on the experimental conditions, i.e., on the concentration of doxycycline which controls the degree of positive feedback. So far, it has been possible to observe the conversion of cells that have low fluorescence intensity into cells endowed with high fluorescence intensity, but never the converse. Although bistability can, in principle, be obtained from a simple positive feedback, it is probably more often obtained through the association of two negative feedbacks. This is not surprising, since we have already seen that two negative feedbacks generate a positive feedback circuit [30–32]. This situation has been experimentally tested with an artificial engineered network called the "genetic toggle switch" [33]. The system is composed of two repressors, LacI and CI, products of the genes *lacI* and *cits*, and two promoters, PLs1con and Ptrc-2 (Fig. 12). PLs1con controls gene *lacI* and Ptrc-2 gene *cits*. Each promoter is inhibited by the repressor expressed by the opposite promoter. Last, two inducers can block the interactions between repressor 1 and promoter 1,

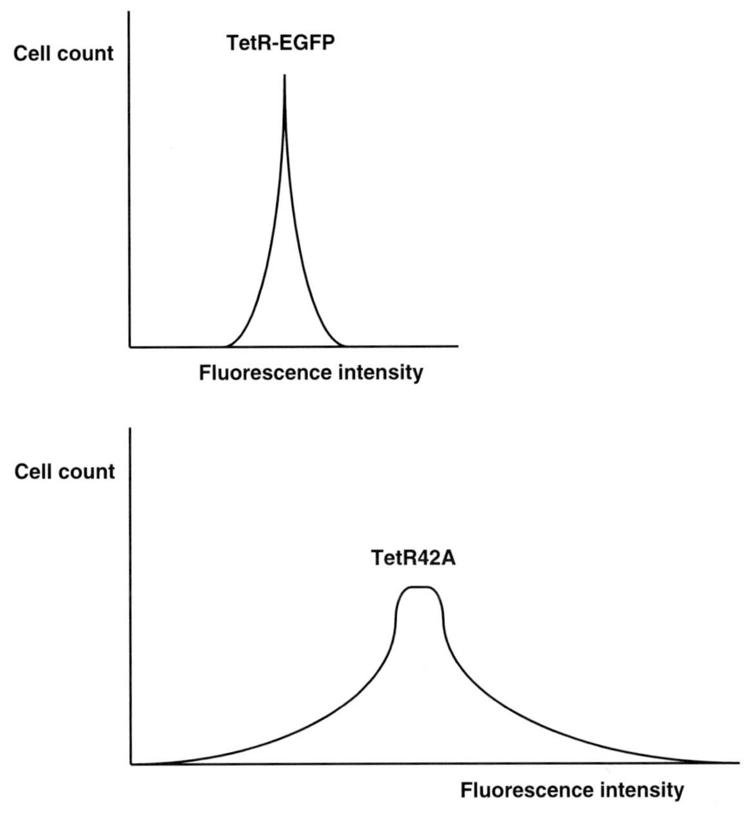


Fig. 10. Stability and negative feedback loops. One can measure the fluorescence of green fluorescent protein (GFP) in the cells through fluorescence microscopy and study the distribution of fluorescence intensity. Top: EGFP is coupled to TetR (TetR-EGFP). The standard deviation of the distribution is small. Bottom: EGFP is coupled to TetRY42A (TetRY42A-EGFP). The standard deviation of the distribution is large. Hence a system displaying negative feedback loop is more stable that of an unregulated one. Adapted from [27].

repressor 2 and promoter 2. Hence one can switch from a state where Ptrc-2 is very active to another state where it is PLs1con which is maximally active. Such a situation is also found in nature in the case of $\lambda$ phage. Gene *cro*, controlled by promoter PR, is expressed as repressor Cro that represses another promoter PRM and conversely CI, controlled by promoter PRM, represses promoter PR (Fig. 13). As shown below, one can jump from one state to the other by acting on either a chemical, or a physical factor that can modulate the inhibition of the promoters.

An example of the reality of this switch has been shown by using, with the artificial system described in Fig. 12, isopropyl-$\beta$-D-thiogalactoside (IPTG) which binds to the repressor lacI, thus preventing the inhibition of Ptrc-2. Alternatively, a temperature of 42°C promotes the denaturation of protein CI and therefore abolishes the inhibition of PLs1con. Hence IPTG and temperature can be used as switches of the system. If one follows experimentally the expression of green

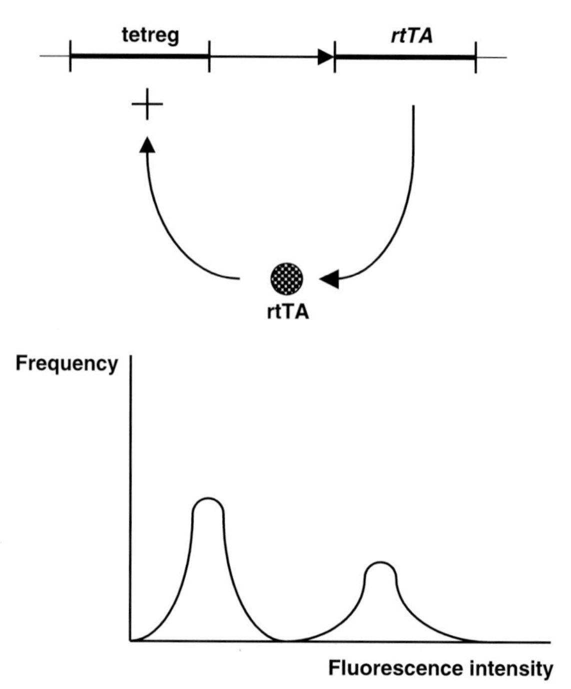


Fig. 11. Positive feedback and multistability. Top: The promoter tetreg controls the expression of the gene *rtTA*. This gene is expressed as a protein rtTA that activates promoter tetreg. Such a system constitutes a positive feedback loop. Bottom: If rtTA is linked with the green fluorescent protein one can measure, by fluorescence microscopy, the distribution of fluorescence intensity in the cells. One can obtain under certain conditions a bimodal distribution. This result leads to the conclusion that bistability requires positive feedback. Adapted from [32].

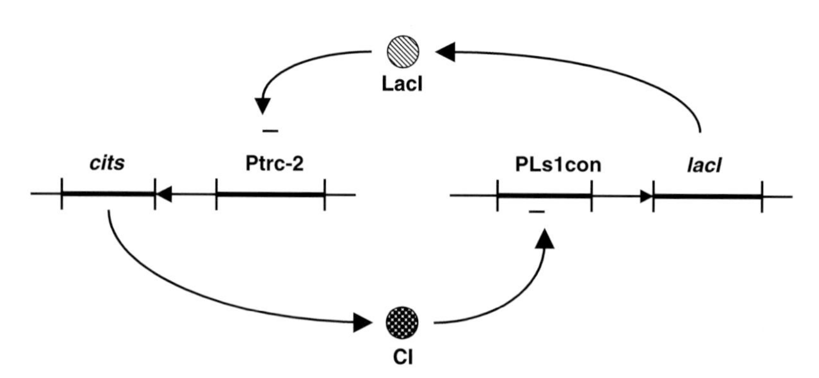


Fig. 12. The artificial genetic toggle switch. In this artificial system, two promoters, Ptrc-2 and PLs1con, control the transcription of two genes *cits* and *lacI*, respectively. Moreover Ptrc-2 is repressed by the product of *lacI*, LacI. Similarly, PLs1con is repressed by the product of *cits* called CI. The switch is therefore made up of two negative feedback loops. Adapted from [33].

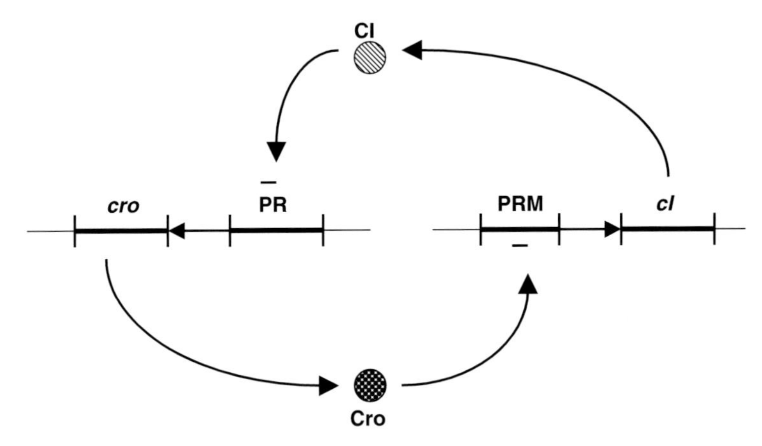


Fig. 13. A natural switch. In this natural system, two promoters, PR and PRM, control the transcription of two genes *cro* and *cI*, respectively. PR is repressed by CI, the product of *cI*. Similarly, PRM is repressed by Cro, the product of *cro*. Adapted from [35].

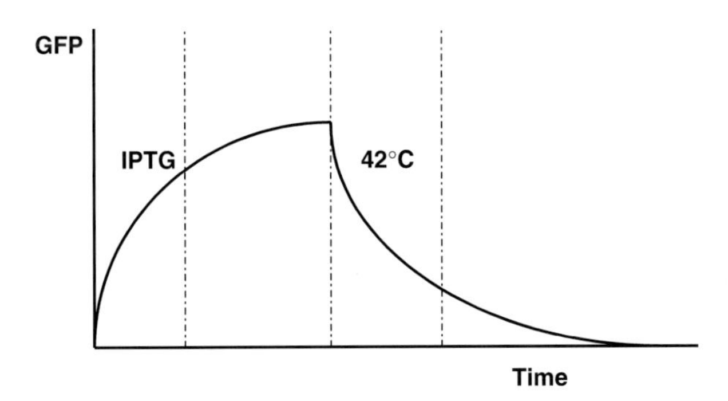


Fig. 14. Roles of IPTG and "high" temperature as switches. In the presence of IPTG, which binds to repressor LacI, repression of Ptrc-2 by this protein is not possible anymore. Hence the fluorescence of cells increases. Increasing the temperature to 42°C leads to denaturation of CI, PLs1con is derepressed and the green fluorescence declines to its initial value. Adapted from [33].

fluorescent protein in *E. coli* cells, one can observe that IPTG treatment promotes synthesis of the green fluorescent protein. This process remains at a high value even if the inducer is removed (Fig. 14). But if the temperature is then brought to 42°C, the cell population is forced to the low green fluorescent protein level (Fig. 14).

### *4.2. Periodic oscillations of gene networks*

Recently, attempts have been made at assembling artificial gene networks that can display sustained periodic oscillations of the synthesis of cellular proteins. A simple synthetic system called repressilator [34–36] is made up of three genes, *tetR*, *cI*, and *lacI* as well as their corresponding promoters PLlac01, PLtct01, and PR. In this system, the lacI protein represses the promoter for the *tetR* gene, the TetR protein represses the promoter for the *CI* gene, and the CI protein represses the promoter for the *lac* gene. Elowitz and Leibler [34] have been looking for experimental conditions that generate oscillations of the system and they found that periodic dynamics takes place if repression is both tight and cooperative and if the rates of decay of mRNA and proteins are similar. This last condition can be obtained if the proteins bear tags that markedly increase their degradation rates.

### *4.3. The logic of gene networks*

We have previously mentioned that engineered gene circuits can be described in the language of electronic circuits. If a gene network requires for its activity RNA polymerase, as usual, and also an activator, such that RNA polymerase binds to the promoter and the activator to the operator (Fig. 15), this situation is equivalent in electronic circuits to an AND gate. It is then possible to establish the following Truth Table [35].

| RNA pol. | Activator | Output |
|---|---|---|
| 0 | 0 | 0 |
| 1 | 0 | 0 |
| 0 | 1 | 0 |
| 1 | 1 | 1 |

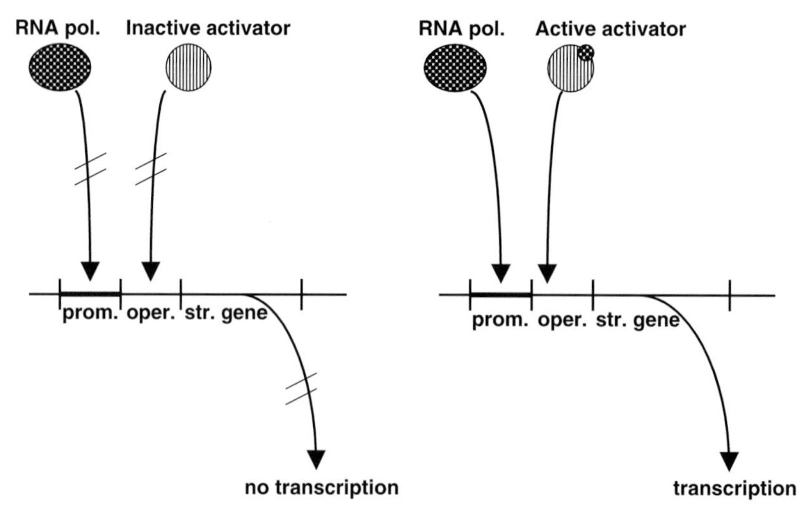


Fig. 15. Genetic networks as logic gates. If a genetic network requires for its activity, in addition to RNA polymerase, an active activator it is equivalent to an AND gate. The small "black" circle is a ligand that activates the previously inactive activator. See text. Adapted from [35].

An equivalent situation is obtained, for instance, in the following engineered gene circuit. A first promoter directs the transcription of *lacI* and *tetR* genes [35]. The second promoter is repressed by either the Lac, or the Tet, protein. If the two chemicals IPTG and anhydrotetracycline (aTc) are both present, the Lac and Tet repressors are inactivated and *gfp* gene located downstream of promoter 2 is expressed, leading in turn to the expression of the green fluorescent protein which can be detected experimentally. This gene circuit can be considered an AND gate because the two chemicals IPTG and aTc must be present in order to obtain the expression of the green fluorescent protein (Fig. 16). This situation can be complicated slightly by "adding memory" to the AND circuit. The network is the same as that presented above, except that an additional gene (*cI*) is inserted between promoter 2 and *gfp* (Fig. 16). If protein CI represses promoter 1 once the system is switched to the ON state it will remain in this state regardless of whether the inducers are present or not, for the expression of protein CI will repress the production of both LacI and TetR.

These studies on the dynamics of engineered gene networks are interesting for several reasons. First, these studies allow us to fill the gap between theoretical and experimental work. Second, they show that the internal logic that governs gene

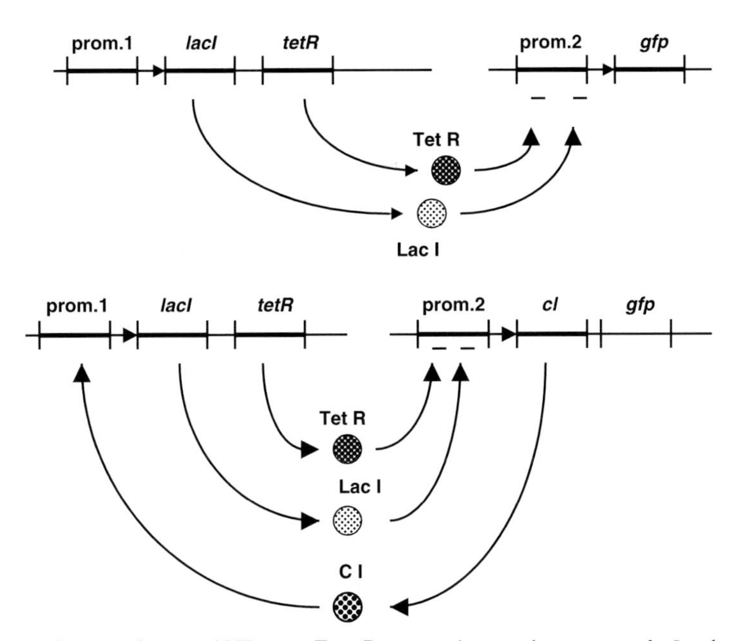


Fig. 16. A genetic network as an AND gate. Top: Promoter 1 controls two genes *lacI* and *tetR* that are expressed as proteins LacI and TetR, respectively. A second promoter that controls gene *gfp* is repressed by either LacI or TetR. In the presence of IPTG and aTc the two repressors are inactivated and the gene *gfp* can be expressed. The system constitutes an AND gate. Bottom: Memory can be introduced in the system by adding in the system another gene, *cI*, under the control of promoter 2 and making use of a promoter 1 repressed by the product of *cI*, CI. Hence when the system is switched ON by both ITPG and aTc, it will remain in this state. The AND gate possesses a memory. Adapted from [35].

circuits is basically the same as that which applies to any network. Third, they leave open the possibility that different gene networks can display synchronization of their dynamic behavior. As a matter of fact, several recent experimental results strongly suggest this may be so [30]. Whatever that may be, these studies on engineered gene networks offer a new example of the potential efficiency of the biomimetic approach of biological phenomena.

## *References*


1. Jacob, F. and Monod, J. (1961) Genetic regulatory mechanisms in the synthesis of proteins. J. Mol. Biol. 3, 318–356.
2. Jacob, F. and Monod, J. (1961) On the regulation of gene activity. Cold Spring Harbor Symp. Quant. Biol. 26, 193–211.
3. Monod, J. and Jacob, F. (1961) General conclusions: teleonomic mechanisms in cellular metabolism, growth and differentiation. Cold Spring Harbor Symp. Quant. Biol. 26, 389–401.
4. Lewin, B. (1997) Genes (6th edition). Oxford University Press, Oxford.
5. Gilbert, W. and Muller-Hill, B. (1966) Isolation of the *lac* repressor. Proc. Natl. Acad. Sci. USA 56, 1891–1898.
6. Friedman, A.M., Fischmann, T., and Steitz, T.A. (1995) Crystal structure of the *lac* repressor core tetramer and its implications for DNA looping. Science 268, 1721–1727.
7. Lewis, M., Chang, G., Horton, N.C., Kercher, M.A., Pace, H.C., Schumacher, M.A., Brennan, R.G., and Lu, P. (1996) Crystal structure of the lactose operon repressor and its complexes with DNA and inducer. Science 271, 1247–1254.
8. Waddington, C.H. (1949) The genetic control of development. Symp. Soc. Exp. Biol. 2, 145–154.15.
9. Delbrück, M. (1949) Unités biologiques douées de continuité génétique. Colloq. Int. CNRS 8, 33–35.
10. Novick, A. and Weiner, M. (1957) Enzyme induction as an all-or-none phenomenon. Proc. Natl. Acad. Sci. USA 43, 553–566.
11. Cohn, M. and Horibata, K. (1959a) Inhibition by glucose of the induced synthesis of the beta-galactosidase-enzyme system of *Escherichia coli.* Analysis of maintenance. J. Bact. 78, 601–612.
12. Cohn, M. and Horibata, K. (1959b) Analysis of the differentiation and of the heterogeneity within a population of *Escherichia coli* undergoing induced beta-galactosidase synthesis. J. Bact. 78, 613–623.
13. Cohn, M. and Horibata, K. (1959c) Physiology of the inhibition by glucose of the induced synthesis of the beta-galactosidase-enzyme system of *Escherichia coli.* J. Bact. 78, 624–635.
14. Thomas, R. (1973) Boolean formalization of genetic control circuits. J. Theor. Biol. 42, 563–585.
15. Thomas, R. (1978) Logical analysis of systems comprising feedback loops. J. Theor. Biol. 73, 631–656.
16. Thomas, R. (1991) Regulatory networks seen as asynchronous automata: a logical description. J. Theor. Biol. 153, 1–23.
17. Thomas, R. and D'Ari, R. (1990) Biological feedback. CRC Press, Boca Raton, Florida.
18. Thomas, R., Thieffry, D., and Kaufman, M. (1995) Dynamical behaviour of biological regulatory networks. I. Biological role of feedback loops; practical use of the concept of loop-characteristic state. Bull. Math. Biol. 57, 247–276.
19. Thomas, R. (1998) Laws for the dynamics of regulatory networks. Int. J. Dev. Biol. 42, 479–485.
20. Thomas, R. (1999) Deterministic chaos seen in terms of feedback circuits: analysis, synthesis, "labyrinth chaos". Int; J. Bifurcation Chaos Appl. Sci. Eng. 9, 1889–1905.
21. Thomas, R. and Kaufman, M. (2001) Multistationarity, the basis of cell differentiation and memory. I. Structural conditions of multistationarity and other nontrivial behavior. Chaos 11, 170–179.

22. Thomas, R. and Kaufman, M. (2001) Multistationarity, the basis of cell differentiation and memory. II. Logical analysis of regulatory networks in terms of feedback circuits. Chaos 11, 180–195.
23. Minorsky, N. (1964) Nonlinear problems in physics and engineering pp. 321–390. In The Mathematics of Physics and Chemistry, H. Margenau and G.M. Murphy eds. Van Nostrand, New York.
24. Ricard, J. (1999) Biological Complexity and the Dynamics of Life Processes. Elsevier, Amsterdam.
25. Nicolis, G. and Prigogine, I. (1997) Self-Organization in Nonequilibrium Systems. From Dissipative Structure to Order through Fluctuations. John Wiley and Sons, New York.
26. Pavlidis, T. (1973) Biological Oscillators: Their Mathematical Analysis. Academic Press, New York.
27. Becskei, A. and Serrano, L. (2000) Engineering stability in gene networks by autoregulation. Nature 405, 590–593.
28. Ferrell, J.E. Jr. (2002) Self-perpetuating states in signal transduction: positive feedback, double negative feedback and bistability. Curr. Opin. Cell. Biol. 14, 140–148.
29. Keller, A. (1995) Model genetic circuits encoding autoregulatory transcription factors. J. Theor. Biol. 172, 169–185.
30. Ferrell, J.E. Jr. (1999) Xenopus oocyte maturation: new lessons from a good egg. BioEssays 21, 833–842.
31. Ferrell, J.E. Jr. (1999) Building a cellular switch: more lessons from a good egg. BioEssays 21, 866–870.
32. Becskei, A., Seraphin, B., and Serrano, L. (2001) Positive feedback in eukaryotic gene networks: cell differention by graded to binary response conversion. EMBO J. 20, 2528–2535.
33. Gardner, T.S., Cantor, C.R., and Collins, J.J. (2000) Construction of a genetic toggle switch in *Escherichia coli*. Nature 403, 339–342.
34. Elowitz, M.B., and Leibler, S.A. (2000) A synthetic oscillatory network of transcriptional regulators. Nature 403, 335–338.
35. Hasty, J., McMillen, D., and Collins, J.J. (2002) Engineered gene circuits. Nature 420, 224–230.
36. Hasty, J., Dolnik, M., Rottschafer, V., and Collins, J.J. (2002) A synthetic gene network for entraining and amplifying cellular oscillations. Phys. Rev. Lett. 88, 148101-1–148101-4.

J. Ricard *Emergent Collective Properties, Networks and Information in Biology*

DOI: 10.1016/S0167-7306(05)40011-3

CHAPTER 11

# Stochastic fluctuations and network dynamics

J. Ricard

In a living cell the number of molecules that can interact is usually so small that molecular noise, i.e., collisions and thermal motion, should explicitly be taken into account if one wishes to understand the internal logic of molecular processes that take place in simple living systems. Moreover how is it possible to explain that, within the cell, the elementary processes are subjected to molecular noise whereas the biological functions that rely upon these processes appear strictly deterministic and devoid of any noise? The aim of this chapter is an attempt at answering these questions. The physics of intracellular noise is briefly discussed first, in particular the random walk of a molecule, the principle of detailed balance, Langevin and Fokker–Planck equations. Then is presented a model that allows one to understand, on a rather intuitive basis, how molecular noise may drive periodic oscillations in a model of genetic network.

*Keywords:* control of intracellular noise, detailed balance, Fokker–Planck equation, generalized Fokker–Planck equation, intracellular noise, Langevin equation, master equation, noise-driven properties.

The mechanism of chemical and biochemical reactions has been usually studied under experimental conditions in which the classical concepts of activity and concentration are meaningful. Thus, for instance, in free solution, the number of molecules that can react or interfere with each other is so large that molecular noise cannot play any part in the overall dynamics of the chemical reaction. In a living cell, however, the number of molecules that can interact in a chemical process is small so that collisions and thermal motion, i.e., molecular noise, must explicitly be taken into account if one wishes to study the dynamics of the chemical reaction within the cell. Under these conditions, the classical concepts of activity and concentration become meaningless and should be replaced by that of probability of occurrence of a molecule at a given place and at a given time. So far, in the previous chapters, we have neglected the influence of molecular noise. It is now time to consider this important matter.

Several questions can be raised. A molecule that undergoes a chemical reaction can be compared to a ball dribbled over the grass by several soccer players [1]. It receives kicks which, for an unacquainted spectator, look random. But, on the other hand, its velocity is slowed down by the grass of the soccer ground. Is it possible to describe the trajectory of the ball in quantitative terms? If we come back to our molecule that receives random pushes from its neighbors and nevertheless undergoes chemical reaction, or is involved in catalysis of a chemical reaction, one can wonder whether it is possible to describe its behavior in physical terms. Another problem arises which is as follows. How is it possible to explain that, within the cell, the elementary processes are subjected to molecular noise whereas the biological functions that rely upon these processes appear strictly deterministic and devoid of any noise? Last but not least, one may wonder whether cell fate and population diversity is not driven in part by molecular noise.

## *1. The physics of intracellular noise*

As outlined above, the classical concept of concentration cannot apply to small systems that contain a small number of reagent molecules. Under these conditions the concepts of probability of occurrence and of random walk have to be used to describe the behavior of molecules in small systems.

### *1.1. Random walk and master equation*

Let us consider a small particle of matter, for instance a macromolecule that receives pushes from other molecules in such a way that it can occupy successive positions $m$ $(0, \pm1, \pm2, \pm3, \ldots)$ in one dimension (see Fig. 1). More precisely, let us assume that a macromolecule that has received, at time $t$, $n+1$ pushes hops and occupies the position $m$. The corresponding probability of occurrence is $p(m; n+1)$. This implies that immediately before the macromolecule hops to occupy the position $m$, its probability of occurrence was $p(m-1; n)$. The corresponding process is shown in Fig. 1 and the transition probability between positions $m-1$ and $m$ is $\tau(m, m-1)$. One has

$$p(m; n+1) = \tau(m, m-1)p(m-1; n) \tag{1}$$

But, in one dimension, the macromolecule may also possibly, at the same time $t$, hop from position $m+1$ to position $m$. Then one has to use the transition probability, $\tau(m, m+1)$, between positions $m$ and $m+1$. In order to take account of the existence

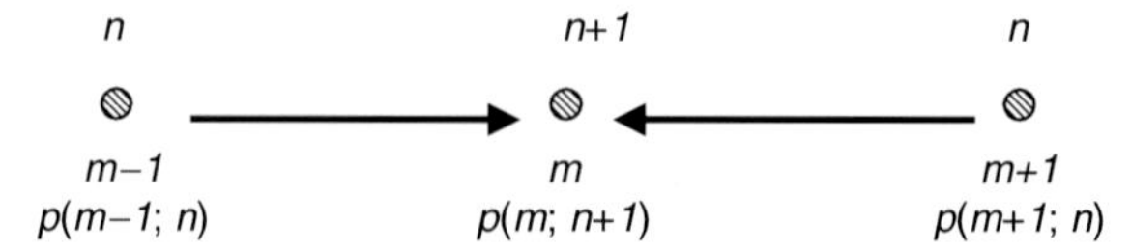


Fig. 1. Definition of a transition probability in one dimension. A particle, or a molecule, has received at time $t$, $n+1$ pushes and occupies position $m$. At time $t-1$, the particle was either in position $m-1$ or $m+1$ and, in any case, has so far received $n$ pushes. The transition probabilities between positions $m-1$ and $m$ or $m+1$ and $m$ are designated $\tau(m, m-1)$ and $\tau(m, m+1)$, respectively. See text.

of these two pathways, Eq. (1) has to be rewritten as

$$p(m; n+1) = \tau(m, m-1)\,p(m-1; n) + \tau(m, m+1)\,p(m+1; n) \tag{2}$$

The definition of a probability requires that

$$\tau(m, m-1) + \tau(m, m+1) = 1 \tag{3}$$

Equations (2) and (3) could easily be generalized if the successive positions of the molecule were defined in two, or three, dimensions instead of one only.

The definition of the probability of occurrence of a molecule at a point $m$ and immediately after step $n+1$ can also be generalized. Thus, for instance, $p(m, n+1; m', n)$ means that the molecule is at a point $m'$ at step $n$ and at a point $m$ at step $n+1$. This probability can be expressed as the probability of occurrence of the molecule at position $m'$ after step $n$, i.e., $p(m'; n)$ times the transition probability that the molecule passes from position $m'$ to $m$, i.e., $\tau(m, m')$. Hence one has

$$p(m, n+1; m', n) = \tau(m, m')p(m'; n) \tag{4}$$

Moreover as the molecule jumps from position $m$ to position $m'$ it travels an "elementary distance" of one, which implies that

$$m' = m \pm 1 \tag{5}$$

Now, if we wish to express the transition time from position $m$ to $m \pm 1$ per unit time, one has

$$\frac{\tau(m, m \pm 1)}{t} = \tau^*(m, m \pm 1) \tag{6}$$

Moreover taking advantage of Eqs. (2), (3), and (6), one can write

$$\begin{aligned} \frac{p(m; n+1) - p(m; n)}{t} = {} & \frac{\tau(m, m-1)}{t} p(m-1; n) \\ & + \frac{\tau(m, m+1)}{t} p(m+1; n) - \frac{p(m; n)}{t} \end{aligned} \tag{7}$$

This expression can be rearranged to

$$\begin{aligned} \frac{p(m; n+1) - p(m; n)}{t} = {} & \tau^*(m, m-1)p(m-1; n) + \tau^*(m, m+1)p(m+1; n) \\ & - \left\{\tau^*(m-1, m) + \tau^*(m+1, m)\right\}p(m; n) \end{aligned} \tag{8}$$

If a large number of steps $n$ are jumped over after time $t$, Eq. (8) can be approximated to

$$\frac{dp(m,t)}{dt} = \tau^*(m,m-1)\,p(m-1;n) + \tau^*(m,m+1)\,p(m+1;n) - \left\{\tau^*(m-1,m) + \tau^*(m+1,m)\right\} p(m;n) \qquad (9)$$

This differential equation is termed the *master equation* and reaches a steady state when

$$\tau^*(m,m-1)p(m-1;n) + \tau^*(m,m+1)p(m+1;n) = \left\{\tau^*(m-1,m) + \tau^*(m+1,m)\right\}p(m,n) \qquad (10)$$

If this relationship is fulfilled, the system is under *detailed balance* conditions.

### *1.2. Detailed balance*

One can give a more general expression of the master equation and of the principle of detailed balance. Let us consider a state, $m_i$, of the system which is connected to other states, $m_j (\forall j = 1,2,3\ldots)$. At time $t$ the probability of occurrence of state $m_i$ is $p(m_i, t)$ and the corresponding transition probabilities per unit time are $\tau^*(m_j, m_i)$. But the states $m_j$ are also connected to the state $m_i$ through transition probabilities per unit time, $\tau^*(m_i, m_j)$. Moreover, the probabilities of occurrence of the states $m_j$ at time $t$ are $p(m_i, t)$ (Fig. 2). Hence the differential equation that describes the variation of $p(m_i, t)$ is

$$\frac{dp(m_i,t)}{dt} = \sum_j \tau^*(m_i,m_j)p(m_j,t) - p(m_i,t)\sum_j \tau^*(m_j,m_i) \qquad (11)$$

The principle of detailed balance requires that there be as many transitions per second from states $i$ to $j$ as from states $j$ to $i$, i.e.,

$$p(m_i)\tau^*(m_j,m_i) = p(m_j)\tau^*(m_i,m_j) \qquad (\forall i,j \in N) \qquad (12)$$

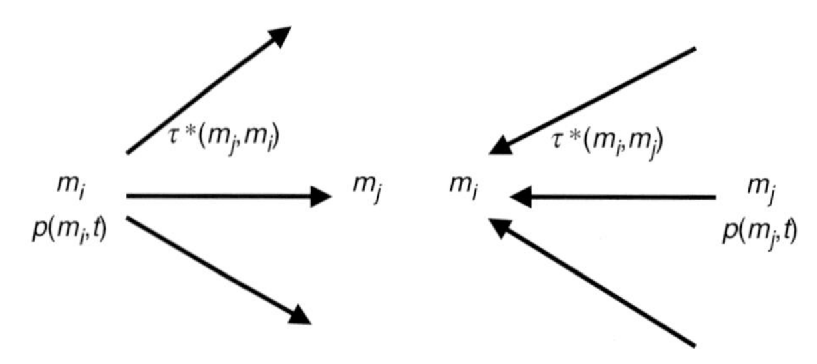


Fig. 2. The principle of detailed balance. Under the conditions of detailed balance one has the necessary (but not sufficient) condition $dp(m_i,t)/dt = 0$. See text.

If the system is under thermodynamic equilibrium, detailed balance of necessity applies for it is then equivalent to classical microscopic reversibility. If, from a thermodynamic viewpoint, the system is open, or if it collects a small number of reacting molecules, detailed balance will not, in general, be fulfilled. In the case of small systems involving a small number of molecules, molecular noise, which is responsible for the departure from detailed balance, generates unexpected behavior, as we shall see later.

Detailed balance is not the only condition that results in the expression

$$\frac{dp(m_i, t)}{dt} = 0 \tag{13}$$

As a matter of fact, if Eq. (11) applies but if

$$\tau^*(m_i, m_j)\, p(m_j) \neq p(m_i)\tau^*(m_j, m_i) \tag{14}$$

the system will not display detailed balance but will be in steady state.

It is easy to illustrate these ideas on the simple enzyme network of Fig. 3. Steady state of the system requires that

$$\begin{aligned}
\tau^*(m_1, m_2)p(m_2) + \tau^*(m_1, m_3)p(m_3) &= \tau^*(m_2, m_1)p(m_1) \\
\tau^*(m_2, m_1)p(m_1) + \tau^*(m_2, m_3)p(m_3) &= \left\{\tau^*(m_1, m_2) + \tau^*(m_3, m_2)\right\}p(m_2) \\
\tau^*(m_3, m_2)p(m_2) &= \left\{\tau^*(m_2, m_3) + \tau^*(m_1, m_3)\right\}p(m_3)
\end{aligned} \tag{15}$$

These expressions may well all be fulfilled even if the three equations

$$\begin{aligned}
\tau^*(m_2, m_1)p(m_2) &= \tau^*(m_1, m_2)p(m_1) \\
\tau^*(m_3, m_2)p(m_2) &= \tau^*(m_2, m_3)p(m_3) \\
\tau^*(m_3, m_1)p(m_1) &= \tau^*(m_1, m_3)p(m_3)
\end{aligned} \tag{16}$$

are not, for the step $EP(m_3) \rightarrow E(m_1)$ is irreversible (Fig. 3). Hence detailed balance cannot apply to this system, which is away from thermodynamic equilibrium.

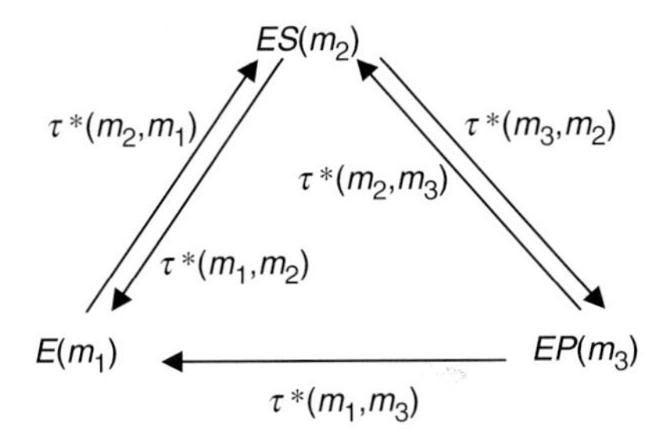


Fig. 3. A system in which detailed balance does not apply. This system could represent a simple one-substrate, one-product enzyme reaction under initial steady state conditions. The system cannot display detailed balance for one step is nearly irreversible. See text.

If detailed balance applies and if the nodes $i$ and $j$ are directly connected, one has

$$\frac{p(m_i)}{p(m_j)} = \frac{\tau^*(m_i, m_j)}{\tau^*(m_j, m_i)} \tag{17}$$

If, however, $n$ steps separate nodes $i$ and $j$, the ratio $p(m_i)/p(m_j)$ is expressed by the ratio of the products of transition probabilities pertaining to the $n$ steps. From an energy viewpoint, this ratio is distributed according to Gibbs–Boltzmann statistics. This point, which has already been outlined, need not be discussed here.

If the system is in steady state, the probabilities $p(m_j)$ can be determined from transition probabilities using Eq. (13) and from the axiomatic definition of a probability, i.e.,

$$\sum_j p(m_j) = 1 \tag{18}$$

In fact, one of the steady state Eq. (13) should be replaced by Eq. (18) so as to solve the Cramer system thus formed. This is in fact the very basis of the method of King and Altman [2] as well as that of Hill [3–8]. These methods are now widely used in scientific literature. Both are in fact derived from the original study of electric circuits by Kirchoff. The language of graph theory is probably best suited to describe how to handle these methods schematically. A network can be considered a graph containing all nodes and edges (see Fig. 4 for a simple graph with three vertices and edges). Expression of the probabilities can be obtained from all the directed trees leading to any of the three vertices (Fig. 4). The directed trees should meet the following requirements:

All the edges and vertices should be edges and vertices of the graph;
They should be devoid of any circuit, i.e., they should not possess any closed loop;
They should be connected.

Hence, in the case of the graph of Fig. 4, there are six directed trees that allow one to derive the expressions of the probabilities $p(m_1)$, $p(m_2)$, and $p(m_3)$. One finds

$$\begin{aligned} p(m_1) &= \frac{\tau^*(1,2)\tau^*(1,3) + \tau^*(1,3)\tau^*(3,2) + \tau^*(1,2)\tau^*(2,3)}{\Delta} \\ p(m_2) &= \frac{\tau^*(2,1)\tau^*(2,3) + \tau^*(2,1)\tau^*(1,3)}{\Delta} \\ p(m_3) &= \frac{\tau^*(2,1)\tau^*(3,2)}{\Delta} \end{aligned} \tag{19}$$

In these expressions $\tau^*(i, j)$ is the transition probability in going from state $j$ to state $i$, moreover $\Delta$ is defined as

$$\begin{aligned} \Delta = {} & \tau^*(1,2)\tau^*(1,3) + \tau^*(1,3)\tau^*(3,2) + \tau^*(1,2)\tau^*(2,3) + \tau^*(2,1)\tau^*(2,3) \\ & + \tau^*(2,1)\tau^*(1,3) + \tau^*(2,1)\tau^*(3,2) \end{aligned} \tag{20}$$

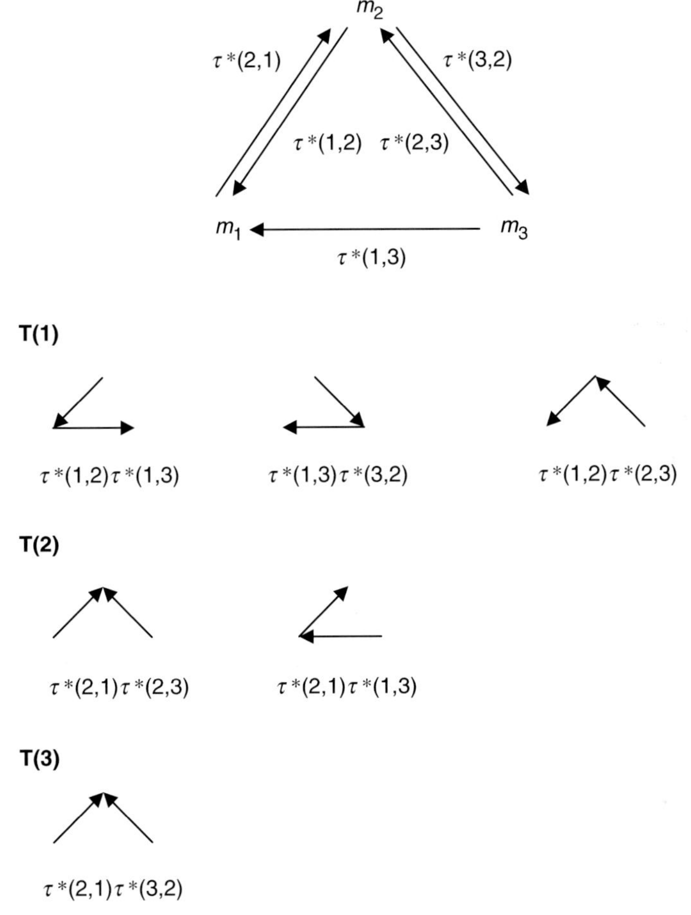


Fig. 4. The principle of the method of King, Altman, and Hill for obtaining the probabilities of occurrence of the nodes. The probabilities of occurrence of the nodes can be obtained from the transition probabilities. See text.

### *1.3. Intracellular noise and the Langevin equation*

If a chemical, or a biochemical reaction is submitted to molecular noise, molecular fluctuations should be added to the deterministic expression of the rate process. This is usually done by adding stochastic processes to the deterministic component of the chemical reaction. Thus, in the case of a chemical process, which is indeed a function of the concentration $C$, one should have

$$\frac{dC}{dt} = f(C) + F(t) \tag{21}$$

This equation, referred to as the chemical Langevin equation, implies that the reaction rate, $dC/dt$, is a function of the classical deterministic component, $f(C)$, plus a stochastic component, $F(t)$, that should be both time-dependent and dependent upon the number of molecules present in the system. The size and evolution with time of the stochastic component $F(t)$ express how molecular noise affects the rate of the process. In this perspective, it is assumed that noise solely originates from an exogenous source. There are many different algorithms that attempt to simulate Langevin equation [9–14]. We do not intend to discuss this matter, but rather we are going to study, on a simple model, the physical nature of molecular noise, the quantitative expression of the role exerted by molecular fluctuations on a rate process, and the conditions required for the emergence of noise-driven novel dynamic behavior of the system.

*1.3.1. The Langevin equation of a macromolecule subjected to random collisions*

Let us consider a macromolecule subjected to a force in a viscous medium, a living cell for instance. Friction is constantly exerted on the macromolecule and slows down its motion in such a way that the friction force is $-\gamma v$. But the macromolecule is pushed back and forth by collisions with other molecules, or small particles. Let us call $\Phi(t)$ the contribution of the collisions to the force exerted on the macromolecule during its migration. The expression for the force reads

$$F = -\gamma v + \Phi(t) \tag{22}$$

As, according to Newton's law,

$$F = m\frac{dv}{dt} \tag{23}$$

it follows that

$$\frac{dv}{dt} = -\frac{\gamma}{m}v + \frac{1}{m}\Phi(t) \tag{24}$$

This equation is a Langevin equation analogous to expression (21). When the macromolecule receives a kick from another molecule, its velocity suddenly increases and the macromolecule changes its trajectory. Probably the best way to describe the kick impulse quantitatively, is to use a Dirac $\delta$-function of strength $\varphi$, thus [1,15]

$$\Phi_i(t_i) = \varphi\delta(t - t_i) \tag{25}$$

The main property of $\delta(t - t_i)$ is

$$\delta(t - t_i) = \begin{cases} 0(t \neq t_i) \\ \infty(t = t_i) \end{cases} \tag{26}$$

Moreover one has

$$\int_{t_i-\varepsilon}^{t_i+\varepsilon} \delta(t-t_i)\,dt = 1 \tag{27}$$

As the macromolecule is subjected to many pushes back and forth, the contribution, $\Phi(t)$, of all these collisions is

$$\Phi(t) = \varphi \sum_i \pm\delta(t-t_i) \tag{28}$$

Here, the plus and minus signs refer to the possibility that the macromolecule is pushed in opposite directions by the collisions. Setting

$$\alpha = \frac{\gamma}{m} \tag{29}$$

and

$$\frac{1}{m}\Phi(t) = F(t) \tag{30}$$

Eq. (24) becomes

$$\frac{dv}{dt} = -\alpha v + F(t) \tag{31}$$

Using the $\delta$-function allows us to understand the effect of a kick on the macromolecule. Combining Eqs. (23) and (25) one finds

$$m\frac{dv}{dt} = \varphi\delta(t-t_i) \tag{32}$$

and integrating both sides of this equation over a short time interval around $t=t_i$ yields

$$m\int_{t_i-\varepsilon}^{t_i+\varepsilon} \frac{dv}{dt}\,d\tau = \varphi \int_{t_i-\varepsilon}^{t_i+\varepsilon} \delta(t-t_i)\,d\tau \tag{33}$$

where $d\tau$ is a differential increment about $t=t_i$. The left-hand member of Eq. (33) is

$$mv(t_i+\varepsilon) - mv(t_i-\varepsilon) = m\Delta v \tag{34}$$

where $\Delta v$ expresses the sudden change of velocity of the macromolecule when it receives a kick from another molecule or a small particle. After integration, the right-hand side member of Eq. (33) becomes

$$\int_{t_i-\varepsilon}^{t_i+\varepsilon} \varphi\delta(t - t_i)\, d\tau = \varphi \tag{35}$$

and comparison of expressions (34) and (35) shows that the sudden change of velocity is equal to $\varphi/m$.

*1.3.2. The function $F(t)$ as the expression of the noise-driven properties*
If one considers the successive discrete, positive and negative, values $u_1, u_2, \ldots u_n$ of the function $F(t)$, which can be obtained from appropriate experiments, these values form a time-series [16,17]. One can define from this time-series monovariate and bivariate moments [16]. Let us consider first the monovariate moments of degree one

$$\begin{aligned} \mu_1(u_i) &= \frac{1}{n-k}\sum_{i=1}^{n-k} u_i \\ \mu_1(u_{i+k}) &= \frac{1}{n-k}\sum_{i=1}^{n-k} u_{i+k} \end{aligned} \tag{36}$$

where $n$ and $k$ are constants. These expressions mean that we split the time-series into two time-series that are shifted by a value $k$ one with respect to the other. The first time-series spans the values

$$u_1, u_2, \ldots, u_i, \ldots, u_{n-k} \tag{37}$$

and the second one the values

$$u_{1+k}, u_{2+k}, \ldots, u_{i+k}, \ldots, u_n \tag{38}$$

Depending on the value of $k$ the two sub-series may, or may not, overlap. If the value of $k$ is such that

$$n < i + 2k \tag{39}$$

then the two sub-series do not overlap. But, for certain values of $i$, it may occur that

$$n > i + 2k \tag{40}$$

and the two sub-series overlap. The two monovariate moments defined by Eqs. (36) are the arithmetic means of the two sub-series. If the number of terms of each sub-series is very large, if the complete time-series is stationary, i.e., has no trend [16], and if the $u$ terms are random, then the arithmetic means defined by Eqs. (36) are equal to zero. This situation can arise if the macromolecule moves in either direction as a consequence of the pushes it receives.

Now it is possible to determine whether the sub-series are correlated or not. This can be tested thanks to the bivariate moment of degree one

$$\mu_{1,1}(u_i, u_{i+k}) = \frac{1}{n-k}\sum_{i=1}^{n-k}(u_i\, u_{i+k}) - \frac{1}{(n-k)^2}\left(\sum_{i=1}^{n-k} u_i\right)\left(\sum_{i=1}^{n-k} u_{i+k}\right) \tag{41}$$

In fact this function is a correlation, or rather a covariance, function. If each function of $u_i$, $F(t)$, is a Dirac $\delta$-function and if the $u$, or $F(t)$ values are distributed according to the Poisson law, one has

$$\langle F(t)F(t+k)\rangle = Q\delta(t-(t+k)) \tag{42}$$

where $Q$ is a parameter that we shall refer to later. The right-hand side member which is the product of two $\delta$-functions is itself a $\delta$-function. If $t \neq t+k$ this $\delta$-function is equal to zero (Eq. (26)). Hence the simplifying assumption that the pushes are expressed by $\delta$-functions implies that the autocovariance function be equal to zero which, as we shall see below, is not necessarily the case for Eq. (42). If the two sub-series $u_i$ and $u_{i+k}$ are made up of random terms, the bivariate moment $\mu_{1,1}(u_i, u_{i+k})$ will be different from zero if $k$ is small in such a way that the two sub-series have many $u$ terms in common. Then relationship (40) applies. But, as soon as $k$ is significantly increased, in such a way that relationship (39) is fulfilled, then the bivariate moment becomes nil. This situation can be best analyzed with the concept of serial correlation [16,17]. Let us first define the two monovariate moments of degree two, viz.

$$\mu_2(u_i) = \frac{1}{n-k}\sum_{i=1}^{n-k} u_i^2 - \frac{1}{(n-k)^2}\left(\sum_{i=1}^{n-k} u_i\right)^2 \tag{43}$$

and

$$\mu_2(u_{i+k}) = \frac{1}{n-k}\sum_{i=1}^{n-k} u_{i+k}^2 - \frac{1}{(n-k)^2}\left(\sum_{i=1}^{n-k} u_{i+k}\right)^2 \tag{44}$$

The serial correlation coefficient can then be defined as

$$r_k = \frac{\mu_{1,1}(u_i, u_{i+k})}{\sqrt{\mu_2(u_i)\mu_2(u_{i+k})}} \tag{45}$$

Hence relation (41) is the counterpart, for a time-series, of a covariance, relations (43), (44) the counterpart of two variances, and relation (45) is equivalent to a correlation coefficient.

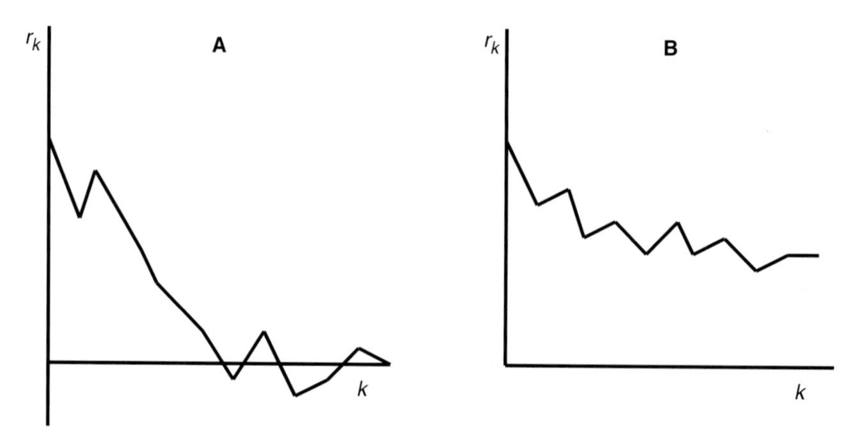


Fig. 5. Variation of the serial correlation, $r_k$, of two sub-series shifted by a value $k$ one with respect to the other, as a function of the shift, $k$. (A) If the two sub-series do not overlap (for $\forall i$, one has $n<i+2k$), there does not exist any correlation between them. Hence the overall time-series that collects the two sub-series is made up of random terms. (B) If the two sub-series do not overlap (for $\forall i$, one has $n<i+2k$)) there is still a correlation between them for $r_k \neq 0$. Molecular noise has driven some sort of organized dynamic behavior.

As expected, it is possible to derive interesting information from the analysis of the variation of $r_k$ as a function of the shift $k$ (Fig. 5). If $k$ is small, $r_k$ is positive, even if the time-series is stationary and made up of random terms. But as $k$ increases, the value of $r_k$ declines, reaches negative values and finally approaches zero. In other words, the curve $r_k = f(k)$ displays damped oscillations (Fig. 5). But it can also happen that the curve does not reach zero even when relationship (39) is fulfilled. This means that the time-series displays some sort of intra-series correlation. Put in other words, this implies that the $u$ terms are not completely random, even if they look so, and therefore that molecular noise has driven organized dynamic behavior.

### *1.4. Intracellular noise and the Fokker–Planck equation*

#### *1.4.1. The Fokker–Planck equation for one-dimensional motion*

Let us consider a macromolecule, or a small particle that moves in the $q$–$t$ plane (Fig. 6). We wish to determine the probability, $p(q,t)$, of finding at time $t$ the macromolecule at a distance $q$. This probability is indeed one if $q = q(t)$, and zero otherwise. Hence the probability density can be expressed by the $\delta$-function

$$p(q, t) = \delta(q - q(t)) \tag{46}$$

Let us consider the integral of this $\delta$-function in the interval $q(t) - \varepsilon$, $q(t) + \varepsilon$

$$\int_{q(t)-\varepsilon}^{q(t)+\varepsilon} \delta(q - q(t))dq = \begin{cases} 1 \\ 0 \end{cases} \tag{47}$$

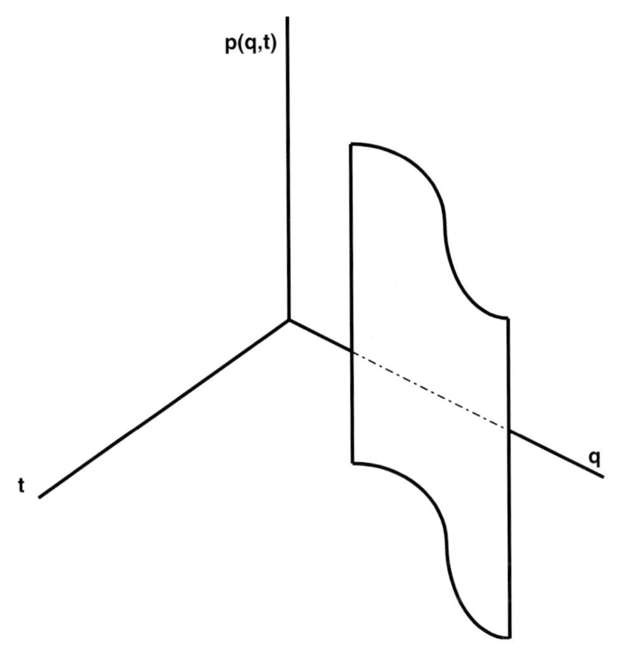


Fig. 6. A probability expressed by a Dirac $\delta$-function. In the course of time, a macromolecule moves along a trajectory in the $q$–$t$ plane. If we want to know the probability that the molecule has travelled distance $q$ at time $t$, this probability will be 1 for $q=q(t)$ and 0 otherwise. Put in other words this means that this probability will be 1 on the surface and 0 away from it. Only a Dirac $\delta$-function can meet this requirement.

This integral vanishes if the integration interval does not contain $q(t)$, and is equal to one otherwise.

If we consider several paths in the $q$–$t$ plane, it is possible to define the mean probability density of finding the macromolecule at distance $q$ and time $t$ as

$$\langle p(q,t)\rangle = f(q,t) = \langle \delta(q-q(t))\rangle \tag{48}$$

The Fokker–Planck equation is the equation that describes the variation of the probability density as a function of time and distance. In order to derive this equation we approximate $df(q,t)$ to [1]

$$\Delta f(q,t) = f(q,t+\Delta t) - f(q,t) \tag{49}$$

Making use of Eq. (48) yields

$$\Delta f(q,t) = \langle \delta(q-q(t+\Delta t))\rangle - \langle \delta(q-q(t))\rangle \tag{50}$$

Setting

$$q(t+\Delta t) \approx q(t) + \Delta q(t) \tag{51}$$

Eq. (50) can be cast into the following form

$$\Delta f(q,t) = \langle \delta(q - q(t) - \Delta q(t)) \rangle - \langle \delta(q - q(t)) \rangle \tag{52}$$

If we compare this expression to a truncated Taylor series

$$f(a+h) = f(a) + hf'(a) + \frac{h^2}{2} f''(a) \tag{53}$$

it follows that

$$\begin{aligned} \langle \delta(q - q(t)) \rangle &= f(a) \\ -\Delta q(t) &= h \end{aligned} \tag{54}$$

Hence one obtains

$$\begin{aligned} hf'(a) &= \left\langle -\frac{d}{dq} \delta(q - q(t)) \Delta q(t) \right\rangle \\ \frac{h^2}{2} f''(a) &= \frac{1}{2} \left\langle \frac{d^2}{dq^2} \delta(q - q(t)) (\Delta q(t))^2 \right\rangle \end{aligned} \tag{55}$$

Therefore Eq. (49) becomes

$$\begin{aligned} \Delta f(q,t) = \langle \delta(q - q(t)) \rangle &- \left\langle \frac{d}{dq} \delta(q - q(t)) \Delta q(t) \right\rangle \\ &+ \frac{1}{2} \left\langle \frac{d^2}{dq^2} \delta(q - q(t)) (\Delta q(t))^2 \right\rangle - \langle \delta(q - q(t)) \rangle \end{aligned} \tag{56}$$

which reduces to

$$\Delta f(q,t) = -\left\langle \frac{d}{dq} \delta(q - q(t)) \Delta q(t) \right\rangle + \frac{1}{2} \left\langle \frac{d^2}{dq^2} \delta(q - q(t)) (\Delta q(t))^2 \right\rangle \tag{57}$$

Now let us consider the Langevin equation

$$\frac{dq(t)}{dt} = -\gamma q(t) + F(t) \tag{58}$$

and let us integrate $dq(t)/dt$ in the interval $\Delta t$. One has

$$\int_{t}^{t+\Delta t} \frac{dq(t')}{dt'} dt' = q(t + \Delta t) - q(t) = \Delta q \tag{59}$$

where $t \leq t' \leq t + \Delta t$. Hence one has

$$\Delta q = -\int_{t}^{t+\Delta t} \gamma q(t')\, dt' + \int_{t}^{t+\Delta t} F(t')\, dt' \tag{60}$$

In order to simplify the mathematics, we assume that, in the interval $\Delta t$, $q(t)'$ has changed very little whereas in the same interval, many pushes have occurred leading to significant changes of $F(t)'$. This means that $q(t)' \approx q$. It follows that

$$\Delta q = -\gamma q \Delta t + F(t + \Delta t) - F(t) = -\gamma q \Delta t + \Delta F(t) \tag{61}$$

Inserting the right-hand member of this expression into the first term of Eq. (57) yields

$$-\left\langle \frac{d}{dq} \delta(q - q(t)) \Delta q(t) \right\rangle = \frac{d}{dq} \{ -\langle \gamma q \Delta t \delta(q - q(t)) \rangle + \langle \delta(q - q(t)) \Delta F \rangle \} \tag{62}$$

In this expression, $\Delta F$ refers to all the kicks that have taken place after time $t$, whereas $q(t)$ is determined by the pushes that have occurred before this time. As these two classes of events are indeed independent, expression (62) becomes

$$-\left\langle \frac{d}{dq} \delta(q - q(t)) \Delta q(t) \right\rangle = \frac{d}{dq} \{ -\langle \gamma q \Delta t \delta(q - q(t)) \rangle + \langle \delta(q - q(t)) \rangle \langle \Delta F \rangle \} \tag{63}$$

Moreover as the kicks take place randomly, $\langle \Delta F \rangle = 0$. Hence it follows that

$$-\left\langle \frac{d}{dq} \delta(q - q(t)) \Delta q(t) \right\rangle = -\gamma \Delta t \frac{d}{dq} \{ \langle \delta(q - q(t)) \rangle q \} \tag{64}$$

The same reasoning allows us to rewrite the second term of Eq. (57) as

$$\frac{1}{2} \left\langle \frac{d^2}{dq^2} \delta(q - q(t)) (\Delta q(t))^2 \right\rangle = \frac{1}{2} \frac{d^2}{dq^2} \langle \delta(q - q(t)) \rangle \langle \Delta q^2 \rangle \tag{65}$$

The expression of $\langle \Delta q \rangle^2$ which appears in Eq. (65) collects the terms $\Delta t^2$, $\Delta F^2$ and $\Delta F \Delta t$. The first of these terms is negligible, the last cancels in such a way that

$$\langle \Delta q^2 \rangle = \langle \Delta F^2 \rangle \tag{66}$$

Moreover as

$$\langle F(t') F(t'') \rangle = Q \delta(t' - t'') \tag{67}$$

it follows that

$$\langle \Delta F \rangle^2 = \int_t^{t+\Delta t} \int_t^{t+\Delta t} Q\delta(t' - t'')dt'dt'' = q\Delta t \tag{68}$$

Hence one has

$$\frac{d^2}{dq^2}\langle\delta(q - q(t))\rangle\langle\Delta q^2\rangle = \frac{d^2}{dq^2}\langle\delta(q - q(t))\rangle Q\Delta t \tag{69}$$

and dividing both members of Eq. (57) by $\Delta t$ yields

$$\frac{\partial\langle p(q,t)\rangle}{\partial t} = -\frac{\partial}{\partial q}\{\gamma q\langle p(q,t)\rangle\} + \frac{1}{2}Q\frac{\partial^2\langle p(q,t)\rangle}{\partial q^2} \tag{70}$$

which is the expression of the Fokker–Planck equation in one dimension. This equation describes the change in the probability distribution of a macromolecule as a function of time. The product $-\gamma q$ is called the drift coefficient and $Q$ is formally equivalent to a diffusion coefficient.

*1.4.2. Generalized Fokker–Planck equation*

The Fokker–Planck equation can be generalized in **q**-space. If we consider the motion of n different types of macromolecules and if we wish to express the joint probability that, at time t, the macromolecules be at fixed positions $q_1, q_2, \ldots, q_n$ this joint probability can be expressed as

$$p(q_1, q_2, \ldots, q_n; t) = p(\mathbf{q}; t) \tag{71}$$

where **q** is the vector of positions. We define a "diffusion coefficient" through the relationship

$$\langle F_i(t)F_j(t')\rangle = Q_{ij}\delta(t - t') \tag{72}$$

and the generalized Fokker–Planck equation assumes the form

$$\frac{\partial}{\partial t}\{p(\mathbf{q}; t)\} = -\nabla\gamma\mathbf{q}p(\mathbf{q}; t) + \frac{1}{2}\sum_i\sum_j Q_{ij}\frac{\partial^2}{\partial q_i \partial q_j}\{p(\mathbf{q}; t)\} \tag{73}$$

where "del" ($\nabla$) is the usual vectorial operator.

## *2. Control and role of intracellular molecular noise*

As already alluded to at the beginning of this chapter, there is little doubt that molecular noise plays an important role in the quantitative and qualitative description of the dynamics of biochemical events as well as in the different fate of cells originating from the division of the same initial cells. Roughly speaking, different problems arise in this context: how is it possible to understand that apparently deterministic biological phenomena are explainable on the basis of elementary stochastic events? How can we explain the robustness of intracellular networks despite molecular noise? What is the origin of noise-driven biological events?

Even though the physics of molecular noise is well understood, the application of physical concepts to biological events is far less satisfactory. It is still probably unavoidable that a number of studies in this domain are empirical and descriptive. Nevertheless, it seems highly probable that the use of physical concepts in the near future will change our views on the logic of the role of molecular noise in cell biology. An example of this kind of promising work is briefly presented below [18]. It seems that the origin of intracellular noise is to be found in the stochastic character of gene expression [19–21]. Proteins appear to be produced in random bursts. This effect is probably amplified by the fact that the same mRNA transcript can be translated several times. The topology of metabolic networks plays a role in the control of noise. Thus it seems that a cascade should display a definite size in order to keep intracellular noise at a minimum value [22,23]. Negative feedback plays an important role in the regulation of metabolism. No doubt it is just as important in the control of noise by providing a simple mechanism that can attenuate random perturbations. Thus for instance the constitutive expression of green fluorescent protein (GFP) by *E. coli* is highly variable. In *E. coli* cells which bear a negative feedback loop such variability is significantly decreased [24]. Probably the most interesting feature of molecular noise is the observation that it can be a functional advantage for the cell. In fact two advantages appear obvious: the emergence of heterogeneity in a formerly homogeneous population of cells and the noise-driven dynamics of biological processes. The first point refers to the problem of development whereas the second refers to the periodicity of many dynamic metabolic processes.

If a dynamic process displays multiple steady states, slight variations of a ligand concentration may result in jumps of the system from one steady state to another (Fig. 7). This is precisely what is occurring with bound enzymes that catalyze chemical reactions. The overall process couples the diffusion of substrate $X$ to membrane-bound enzyme molecules in such a way that the concentration decreases from $X_\alpha$ to $X_\beta$. Substrate $X$ is then transformed into the products of the reaction $Y$ and $Z$ in the immediate vicinity of the enzyme molecules. One has

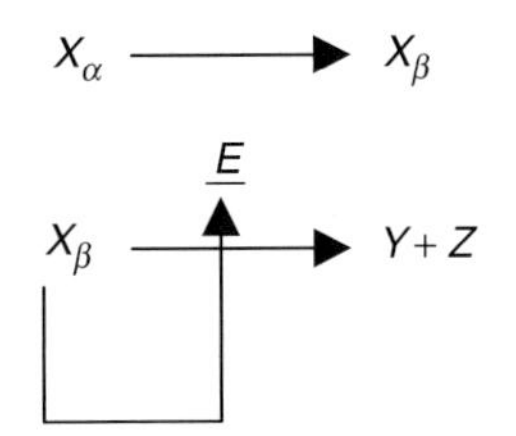

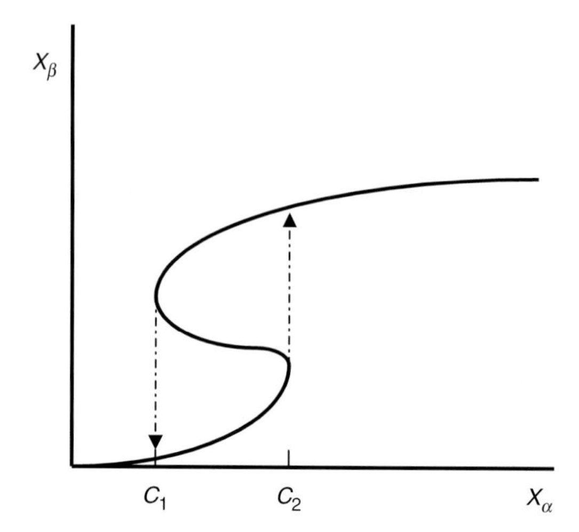


Fig. 7. In the case of a membrane-bound enzyme system, stochastic fluctuations of the bulk number of substrate molecules, $X_\alpha$, in the vicinity of two critical values generate two sets of very different local (close to the membrane) number of substrate molecules, $X_\beta$. When the bulk number of substrate molecules, $X_\alpha$, is very close to either of the critical values $C_1$ or $C_2$, stochastic fluctuations around these values lead to dramatic changes of the local number of substrate molecules, $X_\beta$, which can jump between two sets of values. The local number of substrate molecules should adopt, in space and time, two different sets of values in response to the slight fluctuations of the bulk number of substrate molecules. See text.

If the bound enzyme is inhibited by an excess of substrate $X_\beta$ the coupled system can display three steady states, two of which are stable and one is unstable. Hence the concentration of $X_\beta$ is split into two sets having very different mean values (Fig. 7). If one assumes that the biological effect of $X_\beta$ is different depending on the value of this mean concentration, one can understand that hysteresis has generated biological heterogeneity in a medium that was initially homogeneous. The conditions that are required for hysteresis are nonequilibrium of the system and nonlinearity of the inhibition of the bound enzyme by its substrate $X$.

Probably the most spectacular effect of molecular noise is to drive periodic dynamics of biological systems. A model of particular interest is the system of two genes acting in cooperation [18]. Gene $G_A$ is transcribed as messenger $M_A$ and gene $G_R$ as messenger $M_R$ with rate constants $\kappa_A$ and $\kappa_B$, respectively. The two messengers are subjected to hydrolysis with rate constants $\kappa_{MA}$ and $\kappa_{MR}$ and are also expressed as two proteins $P_A$ and $P_R$ that act as gene activator and repressor, respectively. The activator and repressor are bound to genes $G_A$ and $G_R$ which are then in the states $G_AP_A$ and $G_RP_R$. Under these forms, their transcription as messenger RNA is altered (rate constants $\kappa'_A$ and $\kappa'_R$). The proteins $P_A$ and $P_R$ can associate to form a dimer $P_AP_R$. They can also undergo hydrolysis. Last but not least, the protein dimer $P_AP_R$ can dissociate (Fig. 8).

One can derive the deterministic differential rate equations that describe this model and one can find numerical values of the parameters that lead to periodic oscillations of the concentrations of repressor and activator. However, this reasoning

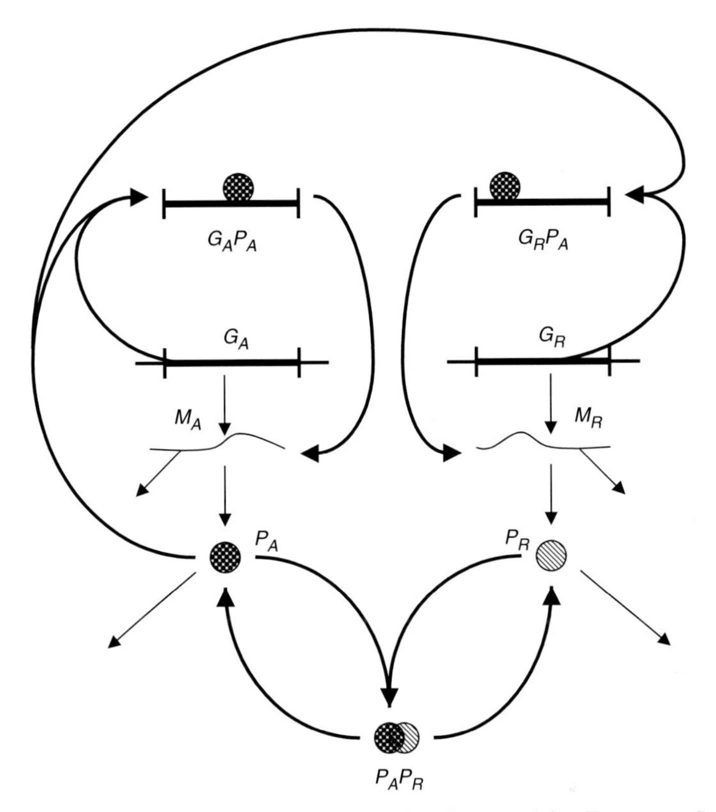


Fig. 8. Stochastic fluctuations can drive periodic dynamics of gene activity. Two genes $G_A$ and $G_R$ are transcribed as messengers $M_A$ and $M_R$ which are subjected to hydrolysis. The corresponding proteins, $P_A$ and $P_R$, form an inactive complex. One of the proteins, $P_A$, interact with the two genes, and the two complexes, $G_AP_A$ and $G_RP_A$, are also transcribed as messengers, $M_A$ and $M_R$. Such a model is indeed hypothetical but its simulation shows that molecular noise can drive periodic dynamics of gene activity. Adapted from [18].

is valid only if the concepts of concentration and rate constants are applicable. This is in fact not the case for small systems such as living cells. Then concentrations should be replaced by probabilities of occurrence and rate constants by transition probabilities. A stochastic simulation of the system can be carried out, as discussed previously, through the Langevin equation or through the master equation and its variant, the Fokker–Planck equation. One will see that the conditions that generate periodicity in the deterministic description of the model also generate periodic behavior, even though these periodicities are less regular. More interesting is a situation in which parameter values can be chosen that do not lead to periodic behavior in the deterministic description, but still lead to sustained oscillations in the stochastic version of the model. In fact, simulations suggest that, owing to a fluctuation, the system can fall in a limit cycle, and displays sustained oscillations.

What is interesting in this model is its ability to be reduced to a two-variable system without losing its main properties [18]. These two variables are the repressor and the complex. The other equations can be assumed to be in quasi-steady state without altering the dynamics of the system. Hence potential variables of the system such as mRNA molecules do not play any role except that of an intermediate in the overall process.

## *References*


1. Haken, H. (1978) Synergetics: An Introduction (Second Edition). Springer-Verlag, Berlin.
2. King, E.L. and Altman, C. (1956) A schematic method of deriving the rate laws for enzyme-catalyzed reactions. J. Phys. Chem. 60, 1375–1378.
3. Hill, T.L. (1988a) Interrelations between random walks on diagrams (graphs) with and without cycles. Proc. Natl. Acad. Sci. USA 85, 2879–2883.
4. Hill, T.L. (1988b) Further properties of random walks on diagrams (walks) with and without cycles. Proc. Natl. Acad. Sci. USA 85, 3271–3275.
5. Hill, T.L. (1988c) Number of visits to a state in a random walk before absorption and related topics. Proc. Natl. Acad. Sci. USA 85, 45–81.
6. Hill, T.L. (1988d) Discrete-time random walks on diagrams (graphs) with cycles. Proc. Natl. Acad. Sci. USA 85, 5345–5349.
7. Hill, T.L. (1968) Thermodynamics for Chemists and Biologists. Addison-Wesley, Reading.
8. Hill, T.L. (1977) Free Energy Transduction in Biology. Academic Press, New York.
9. Gillespie, D.T. (2000) The chemical Langevin equation. J. Chem. Phys. 113, 297–306.
10. Gillespie, D.T. (2002) The chemical Langevin equation and Fokker–Planck equation for the reversible isomerization equation. J. Phys. Chem. A 106, 5063–5071.
11. Gillespie, D.T. (1977) Exact stochastic simulation of coupled chemical reactions. J. Phys. Chem. 81, 2340–2361.
12. Le Novere, N. and Shimizu, T.S. (2001) STOCHSIM: modelling of stochastic biomolecular processes. Bioinformatics 17, 575–576.
13. Gillespie, D.T. (2001) Approximate accelerated stochastic simulation of chemically reacting systems. J. Chem. Phys. 115, 1716–1733.
14. Gibson, M.A. and Bruck, J. (2000) Exact stochastic simulation of chemical systems with many species and many channels. J. Phys. Chem. A 105, 1876–1889.
15. Kraut, E.A. (1967) Fundamentals of Mathematical Physics. McGraw-Hill, New York.
16. Kendall, M.G. (1951) The Advanced Theory of Statistics (Third Edition). Vol. 2 Charles Griffin and Company, London.
17. Sprott, J.C. (2003) Chaos and Time Series Analysis. Oxford University Press, New York.
18. Vilar, J.M.J., Kueh, H.Y., Barkai, N., and Leibler, S. (2002) Mechanisms of noise-resistance in genetic oscillators. Proc. Natl. Acad. Sci. USA 99, 5988–5992.
19. McAdams, H.H. and Arkin, A. (1997) Stochastic mechanisms in gene expression. Proc. Natl. Acad. Sci. USA 94, 814–819.
20. Arkin, A., Ross, J., and McAdams, H.H. (1998) Stochastic kinetic analysis of developmental pathway bifurcation in phage lambda-infected *Escherichia coli* cells. Genetics 149, 1633–1648.
21. McAdams, H.H. and Arkin, A. It's a noisy business! Genetic regulation at the nanomolar scale. Trends Genet. 15, 65–69.
22. Rao, C.V., Wolf, D.M., and Arkin, A. (2002) Control, exploitation and tolerance of intracellular noise. Nature 420, 231–237.
23. Thattai, M. and Van Oudenaarden, A. (2002) Attenuation of noise in ultrasensitive signaling cascades. Biophys. J. 82, 2943–2950.
24. Becskei, A. and Serrano, L. (2000) Engineering stability in gene networks by autoregulation. Nature 405, 590–593.

# Subject Index


Actin filaments 19–22
Additivity and monotonicity of uncertainty functions 85
Affinity of a diffusion process 13, 14
ATP-driven migration of ions 16–18
ATP synthesis 18
ATP–ADP exchange 14
Attack tolerance of networks 75–77

Bacterial operons 227–232
$\beta$-galactosidase 227–230
Binary logic and gene networks 227–232
Binary matrix of a graph 70–75
Binomial distribution 39–42
Bivariate moments 48–51
Boltzmann statistics and emergence 124, 125

Carriers 266–272
Cayley tree (Bethe lattice) 68–70
Central Dogma 102–105
Characteristic function 51, 52
Cis-acting sequences 228
Cluster size distribution 69, 70
Clustering coefficient 63
Code word 98–102
Coding 97–102
Codons 104
Communication and mapping 92, 93
Communication and organization 129–131
Communication channel between DNA and proteins 105–107
Communication system 84, 85
Competition between ligands and mutual information of integration of a protein network 120–124
Conditional entropy 90, 91
Conformational spread and information landscape 223, 224
Connectedness 65
Control of intracellular noise 270–274
Cooperativity 185–199
Coupled scalar–vectorial processes 13–18
Covariance 50, 51
Cumulants 51, 52

Detailed balance 232–236
Diameter of metabolic networks 70–72
Differential activity of a gene 233, 234
Distribution function 36–39
Dyneins 22

Electrochemical gradient 18
Emergence 2, 3, 119–124
Emergence in protein networks 119–121
Energy contribution of subunit arrangement 190–193
Engineered gene circuits 245–252
Entropy 85–92
Evolution of random graphs 61–63

Feedback loops in gene circuits 245–249
Fokker–Planck equation 266–270
Fractionation factor 16, 155
Full-circuit 236–242
Function of connection 171–176
Fuzzy-organized networks 161–163

Generalized Fokker–Planck equation 270
Generalized microscopic reversibility 176–178
Genetic code 102–105
Graph 58

High-level theory 2

Image 32, 110
Induction 230–232
Information and conformational constraints in protein lattices 207–209
Information and organization 110–117
Information and physical interactions between two events 133–136
Information of integration and departure from pseudo-equilibrium 151–153
Integration 2, 3
Integration and emergence in a protein lattice 210–215
Intracellular noise 255–274

Jacobian matrix 235, 236
Joint entropy 89, 90

Kinesins 22
King–Altman–Hill rules 15, 16

Langevin equation 236–238
Laplace–Gauss distribution 44, 45
Logic of gene networks 250–252
Low-affinity sites for repressor protein 232
Low-level theory 2

Mapping 32
Mapping into 32
Mapping onto 32, 97
Markov process 52–54
Master equation 227–233
Mean of the binomial distribution 40
Mean of the Laplace–Gauss distribution 44, 45
Mean of the Poisson distribution 43
Mean uncertainty functions and Stieltjes integrals 87
Metabolic networks 70–72
Metabolic networks as networks of networks 154, 155
Microtubules 19–22
Mitotic spindle 22, 23
Moments 45–51
Monovariate moments 46–48
Multienzyme complexes 167–180
Mutual information 91, 92
Mutual information and nonequilibrium 151–153, 175, 176
Mutual information of a protein unit in a lattice 201–204
Mutual information of an enzyme reaction 146–153
Mutual information of integration 114–117

Negative control 232, 233
Negative feedback 233
Network clustering coefficient 63, 73
Network connections and mutual information 168–176
Network diameter 63
Networks 57–79
Noise-driven properties 236–239
Nonequilibrium cellular structures 19–22
Nonextensive entropies 96, 97

One-to-one mapping 32
Organization and enzyme catalytic efficiency 148–150
Organization of a protein network 111–114
Oscillations of gene networks 249, 250

Path-length of a network 73
Pauling's principle 12, 13, 191, 197
Percolation 63–70
Percolation probability 65
Percolation threshold 69
Permease 227
Phase transition in random graphs 61
Poisson distribution 42, 43
Positive control 232, 233
Positive feedback 233
Power law 74
Pre-equilibrium 11, 12
Prefix condition 99, 100
Probability 32–52
Probability spaces and sub-spaces 84, 111–114
Promoter 228
Protein lattices 199–223

Quaternary constraint energy contribution 193–198

Random networks 59–63
Reduction 2, 3, 125–129
Regular multienzyme networks 157–160
Relations and graphs 30–32
Repressor 230–232
Robustness of multienzyme networks 155–157

Scale-free networks 74, 75
Self-information of integration of a network 125
Sets 27–32
Small-world networks 70, 72–74
Spontaneous conformational transitions in quasi-linear lattices 216–223
Steady state 5–11
Steady state in branched multienzyme networks 171–176
Steady state in linear multienzyme networks 168–171
Stereospecific recognition 1–24
Stieltjes integrals 36–39
Subadditivity in protein networks 117–119
Subadditivity principle 93–96
Szilard–Kraft inequalities 99, 101

Thermodynamic basis for induced conformational spread in a lattice 202–204
Thermodynamic equilibrium 5
Thermodynamics of spontaneous conformational transition in a quasi-linear lattice 216–223

Time hierarchy of steps 15, 155
Topological information 136–140, 162, 163
Transacetylase 227
Trans-acting products 228
Treadmilling 19
Triplet 102–104

Uncertainty functions 87, 89

Variance of the binomial distribution 40, 42
Variance of the Laplace–Gauss distribution 44, 45
Variance of the Poisson distribution 43
Vectors of conformational constraints 207–210, 219, 221, 223

Wavy rate curves 8, 9